Zero to Infinity:
A History of Numbers

Edward B. Burger, Ph.D.

THE
GREAT
COURSES™

PUBLISHED BY:

THE GREAT COURSES
Corporate Headquarters
4840 Westfields Boulevard, Suite 500
Chantilly, Virginia 20151-2299
Phone: 1-800-832-2412
Fax: 703-378-3819
www.thegreatcourses.com

Edward B. Burger, Ph.D.

Professor of Mathematics
Department of Mathematics and Statistics
Williams College

Edward B. Burger is Professor of Mathematics in the Department of Mathematics and Statistics at Williams College. He graduated *summa cum laude* from Connecticut College in 1985, earning a B.A. with distinction in Mathematics. In 1990 he received his Ph.D. in Mathematics from The University of Texas at Austin and joined the faculty at Williams College. For the academic year 1990–91, he was a postdoctoral fellow at the University of Waterloo in Canada. During three of his sabbaticals, he was the Stanislaw M. Ulam Visiting Professor of Mathematics at the University of Colorado at Boulder.

Professor Burger's teaching and scholarly works have been recognized with numerous prizes and awards. In 1987 he received the Le Fevere Teaching Award at The University of Texas at Austin. Professor Burger received the 2000 Northeastern Section of the Mathematical Association of America Award for Distinguished Teaching of Mathematics, and in 2001 he received the Mathematical Association of America's Deborah and Franklin Tepper Haimo National Award for Distinguished College or University Teaching of Mathematics. In 2003 he received the Residence Life Academic Teaching Award at the University of Colorado. Professor Burger was named the 2001–03 George Polya Lecturer by the Mathematical Association of America. In 2004 he was awarded the Chauvenet Prize—the oldest and most prestigious prize awarded by the Mathematical Association of America. In 2006 the Mathematical Association of America presented him with the Lester R. Ford Prize. In 2007 Williams College awarded him the Nelson Bushnell Prize for Scholarship and Teaching; that same year he received the Distinguished Achievement Award for Educational Video Technology by The Association of Educational Publishers. In 2006 Professor Burger was listed in the *Reader's Digest* annual "100 Best of America" special issue as "Best Math Teacher."

Professor Burger's research interests are in number theory, and he is the author of 12 books and more than 30 papers appearing in scholarly journals. With Michael Starbird he coauthored *The Heart of Mathematics: An invitation to effective thinking* , which won a 2001 Robert W. Hamilton Book Award. They also coauthored a general audience trade book titled *Coincidences, Chaos, and All That Math Jazz*.

In addition, he has written seven virtual video CD-ROM textbooks with Thinkwell and has starred in a series of nearly 2,000 videos that accompany the middle school and high school mathematics programs published by Holt, Rinehart and Winston.

Professor Burger has served as chair of various national program committees for the Mathematical Association of America; he serves as Associate Editor of the *American Mathematical Monthly,* and he is a member of the board of trustees of the Educational Advancement Foundation.

Professor Burger is a renowned speaker and has given more than 400 lectures around the world. His lectures include keynote addresses at international mathematical conferences in Canada, France, Hungary, Japan, and the United States; mathematical colloquia and seminars at colleges and universities; presentations at primary and secondary schools; entertaining performances for general audiences; and television and radio appearances including WABC-TV, the Discovery Channel, and National Public Radio. ■

Acknowledgments

Lucinda Robb has encouraged me to return to The Teaching Company classroom for nearly two years. I want to sincerely thank her for her cheerful patience and constant enthusiasm. If it were not for her encouragement and support, I would not have had the wonderful opportunity to create this course. I wish to express my sincere appreciation to the *Zero to Infinity* team at The Teaching Company who made the entire process—from preproduction through postproduction—so pleasurable. Marcy McDonald provided excellent editorial suggestions and comments about the course structure. Zach "Zax" Rhoades was an outstanding producer who beautifully integrated the lectures with visual elements. Tom Dooley and Jim Allen were the technical masterminds performing magic in the control room. Megan Herron and Alisha Reay artfully worked on various pieces of the preproduction process.

Within the world of mathematics, I wish to express my deepest gratitude first and foremost to Professor Deborah J. Bergstrand from Swarthmore College. Professor Bergstrand provided invaluable research on many historical elements. She also offered her outstanding expertise and insightful suggestions—both of which enhanced these lectures. Her contributions and dedication to this project were truly spectacular. I thank her for all her tremendous efforts and talents and for her friendship.

It is a great pleasure to thank my mother, Florence Burger, who suffers from "math-phobia" yet was willing to listen to an extended explanation of binary and ternary expansions of the real numbers during her 78th birthday celebration. I thank her for her helpful suggestions on how to make those abstract notions a bit more palatable for other "math-phobic" individuals. From The University of Texas at Austin, I wish to thank my collaborator, Professor Michael Starbird, for his encouragement and guidance and my Ph.D. advisor, Professor Jeffrey D. Vaaler, who was the first to show me the beauty, wonder, and mystery hidden within the world of number. ■

Table of Contents

Table of Contents

Table of Contents

SUPPLEMENTAL MATERIAL

Zero to Infinity: A History of Numbers

Scope:

In this course, we will paint a picture of an ever-evolving notion of number—coming to the realization that the notion of number is a difficult if not impossible idea to define precisely. Our course will have two main points of focus: the historical evolution of the representation of numbers for communication and manipulation—"the numbers in life"—and the intrinsic structure of those numbers—"the life of numbers." We will discover throughout the course that these two perspectives synergistically inform each other and allow our understanding to grow and evolve with the numbers themselves. Although we will explore these two vantage points and their interconnectedness, we will also offer the recurring historical theme of number as a means to count, quantify, measure, and compare—and surprisingly, all of our intuitions about these basic notions will be challenged.

The Numbers within Our Lives (four lectures)

Starting with the fundamental desire to better understand our world by quantifying it, we will study some of the early attempts to determine "how many?" We will explore clever methods that were used before any abstract notion of number existed. Curiously, as we will discover at the end of this course, these early non-numerical methods hold the keys to unlocking infinity—a realization that set the mathematical world ablaze.

We will then explore the dawn of numerate humanity and celebrate the intellectual triumph of moving number from an adjective ("three" apples) to an abstract noun ("three"). This shift led to a desire to express these new abstract objects in useful and meaningful ways; thus, we will also study the evolution of humankind's struggle with representing and naming numbers. It is here that we will see the dawn of zero, a surprisingly subtle and important concept. We will close this part of the course with a discussion of how numbers have captured the imagination of curious individuals throughout the ages. The allure of numbers has led not only to recreation but also to the beliefs that numbers have personalities and possess spiritual and magical powers.

The Lives within Our Numbers (six lectures)

We will open this part of the course with the discovery that number patterns existed long before our desire to represent them—thus, we see that numbers and the patterns they hold transcend our imagination and are a foundational part of nature and our universe. Once we name the abstract objects known as numbers, however, they are born, take on a life of their own, and attract the interest and capture the imagination of curious minds throughout human history. In this part of our course, we will explore that historical journey to discover and appreciate the structure and beauty of numbers in their own right.

We will look at famous collections of numbers, such as the Fibonacci numbers and prime numbers. We will then move into the fractional world of rational numbers and the much more mysterious and confounding universe of irrational numbers. Even though the ancient Greeks proved that these objects exist, they were unable to view them as numbers. Armed with both rational and irrational numbers, we will explore how these two different forms fit together within the real number line. Finally, we will consider another famous collection of real numbers that exhibits some extremely beautiful albeit counterintuitive properties.

Transcendental Meditation—The π and e Stories (four lectures)

We will open this part of the course by briefly reviewing the historical underpinnings of arithmetic and algebra and their influence on the theory of numbers. We will then share the wonderful stories of the celebrated numbers π and e, discovering not only the histories of these important constants but also their significance in mathematics and beyond.

With these two famous numbers as our inspiration, we will walk the intellectual path of mathematicians toward a better understanding of such numbers as π and e. This exploration will lead us to the subtle concept of transcendental numbers. Here, we will see an illustration of an important recurring phenomenon: That which first appears to be bizarre and strange is, in actuality, normal and commonplace.

Algebraic and Analytic Evolutions of Number (four lectures)

With our initial understanding of numbers in place, we will offer two elegant mathematical perspectives on how numbers have evolved. The algebraic point of view will lead us naturally to "imaginary" numbers—which will then appear far less imaginary. Armed with the imaginary number i, we will discover one of the most amazing and beautiful formulas in mathematics—one that connects the five most important numbers together into one incredible equality.

The analytical evolution of number—based on measuring distances and closeness—will lead us to new numbers that defy our intuition of what "number" should mean or how numbers should look and behave. Paradoxically, these strange and foreign modern numbers have provided physicists with new insights into quantum physics and our not-so-foreign universe.

Infinity—"Numbers" beyond Numbers (five lectures)

In this final part of the course, we will travel beyond numbers and contemplate the enormous question, what comes after we have exhausted *all* numbers? We will tame infinity and discover that, contrary to our initial intuition, infinity shares some basic properties with number—in particular, they both come in different sizes.

The wonderful and surprising realization in these lectures will be that the dramatic and at-first highly controversial theory of infinity arises from the same principle humankind first employed to "count." Thus, we will come full circle, discovering that although the ideas of infinity carry us to levels of abstraction and imagination that most individuals dare not reach even today, the nucleus of that incredible theory relies on an insight of our ancestors from 30,000 years ago. This journey to infinity and beyond will be the final brushstroke on our painting of the endless frontier of the notion of number. ∎

The Ever-Evolving Notion of Number
Lecture 1

Number is not a fixed, rigid idea, but instead an ever-evolving notion.

In this lecture, we will introduce the concept of number and foreshadow an interesting paradox: Although numbers are precision personified, a precise definition of number still eludes us. As our understanding of the world expands and our capacity for abstract thinking grows throughout history, so too does our view of what number means. We will see numbers move from useful tools for measuring quantities to abstract objects of independent interest. This lecture previews the main themes of the course, from an exploration of the life of numbers to the endless world of the transfinite.

Welcome to a world of number. What is your definition of *number*? The distinction between *number* and *numbers* is subtle. Numbers are at once practical notions in our everyday world and abstract objects from our imagination. Before our ancestors could write, they contemplated quantities. Historically, the study of numbers was a central component of one's education—one of the original liberal arts.

Many people incorrectly believe that mathematics is completely understood; most of mathematics, in fact, remains mysterious. Forward progress is extremely slow moving. New discoveries in mathematics are made by building on the work of others who came before.

Our knowledge of the early origins of number is vague; we must depend on relics that archaeologists uncover. Some ancient civilizations recorded their work on materials that stood the test of time. Others employed materials that, over time, disintegrated; thus, our knowledge is as fragmented as the ancient, broken tablets we try to understand. In this course, we will study moments in time to produce a mosaic of small pieces that, when viewed from afar, will allow us to see how numbers grew in our understanding and sparked our imagination.

This course is a blend of mathematics in a historical framework. Although these lectures offer a fluid conceptual development of the notion of number, at times we will gently glide back and forth through history so that we can appreciate and better understand the allure of number as our story unfolds. The course covers three main themes.

- We will discover that numbers are truly difficult to define precisely, despite what most people believe.

- We will come to appreciate the notion of number as one that is always evolving.

- We will also see the recurring theme that what at first appears familiar and commonplace is, in fact, rare and exotic; conversely, what first appeared exotic will later be viewed as the norm.

The first series of lectures focuses on early attempts to quantify. We will journey back to 30,000 B.C.E. and see some of the earliest attempts to count. Numbers slowly evolved into adjectives (e.g., "three" apples). *Counting numbers* (also known as *natural numbers*) became the most familiar numbers (e.g., 1, 2, 3, 4, 5 ...). We will investigate the challenges of communicating and manipulating numbers. Through these investigations, we will see the notion of number expand further, as our ancient ancestors struggled with zero and negative numbers. We will explore how individuals were moved to associate personalities, magic, and even cosmic significance to numerical notions.

We will explore numbers in nature and discover how Fibonacci strove to make them more natural. We will then focus on the nature of numbers themselves. By the 6th century B.C.E., the Pythagoreans were studying numbers as objects in their own right, rather than using them solely as tools for calculation and recordkeeping. Pythagoras may have been inspired by the religious sect in India known as the Jains, whose members may have been the first number theorists. Today, this exploration into the study of numbers is known as *number theory.* Cultures share their passion for number theory; the more we explore, the more our field of vision of number widens.

We will celebrate two of the most important numbers in our universe, π and e, using these famous quantities as the inspiration to see subtle distinctions between different types of numbers. We will also consider two mathematical views of number evolution that allow us to expand our notion of number in new directions. We will also encounter "numbers" that challenge our very notion of what *number* means.

Infinity can be understood and can hold many surprising features.

We will journey beyond the universe of number and delve into the more abstract world of infinity. Using the very first method for counting, we will discover that, just as with numbers, infinity can be understood and can hold many surprising features. Although our discussions will become a bit technical at some points, those details are not the central focus of this course. Our main goal is the realization that the study of number is a beautiful endeavor that has captured humankind's imagination throughout the ages and continues to inspire us to explore its endless frontier. ■

Questions to Consider

1. What is your definition of *number*? You are encouraged to write down your definition after this lecture to see how it changes throughout the course.

2. For what purpose are numbers used? What types of numbers have you encountered in your life?

The Ever-Evolving Notion of Number
Lecture 1—Transcript

Welcome to *Zero to Infinity*, a course that celebrates the history and evolution of the notion of number. In this opening lecture, we'll introduce the concept of "number" and foreshadow the interesting paradox that while numbers are precision personified, a precise, accurate, and satisfying definition still eludes us. In fact, one of the central themes of this course is that the concept of number is not a fixed, rigid idea, but instead an ever-evolving notion. As these lectures unfold, we'll discover that throughout history, as our understanding of our world expands and our capacity to abstract ideas grows, so does our view of what "number" means.

We open our course with a series of lectures, entitled "The Numbers within Our Lives," that offers the early conceptual underpinnings of numbers and how they were used to express different ideas. Throughout the cultures, not only do we see them happening to one or two, but in fact this multicultural contribution led to our current common notion of number. We'll see numbers move from useful tools for measuring quantities to abstract objects of independent interest and intrigue. This transition brings us to the second part of our course, entitled "The Lives within Our Numbers." This exploration of the life of numbers captures a desire that goes back to our ancient ancestors to understand numbers and their beautiful structure for their own sake. This leads us to Part 3 of our course, entitled "Transcendental Meditation— The π and e Stories," where we'll celebrate two of the most important numbers in human history and use them to inspire the modern notion of transcendental numbers.

In Part 4, entitled "Algebraic and Analytic Evolutions of Number," we consider two mathematical perspectives on how to create different types of numbers. The algebraic view naturally leads us to imaginary numbers, while the analytical view challenges our intuitive sense of what number should even mean. In our last part, entitled "Infinity—'Numbers' Beyond Numbers," we explore the endless world of the transfinite and discover that humankind's first method of counting holds the secret to unlocking infinity. Exploiting that ancient, simple idea leads us to a totally surprising and counterintuitive discovery: Infinity, just as the numbers that came before, comes in different sizes. Thus, we close our journey with startling new insights into the infinite that only came into focus after thousands of years of studying numbers and that paradoxically required us to return to the very dawn of the human notion of number.

Well, welcome to the world of numbers. Now, before I say too much and go any further, I want to pose a conundrum for you to consider, and it's in the form of a question. What is your definition of *number*? I want to make, right here on the onset, a subtle distinction between the word *number* and the word *numbers*. In my mind and throughout the course, "number" will mean a more abstract notion, a more abstract idea of all of numbers and all the ideas of numbers put together in whole, whereas "numbers" will usually refer to a particular collection or group of numbers. This is similar to the distinction between the words "humanity" and "human being." Often, "humanity" refers to all humankind, considered as one abstract collective whole, while "human beings" often refers to a specific group of individuals. I'll make that distinction here between "number" and "numbers" throughout the course.

The question is, what is your definition of the abstract idea of *number*? In fact, if you're so inclined, I actually invite you to stop and jot it down before continuing on—if you so take the challenge. The reason why I raise this challenging question, as we'll discover, is that in reality it is a very difficult task to precisely define "number." In fact, this difficulty was eloquently articulated over 2,500 years ago by the great number enthusiast Pythagoras (an individual we'll later study in some depth) when he said:

> Number is the first principle, a thing which is undefined, incomprehensible, and having in itself all numbers.

First of all, you can see that even Pythagoras made the distinction between "number" and "numbers," and this abstract notion of number, he claims, in fact is undefined and incomprehensible. But within that realm, within that abstract world, resides all the numbers. In some sense, we haven't made an awful lot of progress since Pythagoras, over 2,500 years ago, because we still struggle with the notion of what numbers mean and the notion of number in an abstract sense.

It's interesting to think that numbers are, at once, practical and abstract. It's an interesting dichotomy. They are one of the most important notions in our everyday world and yet remain mysterious objects of our imagination. Throughout this course, we'll see that interplay between the abstract and the practical. Before our ancient ancestors could even write, they were contemplating quantities. Historically, the study of numbers was a central component of one's advanced education. For thousands of years, learned individuals studied four subjects, and these four areas collectively were known as the quadrivium. These subjects were the theory of numbers, geometry, music, and astronomy.

Let me just say a word about geometry here. Geometry, in this ancient setting, really also involved not just the geometric figures that we've seen in our geometry classes a long time ago, but also the measuring of lengths and the computing of areas. In some sense, within the ancient sense of geometry, we really see also reflections of number theory. So in some sense, two of these four tenants were, in fact, number theory; half of them were number theory. Years later, other subjects were included in this collection, and it became known as the liberal arts. We see the theory of numbers was in fact one of the cornerstones of the liberal arts. So you can see this is really central in one's ancient view of the world and certainly central today. It's remarkable how many of our modern customs, including the way we communicate and think, have been influenced by our ancient ancestors' quest to understand and tame numbers. Even the basic act of writing probably evolved from a basic desire to express numbers. The notion of number permeates nearly every aspect of our lives and continues to inspire the life of the mind today.

Of course, the numbers sit within the realm of mathematics, and so I wanted to say a few words about mathematics and its frontiers before we really start our journey. Many people believe, incorrectly, that mathematics is totally understood—that we know it all. In fact, most of mathematics remains totally mysterious to the research mathematicians and to people from around the world. Now, what a research mathematician tries to do is to move the frontiers of mathematical understanding outward and forward. How does one go about this? Well, in mathematics the way we perceive is by actually proving theorems, and the theorem itself is required to have with it this notion of a rigorous proof.

Let me say a word about what those mean. So, in mathematics, we consider statements. A statement is a declarative sentence that's either true or false. If the statement is true, then, in fact, it is given the moniker of "theorem," and that's a great celebration for us in mathematics. But it's not just enough to say that a theorem is true. We have to prove it, and we have to establish it through this idea of rigorous proof, which really goes back to Euclid, as we'll see in a future lecture. The idea here is that we use logical reasoning to write a deductive argument that takes us from points that we already know to a new place. So we build on the previous work, the work of others, and we use previous results to move the forefront of mathematics forward. This process is extremely slow moving, and new discoveries in mathematics are actually built by building on the work of others that came before.

In some sense—here's how I think about it, I'll share how I view my own mathematical work and mathematics in general. Imagine if you will a very large copper orb, and the mathematicians live inside this copper orb, and the copper orb represents the boundary of our mathematical understanding. Every little mathematician is equipped with a ball-peen hammer, and what we're doing is we're just hammering inside this copper orb. Just as you would make a pan out of copper, slowly, slowly, we put these dimples, and dimples upon dimples, and slowly the orb begins to grow. Well, it's that slow growth that expands our understanding and the frontiers of mathematics. In fact, that's how I view moving mathematics forward—slowly making this tiny progress, and slowly, dimple upon dimple, we see the light.

The interesting paradox here is that the more we learn, the more we realize we don't know. In fact—it's funny—every time we take one tiny step forward toward the frontier of mathematics, that insight reveals far more unanswered questions than the question we just answered. In some sense, with every tiny step we take—we look and think that we're closer to the frontier—the frontier, in fact, jumps out another 100 miles, and so it seems like we're losing ground, even though we are, in fact, moving forward.

In fact, moving mathematics forward is an art that requires both imagination and creativity. Mathematicians are, at once, artists and explorers. We're explorers in that we're trying to understand the truths in nature and in the universe based on the logic of mathematics and the nature of mathematics.

But we're also artists, because to carefully construct and create the proofs requires a tremendous amount of creativity, imagination, and originality. We walk a fine line. On the one hand, we're trying to be explorers and discover truth, and on the other hand, we're being artists. Unlike within the arts, if we have a statement that we particularly like, but it turns out it's false, there's no "in fashion" or "out of fashion." The statement is false and then disregarded within the mathematical realm. All we care about in mathematics is the truth.

Let me return now to our journey, our journey through numbers, and let me say a word about the archaeology of numbers. Our knowledge of the early origins of numbers is very spotted and vague. In fact, we must depend, obviously, on the relics that the archaeologists uncover for us. Thus, the medium upon which early cultures used to record their thoughts (and even the climate) dictate what we know and what we don't know. Some ancient civilizations, for example, recorded their work on materials that stood the test of time. For example, in the dry climate of the Middle East, we can find carvings in stone, clay tablets, and even early forms of paper.

Other civilizations employed materials that, over time, disintegrated. For example, the tree bark that was used in ancient writing in Asia was not able to survive the humid climate. Thus, our knowledge of the early history of numbers is as fragmented as the ancient, broken tablets we try to piece together and understand.

As a result, in this course, we offer moments in time in order to produce a mosaic of small pieces that, when viewed from afar, will allow us to see how numbers grew in our understanding and sparked our imagination. So in some paradoxical sense, to really appreciate this, we have to see it all at once. At the last lecture, we can finally step back and see the entire panoramic, which I assure you, will be quite spectacular. In some sense, that mosaic is one that can be best appreciated at the end of the journey, and we'll see it then.

These pieces will represent some of the major advances in our understanding of number. Now, remember that mathematics is a game of very, very tiny steps; however, here in this course, we'll only highlight some of the big and intriguing ones. We're going to make quantum jumps, but we have to understand that behind the scenes what we have are in fact tiny, tiny, tiny steps, but we're only going to highlight the ones that are quite exceptional.

Well, let's take a look at the journey ahead. This course is really a mathematics course within a historical framework. We'll meet many of the individuals who actually hammered within that copper orb that I mentioned (in this case, the copper orb of "number," which is where I live). Let me assure you that many of them are quite interesting characters, and there are quite a number of interesting stories that will be revealed. Our vignette scenes will attempt to offer a fluid conceptual development of the notion of number. At times, we'll travel back and forth through history so that we can appreciate and better understand the concept at hand as our story unfolds. It's important to remember that the main character in our story is number, and our main plot line is to intuitively see the evolution of number from the dawn of human thought to today.

I now want to outline three main themes that will be with us throughout all the lectures. The first one is that, despite what most people believe, we'll discover that numbers are truly difficult to define precisely. Second, we'll also come to appreciate the notion of number as always evolving. It's not a fixed idea, but in fact a moving target. In some very real sense, the fact that this is an ever-moving target for us is what makes this chase so exciting. Finally, we'll see the recurring theme that what at first appears familiar and commonplace is in fact rare and exotic. Conversely, what first appears to be exotic will later be viewed as the norm. This theme will really challenge our intuition and show us that there is more to number than meets the eye. So we'll see quite often that our intuition and our initial guess isn't quite up to the correct idea of number as nature laid it out for us.

In fact, I hope that the themes of this course will inspire a newfound perspective and a lens through which to see your life and entire world in a whole new way—an agile mind-set that is open to new ideas and change. As human beings, we come to every issue with preconceived notions, biases, and prejudices, and that's just the reality. But our job as thinking individuals is to put those biases to the side. When we see reality, and when we see truth, we need to realize that this could be, potentially, a surprise. A surprise is a wonderful moment. It's a moment when our intuition runs counter to reality, and it's a wonderful moment because it's a moment when learning can occur. So we should constantly be searching for surprises. When we see a surprise, our mission as thinking people is to not give in to that, but instead to say, "Now is my opportunity to retrain my intuition," so that what first seems surprising and counterintuitive is now not nonsensical. It makes complete sense—commonsensical.

The other important lesson to keep in mind is that we should never be afraid of making mistakes. Failing is the road to creativity, understanding, and enlightenment. We will see that as a recurring theme throughout this course, because we will see great minds throughout history fail, and fail aggressively, and fail again and again. We celebrate that failure, because through every misstep we understand the truth, and we understand the correct path that will lead us to a new insight. We'll see humankind struggle with both failure and with bias (in the context of number) throughout the whole course, but I hope that it will permeate into our whole lives.

Part 1 of our course is entitled "The Numbers within Our Lives," and in this series of lectures we'll first focus on the very early attempts to quantify. We will journey back to 30,000 B.C.E. and see some of the earliest attempts to count. For example, suppose that we couldn't count very high. Suppose that we couldn't even count to, let's say, five. But the question is, how do we know if, in fact, I have the same number of fingers on this hand as I do on this hand? Well, of course, for us today this is not a big deal. We count: one, two, three, four, five. We count: one, two, three, four, five. They're both five; they're the same.

But suppose we couldn't count that high. What would we do? Well, if we couldn't count that high—and this was the issue that early civilization had, because they had no language for numbers, of course, before they were discovered and invented—how could I compare and see if the number of fingers on this hand is greater than, or smaller than, the number of fingers on this hand? Well, it's so natural that even a child can do it: They would just do *this*. We see, we've paired up each finger from each hand with exactly one finger from the other hand, and there are no fingers left without a partner. We see through this pairing that, in fact, there are the same number of fingers on this hand as this hand, and yet we have no idea of the notion of five.

Comparing collections rather than counting them is the dawn of numbers and, in fact, the key to unlocking infinity, as we'll see at the end of the course. Well, numbers slowly evolved into adjectives, as in "three apples." Different names for the same quantity of different objects actually came into fashion—for example, a "brace of pheasants," a "couple of days," a "pair of dice." Well, they all meant "two." But moving from an adjective to the abstract noun "two" was a dramatic leap in the intellectual development of humanity. We'll see that leap for ourselves and enjoy it.

This leap leads us to the "counting numbers," the familiar numbers that we use to count things, which we'll call in this course the "natural numbers." The natural numbers, which are one, two, three, four, five, six, and so forth, are the numbers that we can really count on. We'll then investigate the challenges of communicating and manipulating these numbers.

It's one thing to count, and it's another thing to communicate, keep records, and perform arithmetic. Through these investigations, we'll see the notion of number expand yet further as we witness our ancient ancestors struggle to understand zero and to understand negative numbers. Well, now you're saying, "Zero's no big deal." Well, in fact, zero is extremely important. Without zero, we face serious numerical conundrums. For example, take a look at this number: a 1 followed by a 1. Is it 11? Or is it 101? Well, without zero, we don't know. Zero, as we'll see, is extremely important. But should it be considered a number? Here we'll see the first instance in which humankind was faced with the challenge of expanding its notion of number and what number should mean. Finally, we'll explore how individuals were moved to associate personalities, magic, and even cosmic significance to numerical notions. How numbers were believed to have warded off the plague, and even appease the gods. Even Benjamin Franklin, as we'll discover, could not resist the magic of numbers.

Once numbers were born within our imagination, they took on a life of their own, following the laws and nature of mathematics. This brings us to the second part of our course, "The Lives within Our Numbers." Here, we move from the practical side of numbers (as a means of counting and commerce) to the study of numbers themselves. This area is known today as number theory. We'll open our explorations by first discovering an amazing pattern in nature that was made famous in a 13th-century mathematician's book written by Leonardo de Pisa (better known today as Fibonacci). From here we'll focus on the nature of numbers themselves.

By the 6th century B.C.E., the Pythagoreans were studying numbers as objects in their own right, rather than solely as tools for calculation and recordkeeping. The Pythagoreans may have been inspired by the religious sect in India known as the Jains. It appears that the Jains were the first number theorists. Different cultures all share their common passion for number theory, and the more we explore, the more our field of vision of number widens. Here, we see ratios of natural numbers and discover the existence of numbers that are not ratios (the exotic, so-called "irrational" numbers)—a notion that was so disturbing that it may have driven the Pythagoreans to murder. That's right. Here you think that this is going to be a humdrum course on numbers, but in fact there's a lot of drama. That's a pretty good cliffhanger right there, don't you think?

Next, we'll celebrate two of the most important numbers in our universe: π, famous in geometric circles, and e, a more modern, extremely important number used to measure the growth and decay of populations and for computing compounding interest rates. Both numbers help us better understand nature and our universe, and we'll explore them and enjoy them in some depth. We'll then use these famous quantities, π and e, as the inspiration to see subtle distinctions between different types of numbers. This will lead us to the mysterious modern world of what's called "transcendental numbers." We'll consider two different mathematical approaches to number: an algebraic view based on solving equations—this leads to new, "imaginary" numbers—and then an analytic view based on using numbers to measure distances and closeness. This point of view will actually lead us to a parallel universe of number unlike anything we've ever seen before. In fact, through these new and strange numbers, we might better be able to empathize with the Pythagoreans who were unable to accept certain disturbing quantities as numbers. (Although I hope it won't drive us to murder.) Along the way, we'll also study the most beautiful and famous formula in all of mathematics, one that elegantly connects the five most important numbers in our universe.

Finally, at the close of the course, we will journey beyond the universe of numbers and delve into the at-first forbidding world of infinity. Using the very first method for counting, we'll discover that, just as with numbers, infinity can be understood and holds many surprising features. This theory was constructed essentially by one man, Georg Cantor, in the late 1800s. It was so controversial and so counterintuitive that his work was not accepted by many of his peers until years after he introduced his ideas. Here, we'll see the very human drama of a wildly imaginative individual, who was also troubled, and who set the mathematical community ablaze with his incredibly original discovery.

I want to say just a couple of words about the overarching objectives of this course together. While at some points our discussions will become a bit mathematically technical, I assure you that those details are not the central focus of this course. No one should be alarmed or overwhelmed by the digits. So when you see big numbers (or every once in a while—and there won't be a lot—I'll show a little something that looks more mathematical than not), the goal here is for us to take these mathematical moments and allow them to wash over us. We can appreciate the creativity within a mathematical context just as we would appreciate a piece of music or a piece of art. We can sit back and take in whatever we can. But I invite you and I encourage you to take in as much as you can. I urge you to push a little bit. If it gets a teeny bit mysterious, that's fine—we can let it go. But always be open. I assure you that we will not be technical per se, and that is certainly not the thrust of what we're going to be doing here. So stay with me, and trust me.

I have to close by saying that I'm a number theorist, and so personally I'm extremely excited to have this opportunity to share with you what I believe are some of the most incredible and stunning ideas of human history. The great 20th-century British number theorist, G. H. Hardy, once wrote:

> A mathematician, like a painter or poet, is a maker of patterns. ... The mathematician's patterns, like the painter's or the poet's must be beautiful; the ideas, like colors or the words, must fit together in a harmonious way.

Thus, while my hope is that our journey will be both uplifting and rewarding, my main goal is for all of us to realize that the study of number is a beautiful one that has captured humankind's imagination throughout the ages and continues to inspire us to explore its endless frontiers.

The Dawn of Numbers
Lecture 2

Long before our ancient ancestors figured out how to count, they used the innate number sense that's within us all to instantly compare small collections of objects.

W hat motivated humans to count? Thousands of years before there were writing, literacy, or even numeral symbols, shepherds tending flocks had to keep track of their sheep. As agricultural societies developed, people needed to measure and divide land, keep track of livestock, record harvests, and take census data. With growing populations and clashing cultures came conflict, requiring armies to face the logistics of arming and feeding their soldiers. Bountiful agricultural fruits of labor required counting days and lunar cycles as part of calendars to better predict the change in seasons, annual floods, or dry spells.

Human beings have an innate number sense. This innate number sense allows us to instantly compare small collections of objects. If a Sumerian shepherd has a very small number of animals, he can keep track of them without the need for counting. It is easy to see the difference between a herd of four sheep and a herd of three without actually counting. With a larger herd, we are unable to determine (by simply looking) whether the collection of

Before writing and counting, primitive calendars, like this Mayan one, were developed to measure lunar cycles.

sheep we have after grazing is the same size as the collection with which we started. This limitation is referred to as the limit of four. This limit of four might underlie the barred-gate system of counting we still employ today.

Other creatures also sense numbers. Studies with goldfinches reveal that when presented with two small piles of seeds, they usually pick the larger of the two piles; crows have also been known to distinguish between collections of different sizes. Evidence suggests that animals do not, however, have a notion of number as an abstract object.

The human concept of number may have developed in a manner similar to the way in which it develops in children. Ordination comes first; that is, the ability to see that one set of objects is larger than another. We learn to order objects according to size before we learn to count them. Learning ordered lists is a classic component of early education; children are taught to recite the alphabet, numbers, and even the days of the week, often well before they understand the meaning of these sequences. Children next begin to grasp the idea of natural numbers (e.g., 1, 2, 3, 4 ...). Finally, children master cardination (or true counting), in which the objects in one collection can be counted or paired up with objects from another collection.

Notched bones from as long ago as 30,000 B.C.E. have been found in Western Europe.

Many societies used various forms of sticks as counting tools to record one-to-one correspondences. Notched bones from as long ago as 30,000 B.C.E. have been found in Western Europe. Notched sticks called tally sticks have been used for millennia and may have inspired the development of Roman numerals. A wooden tally stick could be marked and then split lengthwise so that two parties could keep track of a transaction.

In order to make the one-to-one correspondence physical, the Incas and cultures along the Pacific Rim and in Africa used knotted strings. In the 5th century B.C.E., Herodotus of Greece wrote in his *History* that Darius, the king of Persia, used a knotted cord as a calendar. Catholic, Muslim, and Buddhist rosaries and prayer beads allow the devout to recite the appropriate number of litanies without the need for an abstract counting system. As early as 3500 to 3200 B.C.E., our Sumerian shepherd most likely used a pile of pebbles to "count" his sheep through a one-to-one pairing.

The human hand is a natural counting tool. The limit of four in humans made the five-digit hand particularly useful for counting and led to the use of five as a basic grouping for counting. The continued popularity of the barred-gate tally system may be due to its basis in counting by 5s. The 10 digits on our two hands may have led to the modern-day dominance of a base-10 numeral system. Toes are the obvious extension of the hand as a counting tool. Given the hand's convenience, some cultures extended their counting to include the joints of fingers. Other body parts have been incorporated into counting systems in many cultures.

Sumerian methods of counting have been studied extensively. Sumerians created different clay tokens, called *calculi*, to represent quantities of different items. To record a quantity of goods, such as measures of grain in a storehouse, Sumerians would seal the appropriate collection of tokens in a clay jar. Evidence for this method of counting can be found as early as 3200 B.C.E. The tokens were used to make impressions on the outside of the jar before the clay hardened, indicating the quantity within. The token markings later came to represent the numbers themselves, eliminating the need for the tokens in recordkeeping. Sumerian markings led to one of the first numeral systems and, consequently, to what may have been the first form of written language: cuneiform. ∎

Questions to Consider

1. Without counting, determine whether the collection of @ signs below or the collection of & signs is larger.

    How were you able to perform this task without counting?

Now determine without counting whether the collection of @ signs below or the collection of backslashes is larger.

Discuss why one comparison was easy and the other more difficult.

2. Find examples in your everyday life where you use a one-to-one pairing to compare the sizes of two collections without actually counting.

The Dawn of Numbers
Lecture 2—Transcript

In this lecture, I want us to consider the dawn of numbers. Suppose we were Sumerian shepherds with no knowledge of numbers. How would we keep track of our sheep? How would we know if all our sheep returned after grazing in the pasture? Humans have an innate capacity to accurately compare small quantities. However, to compare large quantities, early civilizations employed the idea of a one-to-one pairing. Notched bones, knotted strings, and piles of pebbles allowed them to keep track of their animals and to conduct commerce. While the human hand is in many ways one of the most fundamental counting tools, studies of primitive cultures reveal subtle use of the entire body in counting practices. Some of these early counting techniques actually led to important number systems, and even persist in modern times.

Humankind has been counting for at least 30,000 years, but are humans the only creatures to possess a number sense? Here we see that even some animals appear to have the capacity for numerical concepts. Next, we'll turn to the development of the abstract notion of number. When did the adjectives "three" sheep or "three" apples become the noun *three*? While there is no precise moment, of course, evidence of abstract numbers in Mesopotamia dates back to 3300 B.C.E. We'll examine some of the early counting tools as a means to determine how humanity's understanding of numbers initially developed. By understanding early cultures' use of one-to-one pairings, we will in fact be laying the foundation for understanding how a 20th-century mathematician named Georg Cantor unlocked the secrets of infinity.

Let's begin with the motivation to count. Thousands of years before there was even writing, or even numerical systems or symbols, shepherds tending flocks had to keep track of their sheep. As agricultural societies developed, people needed to measure and divide land, keep track of livestock, record harvests, and collect census data. Growing populations and clashing cultures led to— of course—inevitable conflict, thus requiring armies to face the logistics of arming and feeding their soldiers. Producing bountiful agricultural harvests required counting days and lunar cycles as part of calendars to better predict the change in seasons, annual floods, and dry spells. These were, in fact, the initial motivations for counting. Long before our ancient ancestors figured out how to count, they used the innate number sense that's within us all to instantly compare small collections of objects.

So, we return to the Sumerian shepherd. If he had a very small flock, he could easily keep track of them without the need for counting. For example, it's easy to see the difference between a herd of four sheep and a herd of three sheep without actually counting. With a larger herd, we're unable to determine, by simply looking at them, whether the collection of sheep we have after they return from grazing is the same size as the collection with which we started. This limitation is often referred to as the "limit of four." We can look at small collections of things, usually up to four, and just see how many are there without actually counting them. But beyond four, we traditionally end up counting. In fact, this limit of four might underlie the "barred gate" system of counting that we still use today. This is where we write a notch, mark one, two, three, and four. Instead of marking five, which would be hard for us just to look at and visualize, we put the gate across and call that a five, and then we'd add more—one, another one, another one. If we stop there, we can easily see this is just three, and here we have five, and so we have a total of eight. So, in fact, we can see that very easily, and that's why we have this barred gate system, because of this limit of four.

Are we the only animals that have this innate sense of number? Well, studies with goldfinches reveal that when presented with two small piles of seeds, they will usually pick the larger of the two piles. So goldfinches have a sense of number in this sort of generic sense. Another illustration comes from the late 1700s, when the scientist Gilbert White performed an experiment involving plover eggs in a nest. He would secretly approach the nest, and he would remove one egg from the plover nest, and the mother bird would then later realize that she was deficit an egg, and she would lay another egg to replace it. She was able to detect that the number was off. If he didn't remove an egg, she would not lay another, but if he removed another egg, she would again produce and lay another one. So this plover actually seemed to have a sense of the number of eggs in her nest.

Crows have been known to distinguish between collections of sizes one, two, three, and even four. I must tell you the wonderful story of a squire who was determined to get rid of a pesky crow who made its nest in the watchtower of his estate. So this crow made a nest at the watchtower. Every time the squire would approach in order to shoot the crow, the crow would fly away, go to a distant tree, and stay there and watch until it saw the squire leave, and then the crow would return back. Finally, the squire had the brilliant idea of bringing a friend along. So two people went into the tower, and of course the crow flew away, and then the squire sent his friend out so the crow would see one person leaving. But the crow realized that two went in and one went out; it wasn't the same. So, the crow waited, until finally the squire gave up, and the squire left, and the crow returned to its nest. Then the squire brought two friends. So three of them went in, and the crow flew away. Two of them came out, and the crow still was able to tell that it wasn't the right number. Well, four came in. So, the squire and three of his friends came in. Three of them left, the crow watched, but the crow was still able to detect the difference between three and four. Well, finally, the squire brought four of his friends. So all five went in. The crow flew away. Four came out, but now the crow was unable to detect the difference between the four and the five. It seemed as though maybe five left, and he sadly returned to his perch and his nest, and—let me just say—that was the end of the crow.

So, sadly, the crow exhibited our limit of four as well. Do animals have a notion of number as an abstract object? Can horses really perform addition by tapping their hooves, as we see in circuses and sideshows? Well, evidence suggests that the answers are really no. A famous example from the 19th century is Clever Hans, a horse whose trainer taught him to count and to add. Instead of actually performing addition by counting with his hooves up and down, experts now believe that the horse was actually responding to subtle and subliminal responses, most likely unintentional visual clues from his overenthusiastic trainer. That was what the horse was responding to, rather than this innate sense of arithmetic.

Let's now move from animals to children. Interestingly, a child's development of the concept of number appears to mirror humankind's historical development of number, which is really quite fascinating. It's a microcosm of the scope of the entire history. The word "ordination" comes first (that is, the ability to see that one collection of objects is larger than another). In other words, we first learn to order objects according to their size before we learn how to actually count them; this is ordination. Learning ordered lists is a standard component of early education. Children are taught to recite the alphabet, numbers, and even the days of the week, often well before they understand the meaning of these sequences. But they get the sequence down—ordination. Then the child begins to grasp the idea of the natural numbers, which are the ones we count with—one, two, three, four, five, and so on. Finally, true counting—or what we call "cardination"—is mastered, in which the objects in a collection can be counted, or paired up with the objects from another collection. In our discussion of infinity, in Lecture Twenty at the end of the course, we'll actually revisit these notions of ordination and cardination.

So now let's return to the first form of counting—counting via comparing. Many societies used various forms of tally sticks as counting tools to record one-to-one correspondences. So this was a stick, a device—in fact, I have one right here—where marks would be made, like this, to denote whatever it is that they were trying to quantify, what they were trying to keep track of. In fact, we see evidence of this early on—notched bones from as long ago as 30,000 B.C.E. have been found in Western Europe. So we see the desire to count dates back tens of thousands of years. It's quite a really old and important idea. Now, notched sticks, such as the one I have here, have been used for millennia, and they may have in fact inspired the development of Roman numerals, really, if you think about it for a second, or maybe even turn it on its side. But this is something we're not sure of.

Now, a wooden tally stick would be marked like this, and let me explain to you how this would work. Suppose that you and I were going to make some kind of business transaction. Maybe we were going to make some deal where I was going to purchase some grains of some kind of substance like wheat, or barley, or something. So we agree upon the amount. We don't know how to speak these numbers yet, but we do so on this tally stick. Well, of course, now who holds the tally stick? If I hold the tally stick, then I might add some more notches while you're not looking, and I say, "Well, look, the deal we made—you actually have to give me this much more." Or if you held the tally stick, then somehow you might try to erase some of these marks and then say, "Well, in fact, I only owe you this much," which was less than the deal.

So, to ensure that, in fact, this was fair, what they would do is they would take the tally stick, and they would actually break it in half, like this. So the marks are made, and they would break it in half, like this, and then you would have one, and I would have one, and that way neither can cheat. Because if I mark mine, when I come back, the one-to-one pairing will not match up perfectly. So it was a wonderful way to keep track of transactions and to make sure that both parties were being honest.

In fact, you can think of these tally sticks as stock, a stock of wood. Really, when you actually hold onto it, you are in fact a stockholder. That's where the word "stockholder" actually comes from: from our ability to try to look at a one-to-one pairing, our ability to want to count and to make sure the transaction is going to be fair and honest. You hold the stock. This is also where the phrase "bank stock" comes from, because banks would actually do the same operation. They would give you one of these stocks, so you would have bank stock. It's amazing how much of the language in our everyday world that we take for granted really, in fact, have origins in our understanding or our desire to understand numbers—really fantastic. So, "stockholder" is a wonderful word to remember and especially to remember its origins.

In fact, from the 1300s all the way up to 1828, the British government made tax demands and gave receipts in the form of tally sticks. This is actually how they did it all the way up until the 1800s. In 1834, after this system was replaced, the government decided to burn the enormous number of sticks that were kept in the vaults of the Houses of Parliament. You can just imagine—after all these generations—there were just enormous piles of these wooden sticks that were no longer of value. They set these on fire and, in fact, generated an enormous inferno, so large that the buildings themselves, the Houses of Parliament, burned down. The great artist William Turner painted several paintings called *The Burning of the Houses of the Lords and Commons, 16th October, 1834*. Turner himself, in fact, witnessed this inferno from a boat on the Thames River. So you can see—if you're going to try to mess with numbers and get rid of them, you are in for some serious havoc.

Another early method of counting that was used was the notion, the idea, and the actual action of knotted strings. In order to make the one-to-one correspondence physical, the Incas, as well as cultures along the Pacific Rim and in Africa, used knotted strings. The idea was that there would be a string, and they would put a knot in it when they wanted to represent one unit. The more knots they put in would designate the more elements they'd have in that collection they're trying to keep track of. Sometimes these became very intricate and quite beautiful. They're now in museums. In the 5th century B.C.E., Herodotus of Greece wrote, in his book called *History*, that Darius, the King of Persia, used knotted cords as calendars. So, we can see evidence even in the literature that these knotted strings were really used as a value before we had a notion of the numbers themselves. It was, again, this one-to-one correspondence, this one-to-one pairing.

This tradition actually continues to this day with the knotted strings. Catholics', Muslims', and Buddhists' rosaries and prayer beads actually allow the devout to recite the appropriate number of litanies without the need for counting. So you see here—for example, with this rosary—that I would actually use my thumb, and that would be the way for me to keep track of the litanies without actually counting them. This is, in some sense, a harbinger of a one-to-one pairing or a one-to-one correspondence.

Let's return now to our Sumerian shepherd and his flock. As early as 3200 B.C.E., he may have kept track of his sheep by having them pass through a narrow passage while he paired them up with a collection of pebbles. So the idea here is that as each sheep would go by, he would take a pebble and put it in a container. All the sheep would be out to pasture, and as the sheep were to return, he would remove, pebble by pebble, one for each sheep. Again, it's a one-to-one pairing, or a one-to-one correspondence: One pebble represented one sheep. So at the end of the day, if there were pebbles left over, well, the shepherd knew he had to go out and try to find the missing sheep. But if they worked out just perfectly and there were no pebbles left over, then he knew he had his entire flock.

In the 8th century B.C.E., Homer wrote of how Odysseus blinded the Cyclops, who then kept track of his sheep with a pile of pebbles he kept at the entrance to his cave. So even in the classic, ancient literature, we see mentions of this method of using pebbles. In fact, if we fast-forward just a little bit, we can even find a poetic joy within the notion of the one-to-one correspondence.

The 19th-century Austrian poet Adalbert Stifter used a one-to-one pairing of apples to represent the days before he was reunited with his great love. Now, he's a poet, so I don't know why he had to go off and leave his great love, but for some reason he did. What he did was he filled a sack of apples with the exact number of days that he was going to be away, and every day he would eat one of these apples. The correspondence here was that he would eat an apple for the day. Therefore, when there were no apples left, that meant he would return back. He wrote to his beloved a wonderful note where he talks about the numbers of apples left. He wrote the following:

> When I wrote my last letter to you there were only 21—tomorrow there will be only 13. Finally, only one apple will be left, and when I have eaten that, I will shout for joy.

Here we can really see the joyful reflection of a one-to-one correspondence as well as that it's a very powerful way to count without numbers. Of course, we carry with us the most convenient device for one-to-one pairings—namely, our bodies. The human hand is a natural counting tool, and we use it all the time to this day. In fact, the limit of four that I mentioned earlier in the lecture made the five-digit hand particularly useful for counting and led to the use of five as a basic grouping for counting. In fact, the popularity of the barred gate tally system (which we continue to use today) that I mentioned before—where we notch one, two, three, four, and then put a gate across it for the five—actually may be due to this basis of counting by fives. Now, of course, the 10 digits on our two hands may have led to the modern dominance of the base-10 numerical system. This is actually what we use today. We'll describe and explore the base-10 system and other such systems in greater detail in the following two lectures, so let's not worry about it now, except that it is based in a 10 system, which seems very natural to correspond to the 10 digits that we have.

Well, what if we want to count more objects? Well, no problem at all, because we can just add in our toes, and in fact this is what people do. So toes are the obvious extension of the hand as a counting tool. But, given the hand's convenience, some cultures actually extended their counting to include the joints of the fingers. You could actually count quite high with the hands. Let me just show you how they would do this. So they'd say, "1, 2, 3, 4, 5," and then they would count at the joints—"6, 7, 8, 9, 10, 11, 12, 13, 14, 15,"—and then even here, the knuckle—"16, 17, 18, 19, 20." So you could actually count to 20 on one hand, and this is actually a device that was used to keep track of larger numbers without the need for using one's toes.

Well, now let's see how the Sumerians took the notion of a one-to-one pairing one step further. So here is the first point where we're going to see a movement in our ability to abstract and to move the notion of number forward. So it's a very exciting moment for us, because soon what we'll see is a departure from the one-to-one pairing into something totally new. Well, let's now go back to the Sumerians. As early as 4000 B.C.E., Sumerians actually created a different clay token, called *calculi*, to represent quantities of different items. Different shapes actually represented different and various things, for example, various grain measures, a unit of labor, a sheep, a unit of oil, a garment, a unit of honey, and so forth. These images, these tokens, were actually used to depict these objects and thus allowed for the ease of transactions or the ease of recordkeeping.

The word *calculi* means "pebbles." From this root, we actually get the words "calculate," "calculus," and "calculator." If you think about it for a second, it's absolutely fantastic that these fundamental words we use for computation—"calculate," "calculus," "calculator"—all come from *calculi*, which means "pebble," which was the very first attempt to try to quantify, humankind's first effort to articulate a quantity through this one-to-one pairing and these differently shaped tokens.

To record a quantity of goods, such as measures of grain in a storehouse, Sumerians, as early as 3200 B.C.E., would do something quite spectacular. Here's what they would do. You see, if we just take these tokens, or these clay pieces, then what we'd see is—suppose that we were going to make a transaction right now. We agree on a certain amount of barley that I'm going to purchase from you. We do this with these tokens. So what would happen now? What would happen is that I would have these tokens to represent the amount that we agreed upon. Well, if I kept the tokens, if I kept these pebbles, these *calculi*, then I might, when you're not looking, add some more and thus say, "Well, you owe me this much barley; this was the deal we made," and this is not so good. If you were going to hold it, not to accuse anyone here of being dishonest, but one could imagine that maybe that person would take some pebbles away and say, well, we only agreed upon this much barley. That's how much grain that we agreed upon.

So you can see a problem with this. So the way they resolved this problem was to take some soft clay, and they would take the clay and form it into a jar, and this was called a pebble jar. So it was shaped into a jar and then they would take the tokens that we agreed upon, these actual *calculi*, and they would put the *calculi* in the jar, and then they'd seal it up. So it would look like a sphere—like a ball, a clay ball, and that clay would then harden. That way there was no doubt about what the deal was because we would just, in fact, know that we couldn't tamper with it because we've hardened clay that's just completely sealed up like a vault—an early, early instance of a vault.

Well, the problem with this is clear. The problem is that if, in fact, we want to now see exactly what we agreed upon, we now have to crack open this hardened clay jar and then look at the pebbles, look at the *calculi* inside, and then we have to reseal the jar, make a new jar, and so forth. So you could see this was a nuisance; this was, in fact, a big problem. Well, here was the ingenious moment where something dramatic happened within the imagination of humankind. Because what was realized was that we could avoid the breaking up of this clay jar, the pebble jar, by simply taking the tokens, taking the *calculi*, and before we put them into this softened jar, let's just take the *calculi* and make an imprint on the side, on the outside, of this jar and then throw it inside. Take the next *calculi*, the next pebble, make the imprint, and throw it inside, and do this for every single pebble that we have. So when we're done, we'll have all the pebbles safely inside the jar. We'll seal the jar up, and it will harden. There's no need for us to actually break the jar open to see what's inside because, in fact, these tokens actually have their impressions on the outside of the jar. Soon, the impressions themselves actually became the unit of measure. There was no need to actually look inside the jar anymore.

If you think about this, this was a dramatic moment in our understanding, because now an icon was used in order to denote a quantity. There's an image now, an abstract image, that's associated with a particular quantity that was pressed into this clay container. Well, the token—in fact, the markings on the side—represented the quantities for themselves. In fact, when you think about it, it actually then eliminated the need for the tokens being put inside. So in fact, they don't even have to put the tokens or the pebbles inside. All they have to do is press the tokens on the outside—that is all that was needed in order to keep track of this. Well this really is a brilliant idea because soon what they realized was, in fact, we don't even need the jar. All we need to do is to take a piece of softened clay, and lay it out flat, and then impress inside the clay the images of these tokens. That will represent the quantity of a transaction, or the quantity of units of grain in storage, or whatever quantity we want. We have an abstract visualization of that quantity, rather than actually counting the stones themselves and putting them in a one-to-one pairing with the quantity itself.

This was a dramatic moment in the life of the mind, for us, because we now move to an abstract symbol that represented the idea and the notion of a quantity. If you think about it, so much of our culture is based on our desire to understand and record numbers. In fact, you can think of the phrases that we saw, like "stockholder" and "bank stock," which now really have some serious significance to us. We understand that they really correspond to this one-to-one pairing idea that was early civilization's answer to counting before we even had numbers. But now we see a movement forward, a step beyond, where in fact now an abstract image—an icon, if you will—could be used to replace the one-to-one pairing.

In the next lecture, we'll see how the Sumerian markings led to one of the first numerical systems and, consequently, to what we may in fact consider the very first form of written language—amazing! In fact, the written language may have its origins in our desire to express quantity. Numbers are even behind our writing. We'll see that in the next lecture.

Speaking the Language of Numbers
Lecture 3

Numbers arose out of utility. However, once the notion of number came alive in our imagination, it took on a life of its own, and this gave birth to number theory.

Number systems evolved at different times throughout the world. Around 3500 B.C.E., Sumerian round jars evolved into flat tablets with symbols known as pictographs, which might be the moment when numbers were first viewed as nouns rather than simply as adjectives. These pictographs, created by using a reed-stem stylus to mark in clay, were developed by the Sumerians and Babylonians (collectively known as Mesopotamians). The reed-stem stylus gave way to a pointed stylus for creating more delicate shapes. This style of creating number symbols evolved into cuneiform and may be the first instance of writing in human history. The Babylonians began with a system that used symbols for powers of 60 (1, 60, 3600 ...), 60 being the smallest number that can be divided evenly by 2, 3, 4, 5, and 6. During the 2nd millennium B.C.E., the Babylonians developed a positional system; by 300 B.C.E., they had a symbol for zero as a placeholder.

Mesopotamians invented the stylus to create pictographs to count, which evolved into cuneiform.

The Egyptians also employed pictographs for numbers as early as 3500 B.C.E. With the invention of papyrus, the Egyptians were able to record more complex bookkeeping methods and calculations. The famous Rhind Papyrus (c. 1650 B.C.E.) contains many calculations and includes hieroglyphs for addition and subtraction. The Chinese used a system involving powers of 10 as early as 1400 B.C.E., with symbols for 1 through 10, in addition to 100 and 1000. Few records survived the humid Chinese climate, making the culture's history of number development and use more uncertain. The ancient Chinese used the words for large numbers in a poetic manner to create new words.

The Mayans of Central America had a numeral system by the 3rd century C.E. Their system involved powers of 20 and used symbols that consisted of dots and lines. Dots and dashes still remain with us today—consider the symbol for division: ÷. The Mayan system, though compact, was still cumbersome for calculation. The Mayans had a symbol for zero but used it solely as a placeholder.

Numbers arose out of utility; thus, some naming conventions are limited. Primitive cultures may function very well with words only for *one*, *two*, and *many*. According to 19th-century anthropologists, the Botocudos from Brazil would count up to four and express *many* by pointing to their hair, implying that, beyond four, things were as countless as the hairs on one's head. For the Lengua people of Paraguay, words for numbers reflect body counting. The word for *ten* literally means "finished both hands." The word for *fifteen* means "finished foot."

The famous Rhind Papyrus (c. 1650 B.C.E.) contains many calculations and includes hieroglyphs for addition and subtraction.

In Indonesia, *lima* means "five" even though it literally means "hand." In some languages, the word for *whole person* also means "twenty." The Spanish word *tres* ("three") and the French word *très* ("very much") derive from the Latin word *trēs*, which means, among other things, "beyond." The word *three* in English is derived from the Anglo-Saxon *thria*, which is related to the word *throp*, meaning "pile" or "heap."

In the days of Romulus (c. 750 B.C.E.), the Roman calendar had 10 months, starting with March. Months 5 through 10 were named Quintilis, Sextilis, September, October, November, and December, designating their position in the year. January and February were added later, and Quintilis and Sextilis were renamed Julius and Augustus in honor of two Caesars.

Number theory was born in the 4th century B.C.E. At around this time, members of a religious community in India called the Jains were exploring mathematics with a new, sophisticated perspective. The Jains classified numbers into different categories, including at least four different types of infinity. In some sense, the Jains were the first number theorists—that is, the first group of individuals to study number as an independent, abstract object of interest.

Pythagoras explored the abstract notion of number in great depth in the 6th century B.C.E. Pythagoras might have been inspired by the work of the Jains. The Pythagoreans were among the first to consider numbers as abstract objects, rather than tools, and the first to prove theorems involving numbers. This period truly represents the birth of the life of numbers.

When did negative numbers enter the scene? In the 1st century B.C.E., the Chinese solved systems of equations simultaneously using negative numbers. There is evidence of calculation with negative numbers in India between 200 B.C.E. and 200 C.E. In the 3rd century C.E., the Greek mathematician Diophantus described an equation equivalent to $4x + 20 = 0$ as "absurd" because its solution is a negative number ($x = -5$). In 7th-century India, negative numbers were used to represent debts. The great Indian astronomer Brahmagupta (598–c. 665 C.E.) was the first to offer a systematic treatment of negative numbers. As late as the 18th century, some important mathematicians discarded negative solutions because they viewed them as unrealistic. Today, we define *integers* as the collection of all natural numbers, their negatives, and zero (e.g., ... $-3, -2, -1, 0, 1, 2, 3$ ...).

Illustrations of ancient arithmetic abound. One of the oldest recorded division calculations dates from 2650 B.C.E. in Sumeria. The challenge was: A granary has a given amount of barley to be distributed so that each man gets 7 *sila* (a measurement unit) of barley. Find the number of men. Historians speculate that this calculation was performed using *calculi*. To multiply, the Egyptians employed a technique called doubling. They also performed calculations involving fractions with numerator 1 (unit fractions). This convention limited their understanding of more complicated numbers. The Babylonians computed square roots as early as 1800 B.C.E. They were able to produce a very good approximation to $\sqrt{2}$. They understood the Pythagorean Theorem by 1600 B.C.E., more than 1000 years before Pythagoras.

What is a calculator? Before the invention of modern calculation devices, the word "calculator" referred to an individual who performed calculations. Many ancient cultures performed calculations on a sand table. Pebbles and other markers were used in columns representing different numbers that were then moved to depict addition or subtraction. Another possible origin for the zero symbol (0) might be the round dimple left in the sand when a pebble was removed, leaving an empty column. The classic Chinese abacus came into use in China during the 14th century C.E. Long before that time, as early as 450 B.C.E., the Chinese, as well as the Egyptians, Romans, Greeks, and Indians, used a simpler calculating board, similar to sand tables. The ancient Greek historian Herodotus (c. 484–c. 425 B.C.E.) is credited with the observation that "the Egyptians move their hands from right to left in calculations while the Greeks from left to right." The abacus allows for fast calculation in an additive numeral system and is still in use today.

Modern Hindu-Arabic numerals arose from a mysterious mixing of traditions from two cultures. The origin of numerals can be traced back to India and evolved from 400 B.C.E. to 400 C.E. They grew out of Brahmi numerals. These symbols were introduced to European nations through the writing of Arab and Persian mathematicians and scientists. Although we have a clear understanding of the origins of some numerical symbols (e.g., 1, 2, 3), historians disagree on the origins of other numerals. Writing these numerals became a form of art, as seen in the work of Albrecht Dürer. The 10 Hindu-Arabic numerals form an efficient basis for our current decimal system. ∎

Questions to Consider

1. Suppose you have an additive numeral system in which 1, 5, and 25 are denoted by |, @, and %, respectively. What numbers do the following represent?

%%@@@|, %@||||, %%%|, @@@@

2. Create a context in which negative numbers would be absurd.

Speaking the Language of Numbers
Lecture 3—Transcript

In the previous lecture, we discovered that our number sense began with the idea of a one-to-one correspondence. Here, in this lecture, we'll examine the words and symbols that early cultures employed to represent numbers as abstract quantities. Many advanced societies, including the Sumerians, the Egyptians, the Mayans, and the Chinese, developed sophisticated numerical systems. The Babylonians used counting tokens that led to one of the earliest forms of writing, called "cuneiform." While each of these different systems offered a means of keeping records, very few provided an efficient means of calculation. The desire to calculate led to the development of arithmetic aids such as counting boards, the precursors to the abacus. Here we'll consider how these important arithmetical tools may have laid the groundwork for a place-based number system that we use today.

Now, where did our Arabic numerals—the 10 symbols "0," "1," "2," "3," all the way out to "9"—come from? The surprising answer is India. But what about cultures whose primitive number language remained primitive? Here, we'll see connections between our modern world and societies whose number language never progressed beyond "one," "two," "many." Even early societies artfully employed numbers to create sophisticated calendars, keep track of agricultural production, and record commercial transactions. But they still struggled with a complex arithmetic idea, especially calculations that involved fractions. The age-old struggle to perform complex arithmetic ran parallel with the age-old struggle to devise an appropriate means to represent numbers.

Well, let's begin our study of early number systems in Mesopotamia. We recall from the previous lecture the pebble jar used by the Sumerians as a means of counting was one where tokens were used to later stamp markings on the jar itself. Then, the tokens were replaced by the stamped images in the soft clay. Around 3500 B.C.E, the round jars evolved into flat tablets with symbols, known as pictographs, embedded in these tablets. Here might be the moment when numbers were first viewed as nouns rather than as adjectives. Even fairly recently, different words were used for the same number of different objects. There's a wonderful quote by the 20th-century British philosopher and logician Bertrand Russell. He wrote:

It must have required many ages to discover that a brace of pheasants and a couple of days were in both instances the number two.

Again, we're moving from the actual counting of things to an actual noun, rather than an attribute. These pictographs created by a reed-stem stylus in clay were developed by the Sumerians and Babylonians, collectively known as Mesopotamians. I want to show you how this actually worked. So, for example, here we see a clay tablet, and they would use a reed stylus similar to this, and they would just make an impression in the clay. So, for example, they would just do this, or this, or this, and these circular impressions you can see. Well, these actually represented the units 10. So, each of these were 10. Then notice something really interesting—if they took the stylus and, instead of imprinting straight in to get a circular image, they were to put it at an angle, a little bit of skew, then in fact they would see an image that looked like this, or this, or this. These askew images actually stood for the number 1. So in fact, this right here would represent—well, 10, 20, 30, 31, 32, 33—this would represent 33, perhaps 33 units of oil, for example. So this is how the symbol is used, and you can see that with one instrument, they were able to produce two different styles of writing, and they each took on a value.

Well, soon what they did was they actually sharpened the reed stylus, and it gave way to a pointed stylus, and therefore we were able to create more delicate shapes. In fact, these number symbols evolved into what is called "cuneiform" and may have been the first instance of writing in human history—all coming from our desire to articulate and express numbers. If we journey to Babylonia, we find that the Babylonians developed a system that used symbols for powers of 60—so 1, then 60, then 3,600, and so forth. So, what's so special about 60? Why were they using 60? Well, it happens to be the smallest number that can be divided evenly by 2 (you can cut it in half), by 3, by 4, by 5, and by 6. So it's a very convenient number if you're going to chop it up into equal pieces. But historians aren't really sure why they actually used 60, although this is a property about the number 60.

Well, in cuneiform we see two different symbols: the so-called "nail," which represented the number 1, and the "dovetail," which represented the number 10. So you can see that the circle was replaced by the more interesting and artistic-looking dovetail, which stood for 10. Well, their number system was both additive, like the Roman numerals—we add up the symbols we see—and also positional, like the decimal that we use today. I wanted to illustrate both of these qualities for you with a couple of quick examples. So in this example here, we see two dovetails and three nails. So we add, like we would with Roman numerals. We see two dovetails—that means 10 and 10 are 20. Three nails would be another three, so we add and we get 23, just as we would with Roman numerals.

In this example, here is a very large number. We see one, two, three, four, five—five dovetails. That's then 50 when we add them all up. We see nine nails. So that's nine, so that's 50—59. But also there was a positional quality in cuneiform, which is very similar to the way we write numbers now. We've got a unit spot and then a 10 spot, and then a 100 spot, and so forth. So in this case, we have to remember that they were using a base of 60, which means that when we moved over to a spot, it wasn't the 10s spot, but it was actually the 60s spot—a little confusing. But let's take a look at this example. Here I see two nails in the 60s spot, so I have two 60s, which would be 120. Then I have one nail in the units spot, and so this number represents 121. With this number written in cuneiform, we see that we have 10 (a dovetail) in the 60s spot, and so that would be 600. Then in the unit spot, I see a 10 and a 1, which is an 11, so this represents the number 611. You'll notice it's really important that we have the appropriate spacing. Spacing here is crucial. In fact, there's actually some ambiguity since there was no symbol for zero; zero wasn't even a number yet.

We have an ambiguity here. Let me show you what I mean with an example. If we take a look at this, what do we see? We see two nails that are spread apart. Well, are they spread apart so that, in fact, that first nail is in the 60s spot (in which case the number would be 60 + 1, or 61), or is that nail actually further over and, in fact, in the 3,600s spot (in which case this number would be 3,601)? So you can see that there is definite ambiguity here without the utility of our zero. While this positional system was developed and put into effect by the 2nd millennium B.C.E., it wasn't until about 300 B.C.E. that they actually had a symbol for zero, but it was just as a placeholder. In fact, it was not considered a number, and we'll talk more about the positional systems and the importance of zero in the next lecture. So we'll hold that off for now.

If we now move to ancient Egypt, we see that the Egyptians also employed pictographs for numbers as early as 3500 B.C.E. But with the invention of papyrus, which is an early form of paper made from the papyrus plant, the Egyptians were able to record more complex bookkeeping and calculations. The famous Rhind Papyrus from around 1650 B.C.E. contains many calculations and includes hieroglyphics for addition and subtraction. This important document is our principle source of information about Egyptians' understanding of numbers. This is really a significant document. The author is conjectured to be Imhotep, an architect and builder of the Sakkara pyramid. The scribe of the Rhind Papyrus, the person who actually put pen to papyrus and wrote it down, was named Ahmes. Now, the work opens in a very dramatic style, which gives you an indication of the import and the magnitude of such a document. So in fact, the Rhind Papyrus opens with the following:

> Accurate reckoning. The entrance into the knowledge of all existing things and obscure secrets. … It is the scribe Ahmes who copies this writing.

So, very dramatic, very important—and he actually includes his name in this document, which just shows you that he was really very proud of this work. In fact, scribes of the time were also considered teachers. They were the equivalent of teachers, in fact, back then. There's a wonderful ancient quote about scribes as teachers that I wanted to share with you now:

> As for the learned scribes … they did not make for themselves pyramids of copper with tombstones of iron. They were unable to leave an heir in the form of children who would pronounce their name, but they made for themselves an heir of the writing and instruction they made.

I mean the truth is, writing is our true link to our past, and indeed a great monument, and perhaps even more significant than the pyramids. I mean, really—I'm an educator, so I'm biased toward this—but education really endures. We'll actually see the impact and the influence of the Rhind Papyrus and its scribe, Ahmes, and that they had on the world of numbers throughout the entire course. So we will continually return to Ahmes and the Rhine Papyrus.

If we now move to ancient China, we discover that the Chinese appeared to have had a number system involving powers of 10 since as early as 1400 B.C.E., with symbols for 1 through 10, in addition to having symbols for 100 and 1,000. Unlike the clay or papyrus that we've seen in the dry Middle East, much less survived the humid Chinese climate, making their historical record of number development and their use much, much sketchier. However, there is one thing that we have evidence of, and that is that the ancient Chinese used the words for large numbers in a poetic manner in order to create new words. They would make compound words using the words for large numbers, and I want to give you some examples of these, which I think are really quite beautiful. For example, they would take the number 100 and then put it together with "thing." So "100 thing" meant "everything." You see how the large number was actually depicting this "everything," this "all things." "One hundred times think," that meant prime minister, right, someone of great wisdom. "One hundred worker" meant "working class." "One thousand direction" meant "versatile," which I think is maybe my favorite thing. It really captures the image. "One thousand direction" really is "versatile"—you're all over. A "100,000 month" meant "antiquity." So again—a very poetic way of using the words of numbers to capture what we mean.

If we move to Central America, we find that the Mayans had a numerical system by the 3rd century C.E. Their system involved powers of 20 and used symbols involving dots and lines. Dots and dashes still remain with us today if you consider the symbol that we use for noting division. If you take a look at these symbols, what you'll actually notice here is certainly this limit of four, because we get to four, and then the fifth thing—you see the symbol actually changes. The dots and the lines begin to come together. Again, we see reflections of the barred gate method of counting that we actually use today. Their system was certainly more compact than some of the others we've seen, but it still was cumbersome for calculation. So, in this direction we haven't made an awful lot of progress. They did have a symbol for zero but used it solely as a placeholder and did not view it as a number.

Well, now, in all these and the other number systems, it's important to note that the utility dictates the level of complexity and the naming conventions of numbers. Primitive cultures may function very well with words only for *one*, *two*, and *many*. They don't have billion-dollar gross national products to record. In their world, they only need to be able to count up to, say, two. In fact, according to a 19th-century anthropologist, the Botocudos from Brazil would count up to four and then express *many* by pointing to their hair. The implication here—what they were saying was, "Beyond four, things are as countless as the hairs on my head." Again, we see this limit of four as a recurring theme. Moving from the hairs on our body to looking at our entire bodies, we consider how the body has influenced the development of the language of number.

For the Lengua people of Paraguay, words for numbers reflected body counting. The word for "ten" literally means "finished both hands," which is a great image—we've counted through both hands. I love this one: the word for "15" means "finished foot." So we've got the two hands and one finished foot, which gives us 15. In Indonesia, *lima* means "five," even though it literally means "hand." So you can see the reflections of our body within the notion of number. In some languages, the word for *whole person* is the same thing as "twenty." As we move toward language, the language of numbers used today, we discover that there are many echoes of early references. The Spanish word *tres*, for "three," and the French word *très*, for "very much," derive from the Latin word *trēs*, which means, among other things, "beyond." So the idea is you have one, two, and then beyond; that's where we see "very much" or "three." The word *three* in English is derived from the Anglo-Saxon *thria*, which is related to the word *throp*, which means "a pile" or "a heaping." We'll see more about these piles in the context of numbers a little bit later.

In the days of Romulus, around 750 B.C.E., the Roman calendar had 10 months, starting with March. The last 5 months of the year, the months 5 through 10, were named in order to describe their distinction in the position of the year. So they were called Quintilis (5), Sextilis (6), and then maybe some that you've heard of—September for 7, October for 8, November for the 9, and December for the 10. It's amazing that September, October, November, and December have within them the reflections of numbers, at least as they were positioned in the calendar year back then. January and February were added later, and Quintilis and Sextilis were renamed Julius and Augustus in honor of the two famous, or infamous, Cesars.

Numbers arose out of utility. However, once the notion of number came alive in our imagination, it took on a life of its own, and this gave birth to number theory. That birth took place in India. By the 6th century B.C.E., members of a religious community in India called the Jains, or the Jaina, were exploring mathematics with a new and sophisticated perspective. The Jains classified numbers into different categories, including at least four different types of infinity. Well, we'll delve into infinity at the beginning of Lecture Twenty, toward the end of the course. Many argue that, in some sense, the Jains were the first number theorists—that is, the first group of individuals to study number as an independent, abstract object of interest. The Jains were really a religious sect, and thus they viewed number and mathematics more as a means toward their religious studies than an end, which is really a shame for the math community, because with their insights, their contributions could have been even more dramatic than they actually were.

During the same period, there was an interesting character trying to find his way in life as well as finding his way in numbers. This character was the influential Pythagoras, who was born in Samos, off the Turkish coast, in 569 B.C.E., and traveled extensively, including to Egypt and to Mesopotamia, before finally settling in the south of what we now know as Italy—so, a quite beautiful part of the world. Now, Pythagoras led the ancient Greek movement in the study of the abstract notion of number. In fact, some believe that Pythagoras, as well as Plato, might have been inspired by the work of the Jains. So again, we can see cultures coming together within the context of number. The Pythagoreans were among the first to consider numbers as abstract objects, rather than as tools, and to prove theorems involving numbers. We'll study the Pythagorean brotherhood and their seminal work in greater detail in Lectures Five and Eight.

This period truly represents the birth of the life of numbers. Up to this point, "number" meant, by the way, the natural numbers, that is, the collection of numbers one, two, three, four, and so forth. So, when did the negative numbers enter the scene? Again, the development came at different moments in history in different cultures.

In the 1st century B.C.E., the Chinese solved systems of equations simultaneously using negative numbers. There's evidence of calculation with negative numbers in India between 200 B.C.E. and 200 C.E. In the 3rd century C.E., the great Greek mathematician Diophantus described an equation, equivalent to an expression, and the answer turned out to be −5. He called this equation absurd, because the answer was a negative number, which to him made no sense. This equation—I can actually say—in modern language it would look like "$4x + 20 = 0$," so here's the first equation we see, but don't worry about it. The point is, there's an unknown value for x, and we have to figure out what value of x would make that equation be true. Well, the way we could actually solve this is to subtract 20 from both sides, and so that would then give us $4x = −20$, and then we could simply just divide both sides by 4 to undo the multiplication by 4 and the x, and I see that $x = −5$. Therefore, since this wasn't a number in the mind of Diophantus, he called this absurd.

In 7th-century India, negative numbers were used to represent debt, so you could really see that was a sophisticated point of view. In fact, the great astronomer Bhramagupta was the first to offer a systematic treatment of negative numbers during this period. Well, while negative numbers have been around for over 2,000 years, as late as the 18th century some important mathematicians discarded negative solutions to equations because they viewed them as unrealistic. Today, we define the collection of "integers." That's the word we're going to use to describe the collection of all the natural numbers, together with their negatives and with zero. So the integers consist of 0, 1, 2, 3, 4, and so on, together with −1, −2, −3, and so on. That's the collection we're going to call the integers. It's interesting to consider that just 200 years ago or so, mathematicians dismissed negative numbers as unrealistic. Today we consider these numbers, of course, very real and almost take them for granted. Here again, we see the challenge of expanding humankind's notion of number, a recurring theme throughout our course.

Given that we mentioned solving equations, I thought it would be fun to take a minute to enjoy some moments in ancient arithmetic from a variety of different cultures. So, from Sumerians, let's consider some early division. One of the oldest recorded division calculations dates from 2650 B.C.E. in Sumeria, and here was the challenge. You have a granary with a given amount of barley that is to be distributed so that each man gets seven *sila* of barley, which was a unit of measurement. The question was how to determine how many men there were. In fact, it's interesting because historians actually conjecture that this division calculation was performed using the *calculi*, the tokens that we mentioned in the previous lecture.

Well, let's now look at the early Egyptians, and we'll consider multiplication. They had a brilliant way of multiplying numbers, and this technique was called "doubling." I want to give you an example. Let's actually multiply 13 by 21, and their method was to continually double the 21. So we start with 1 × 21, which is 21. Then we double—2 × 21 is 42. Then we double again—4 × 21 is 84. Then we double yet again, and we see 8 × 21 is 168. If we were to double again, we'd have 16 × 21, but remember we're trying to multiply 13 × 21, so that would exceed the value we're looking for. So we stop right here at 8 × 21, and then we take the number 13, which is the real thing we multiply by 21, and we write 13 as the sum of these doubled numbers. So, in fact, how could we write 13? There's only one way of writing 13 in terms of 1, 2, 4, and 8, and it turns out it's 8 + 4 + 1. That equals 13. So, what's 13 × 21? It's 8 + 4 + 1, all times 21, which we see is, from our chart, 168 + 84 + 21. We see that the product is 273. Their method really foreshadows the important modern notion of binary expansions that drives our technological age and actually runs our computers. We'll describe this idea in the next lecture.

They also performed calculations involving fractions, but only those with the numerator 1 (only 1 in the top of the fraction). These are called unit fractions. This convention actually limited their understanding of more complicated numbers. If we now hop to the Babylonians, we could actually see the occurrence of an early square root, so the Babylonians were computing square roots as early as 1800 B.C.E. They were able to actually calculate a very good approximation to the number $\sqrt{2}$. In fact, if you look at this image, you see a square, and notice that we see three dovetails. Those three dovetails, if you remember, means we have 30, and so this actually meant the square had size 30. They need to actually calculate $\sqrt{2}$. If you notice in the image here, we see the numbers 1, 24, 51, and 10, which actually gives the expansion for $\sqrt{2}$. But remember the Babylonians used this funny base-60 expansion, and so this would really be read today as $1 + \frac{24}{60} + \frac{51}{60^2} + \frac{10}{60^3}$. If you would actually add that up on a calculator, you would see the number 1.414213, which, notice, actually agrees to the precise value of the number $\sqrt{2}$, as you can see, to six decimal places, which is really an amazing approximation without a calculator.

In fact, what is a calculator? The question really is, *who* is a calculator? Before the onset of modern calculation devices, the word "calculator" referred to an individual who would perform calculations. In terms of computing devices, many ancient cultures performed calculations on sand tables, and this was really the earliest form of the abacus. Pebbles were used inside the sand in various columns to mark the ones and the tens, and so forth, and moved the pebbles back and forth to depict the addition or the subtraction. Perhaps this is the inspiration for our "placed-based" decimal system. By the way, it's another possible origin for the zero symbol, because when we take out a pebble, it leaves an indentation—a round dimple, if you will—in the sand and that, of course, means we have nothing there, and perhaps that's where we get zero.

Well, the sand table led to the abacus. The classic Chinese abacus came to use in China during the 14th century C.E. However, long before that time, the Chinese (as well as the Egyptians, Romans, Greeks, and Indians) used simpler calculating boards, similar to the sand tables, as early as 450 B.C.E. The use of these abaci varied from culture to culture. In fact, the 5th-century B.C.E. Greek historian Herodotus is credited with the observation that "The Egyptians move their hands from right to left in calculations, while the Greeks from left to right." Well, the abacus allows a very fast way of calculating things, and we still use it today.

I want to close this lecture with the evolution of the symbols that we use for numbers today. Our modern 10 Hindu-Arabic numerals, the symbols 0 through 9, arose from a mysterious mixing of traditions from two cultures. The origins of the numerals themselves can be traced back to India and evolved from 400 B.C.E. to 400 C.E., and they grew out of the Brahmi numerals. These symbols were introduced to European nations through the writing of Arab and Persian mathematicians and scientists. Today, we refer to these symbols, these digits, as the Hindu-Arabic numerals. Where did these origins come from? Well, one, we understand, is 1. Two—if you look back to the Chinese, for example, we see those two lines. If you connect those two lines very quickly, you get a 2. The three hashmarks—if you connect them, you actually get a 3. But historians actually disagree on the origins of the remaining numerals. So those still remain a mystery, those actual symbols.

Writing these numerals actually became a form of art as we see in the work of a variety of artists. The 10 Hindu-Arabic numerals actually form the heart of our current decimal system, and that system is one that finally allows humankind to have an efficient means of performing calculations. Now we'll describe this powerful decimal system, an example of what is referred to as a "place-based," or "positional," system in our next lecture. But for now, let's just enjoy the evolution of the abstract idea of number.

The Dramatic Digits—The Power of Zero
Lecture 4

Have you ever tried to multiply using Roman numerals? How would you write the number 10,030 without using zero? A compact, place-based (or positional) number system with a symbol for zero opens the floodgates for arithmetic calculations and the discovery of new numbers.

Ancient numeral systems were mostly additive. Most of the systems we studied in the previous lecture required the repetition of symbols (e.g., XXIII for 23 in Roman numerals). Although computation with additive systems was fast, thanks to such tools as the abacus, those systems required very long lists of symbols to denote larger numbers. Additive systems made it difficult to look at more arithmetically complicated questions, thus slowing the progress of the study of numbers.

For millennia, zero did not count as a number.

For millennia, zero did not count as a number.
In the Rhind Papyrus from 1650 B.C.E., numbers were referred to as "heaps." This tradition continued with the Pythagoreans, who in the 6th century B.C.E. viewed numbers as "a combination or heaping of units." Aristotle defined number as an accumulation or "heap." Because we cannot have a "heap" of zero objects, zero was not viewed as a number. The lack of zero resulted in many challenges. A careless Sumerian scribe could cause ambiguities: In cuneiform, different spacing between symbols led to different numbers. The Egyptian system did not require a placeholder such as zero, but the Egyptians' additive notation was cumbersome; in 2000 years, little progress was made in arithmetic or mathematics.

Though the Babylonians and the Mayans had symbols for zero, they considered it a placeholder rather than a number. As noted earlier, the modern symbol 0 may have arisen from the use of sand tables for calculation. Calculations performed on sand tables may have led to a broader use of place-based number systems. Ptolemy later used the Greek letter o (omicron) to denote "nothing," although he did not view it as a number. In 7th-century India, the astronomer Brahmagupta understood zero as a number, not simply as something left from removing a counter in a sand table. He studied $\frac{0}{0}$ and $\frac{1}{0}$, deciding (erroneously) that $\frac{0}{0} = 0$; he did not know what to conclude about $\frac{1}{0}$.

Positional systems can use different bases. The base-10 or decimal system uses the 10 Hindu-Arabic numerals: 0, 1, 2, 3, 4, 5, 6, 7, 8, and 9 (called digits). The position of a digit indicates by which power of 10 that digit is to be multiplied. We will use exponential notation to denote powers of 10: $10 \times 10 = 10^2$, $10 \times 10 \times 10 = 10^3$, and so forth; $10^0 = 1$.

For example:

$$293 = (2 \times 100) + (9 \times 10) + (3 \times 1)$$

$$= (2 \times 10^2) + (9 \times 10^1) + (3 \times 10^0)$$

$$8105 = (8 \times 1000) + (1 \times 100) + (0 \times 10) + (5 \times 1)$$

$$= (8 \times 10^3) + (1 \times 10^2) + (0 \times 10^1) + (5 \times 10^0).$$

A specific mark, called a radix (or decimal) point, is used to separate the whole-number positions from the fractional positions. In some countries (including the United States), this mark is a period:

$$3.14 = (3 \times 1) + (1 \times \frac{1}{10}) + (4 \times \frac{1}{100})$$

$$= (3 \times 10^0) + (1 \times \frac{1}{10}^1) + (4 \times \frac{1}{10}^2)$$

In some other countries, the separator is a comma:

$$87,05 = (8 \times 10^1) + (7 \times 10^0) + (0 \times \frac{1}{10}^1) + (5 \times \frac{1}{10}^2).$$

The base-2 (or binary) system has only two digits: 0 and 1. In this system, the positions are valued as powers of 2 rather than powers of 10.

Consider this example:

$$1011_2 = (1 \times 8) + (0 \times 4) + (1 \times 2) + (1 \times 1)$$

$$= (1 \times 2^3) + (0 \times 2^2) + (1 \times 2^1) + (1 \times 2^0).$$

The number 1011_2 equals $8 + 2 + 1 = 11$ in base 10. The subscript 2 denotes that the number is expressed in base 2.

Fractional positions are analogous to those in base 10. For example:

$$101.11_2 = (1 \times 4) + (0 \times 2) + (1 \times 1) + (1 \times \frac{1}{2}) + (1 \times \frac{1}{4})$$

$$= (1 \times 2^2) + (0 \times 2^1) + (1 \times 2^0) + (1 \times \frac{1}{2}^1) + (1 \times \frac{1}{2}^2).$$

The base-2 number 101.11_2 equals $4 + 1 + \frac{1}{2} + \frac{1}{4} + 5\frac{3}{4}$ in base 10.

The base-3 (or ternary) system has only three digits: 0, 1, and 2. The positions are valued as powers of 3.

Consider this example:

$$2102_3 = (2 \times 3^3) + (1 \times 3^2) + (0 \times 3^1) + (2 \times 3^0).$$

The number 2102_3 equals $54 + 9 + 2 = 65$ in base 10.

Fractional positions are once again analogous to those in base 10. Consider:

$$12.202_3 = (1 \times 3^1) + (2 \times 3^0) + (2 \times \frac{1}{3}^1) + (0 \times \frac{1}{3}^2) + (2 \times \frac{1}{3}^3).$$

The base-3 number 12.202_3 equals $3 + 2 + \frac{2}{3} + \frac{2}{27} = 5\frac{20}{27}$ in base 10. Any whole number greater than 1 can be used as a base; given any base, any number can be expressed in that base. The Babylonians used base 60. Some computer languages use base 16, called hexidecimal.

From the 13th through the 16th centuries, Europeans disagreed about the advantages of a positional system versus an additive system. An additive system allowed speedy calculation using an abacus and did not require memorizing multiplication tables. Some worried that a positional system was more vulnerable to fraud because someone could radically change the value of a number by adding a single digit at one end or the other. The dominance of the positional system has proved enormously valuable, not just in the efficient writing of numbers but in the advancement of the theory of numbers and mathematics in general.

We find harbingers in ancient times of some modern positional systems, along with holdovers of some very old practices in our lives today. The Chinese *I Ching* from 2800 B.C.E. contains patterns of solid and broken lines (called trigrams and hexagrams) that correspond to binary numbers, but they were not commonly used for computation. In 4th-century-B.C.E. India, the poet and musician Pingala used a binary system to notate musical meters. The modern binary system using 0s and 1s was established by the great German mathematician Gottfried Leibniz in the 1660s. The first computer that used binary addition was built by mathematician George Stibitz, a scientist at Bell Labs, in 1937.

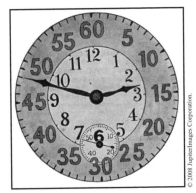

© 2008 JupiterImages Corporation.

The way we divide time is based on the Babylonian base-60 system.

Not all of our numeral notation is positional even today. We designate locations on the Earth's surface using latitude and longitude in degrees, minutes, and seconds. Our division of time also shows the influence of the Babylonian base-60 system. Stock prices traditionally are measured in eighths. The use of this base may result from the practice of hand signaling in the trading pit. ■

Questions to Consider

1. Why do you think it was difficult for so many cultures to consider zero a number?

2. Write the base-10 expansion for 2017. Write the base-2 expansion for 1101_2. Write the base-3 expansion for 212_3. How would the last two numbers be expressed in the shorthand base-10 notation?

The Dramatic Digits—The Power of Zero
Lecture 4—Transcript

In this lecture, we'll celebrate the dramatic impact the digits had on our understanding of number. Have you ever tried to multiply using Roman numerals? How would you write the number 10,030 without using zero? A compact, place-based (or positional) number system with a symbol for zero opens the floodgates for arithmetic calculations and the discovery of new numbers. With only ten symbols, we have the machinery to describe new numbers that grow beyond our imagination. Here, we'll explore the origins of zero and the development of our modern decimal system. With a powerful positional number system in place, humankind was finally equipped with the tools necessary to begin the development of modern mathematics.

Along with the commonly used decimal system, we'll consider the binary number system, a system that has roots in ancient China and India. Binary numbers are the basis for modern computers. Who would have imagined that the basic structure utilized by today's computers is the basis for the same structure used in the ancient *I Ching*? We close our discussion by contemplating other base systems including the base-60 system, employed by the Babylonians, and the base-3 system, known as the "ternary number system."

Well, I want to begin with the downside of the ancient additive systems. Well, most of the systems that we've studied in the previous lecture required the repetition of symbols. For example, the Roman numerals XXIII equal 23, and we'd add up the two Xs (10 each), and then the three Is, and get 23. In fact, also recall the dovetails and nails that the Babylonians used. We'd add them up. Although computation with the additive systems was fast using tools such as the abacus, those systems required very long list of symbols to denote larger and larger numbers, and this was a problem in practice.

Additive systems made it difficult to look at more arithmetically complicated questions and thus slowed the progress of the study of numbers. In order to move to what we call a positional system, we need a new number. This inspires a philosophical question. So, the question is: How many items do you see are in this box? Well, is your answer a number? This is the question. This is the question about zero. In the Rhind Papyrus from 1650 B.C.E., the scribe Ahmes referred to numbers as "heaps." This tradition actually continued through the Pythagoreans, who in the 6th century B.C.E. viewed numbers as "a combination or heaping of units." Pythagoras and the Pythagoreans will be highlighted in the next lecture, and their view of number will be explored further in Lecture Eight.

But here, even Aristotle defined number as an accumulation or "heap"—again we see the word "heap." Also, recall from the last lecture, the word "three" derived from the Anglo-Saxon word *throp*, again meaning "pile" or "heap." Well, because we can't have a "heap" of zero objects—with zero objects, there would be no heaping at all—zero was not viewed as a number. So this notion of having zero be a quantity didn't make any sense at all because we were thinking in terms of "heaps." Well, the lack of zero caused many challenges. Now, as we saw in the previous lecture, a careless Sumerian scribe could cause ambiguities since, in cuneiform, different spacing between symbols can actually represent different numbers. The Egyptian system, on the other hand, did not require a placeholder like zero, but their additive notation was cumbersome. Again, they had all the symbols together, and we had to add them all up. As a result, in the 2,000 years of the Egyptian numeral system, they made very little progress in arithmetic or, more generally, in mathematics. It's interesting to see how the notation really drives our understanding, and our intuition, and our further quest to consider number.

Well, zero first appeared as an empty placeholder rather than a number. As we noted in the previous lecture, the Babylonians had a symbol for zero by 300 B.C.E. It was a placeholder rather than a number because, again, they were thinking heapings, but they needed to distinguish between numbers. For example, if we just see these two nails spaced apart, like this, then we would understand that leftmost nail—now remember, the Babylonians used this peculiar base-60 system, so instead of being in the 10 spot like we would think today, that's actually located in the 60s spot. So that one nail is 60, and then the other nail produces the 1, and so we see 61. That is now different from this number where we actually see the symbol and icon for zero in between, which means that we have no 60s. But we have that nail located in the 3,600 spot. So we see 3,600 and 1 nail, or 3,601. So, you can see the power of this symbol to represent a placeholder.

The Mayans also had a symbol for zero that they also used only as a placeholder, this eye-shaped symbol. Now, the evolution of the symbol for zero is actually very difficult to chart. As we remarked in the previous lecture, the modern symbol "0" may have arisen from the use of those sand tables for calculations. Remember that we had these sand tables that were used to calculate things, where pebbles would be placed in and moved back and forth for addition or subtraction. What we'd see is when a pebble would be removed, we would have an indentation or a dimple in the sand, which reflects the "0" that we see today. In fact, calculations performed on the sand tables may have actually led to the development of the place-based number systems that we're about to describe and consider.

Later, in the 2nd century C.E., Ptolemy used the Greek letter omicron, which looks like an "O" in fact, to denote "nothing." So this is the symbol for zero, the "0" that we see—the circle. But I want to make it very clear that Ptolemy did not view this as a number, but merely as the idea of "nothing." But you can see, again, these things are coming together; they're slowly moving together. Well, zero as a number really occurred in India, most likely. By the 7th century, the Indian astronomer, Bhramagupta, who, as we saw in the previous lecture, offered a treatment of negative numbers, actually understood zero as a number, not just as a placeholder. In fact, he actually studied 0 divided by 0, and 1 divided by 0, and decided erroneously that 0 divided by 0 equals 0, but he just didn't know what to conclude about 1 divided by 0.

Here again we see a couple of things. First of all, we know today that we can't divide by 0. If we divide by 0, in fact, it does not yield a number, so we leave the realm of number. So we can't do that—no dividing by 0—and we've learned that in school. But we also see a wonderful thing. Bhramagupta, this very important, great mind, was making a mistake, again—something that is to be celebrated rather than to feel embarrassed about. He didn't get it quite right. That's okay; his contributions were enormous. So, finally, we see that humankind expanded its view of number to actually include and embrace zero. Well, given the relatively late inclusion of zero in our world of numbers, a more chronologically accurate name for this course would actually be *One, Two, Many, Zero to Infinity*. I mean, that is really how the notions developed together.

Let me say a few words about this "nothing" number in terms of language. From the 6th to the 8th century, in Sanskrit, we actually see "Sun-Yah," which meant "empty," to represent zero as we think of it. By the 9th century, in Arabic, we have "Sigh-Fr." By 13th-century Latin, we have "Zef-Ear-E-Um." From 14th-century Italian we have "Zef-Ear-Row." By 15th-century English, we have "zero." So we can see the evolution of just that word.

Because of zero's power in computation, some viewed it as mysterious and nearly magical. As a result, the word "zero" has the same origins as another word that means "a hidden or mysterious code," and that word, of course, is "cipher." We can see that "cipher" actually came from the mysterious qualities that zero possessed to our ancestors. So with zero firmly planted in our hearts and minds as an actual number, we're ready for at long last the positional number systems.

Let's start with the familiar base-10 numeral systems that we use in our everyday lives. As we previously mentioned in the other lecture, the base-10, or decimal, system uses the 10 Hindu-Arabic numerals: 0, 1, 2, 3, 4, 5, 6, 7, 8, and 9. These are called digits. The word "digit" actually comes from the Latin *digitus*, which means "finger." Again, quite a nice image—because remember this is early counting with our fingers. Coincidently, Bhaskara I, the great 7th-century Indian mathematician, is credited as the first person to write these numerals, including a circle for zero—so, a wonderful contribution there. The position of a digit in our system indicates by which power of 10 that digit is to be multiplied—so, either 1, the unit spot, or the 10 spot, or the 100 spot, or the 1,000 spot—and as we move further and further to the left, we multiply by bigger and bigger powers of 10.

We'll actually employ exponential notation to denote powers of 10. So, instead of writing 10×10, we'll write just 10^2. Instead of writing $10 \times 10 \times 10$, we'll write 10^3, and so forth. We'll declare, as a convention, that 10 to the zero power will represent the units place, or 1. So 10^0 will be a very fancy way of saying 1. As an aside, I want to remark that this notation was invented by René Descartes in the 17th century. We'll explore the evolution of arithmetical symbols in Lecture Twelve.

So let's take a look at an actual example. If we take a look at what we'd write as 293—I'm not going to actually say 293 and give it away—let's just analyze what that means. I first start off in the far right and notice the 3. That 3 is in the units spot, so that's 3×1 or 3×10^0. Then I add to it. I move over to the next position, which is the 10s spot. It tells me what digit I'm to multiply by 10. In this case I see 9. So I see 9×10^1, or 9×10. Then finally, I add the 2, which is in the 100s spot, so I add 2×100 or 2×10^2, and so 293 actually means: $(2 \times 10^2) + (9 \times 10^1) + (3 \times 10^0)$.

If we take a look at another example—let's take a look at this number here [8,105]. I don't want to say it out loud because that will give it away. Again, I start in the unit spot. I see I have five 1s, so 5×10^0. I have zero 10s, so we skip over the 10s—that's 0×10^1. I see I have 1 in the 100s spot, so I see 1×10^2. Now I see an 8 in the 1,000s spot, which is 8×10^3. So, I see the number is actually 8,105: $(8 \times 10^3) + (1 \times 10^2) + (0 \times 10^1) + (5 \times 10^0)$. Well, a specific mark called a "radix," or decimal point, is used to separate the whole number positions from the fractional positions. In some countries, including the U.S., the mark is a period. So for example, 3.14. In the unit spot, so that's 3×10^0. Now, the 1—since it's to the right of that radix—that's now tenths (1 over 10). So I see $1 \times \frac{1}{10}$, a fractional part. Then that 4 is located in the 100ths, and so I see $4 \times \frac{1}{100}$. In some countries, the separator is actually a comma. So for example, 7,05 would actually mean 7 in the unit spot, plus 0 in the 10s spot, plus 5 in the 100s spot. So, I'd see $(7 \times 1) + (0 \times \frac{1}{10}) + (5 \times \frac{1}{100})$.

Of course, there's nothing special about power 10. The important feature in a base-10 is that we require 10 digits. However, by analogy, we could consider writing in other bases besides 10. In some sense the simplest system is base 2. The base-2 system, it's referred to as the "binary system," requires only two digits. It's base 2, so we only need two digits, and those digits would be 0 and 1. In this system, the positions represent powers of 2 rather than powers of 10. So if we just argue by analogy and think formally, then we'll see exactly what this means. So we have to forego the base 10 we grew up with, and now we're going to think of something new.

So every spot—we'll start with the unit spot, but then one over from that will not be the 10s, it will now be the 2 spot. Next to that, we'll have the 4 spot, which is 2^2. And next to that, we'll have the 8 spot, and so forth. So for example, the number 1011_2. Notice, by the way, the little subscript 2 that we see there actually tells us that this is a number that's denoted in base 2. So we don't read it as 1,011 as if it were read in base 10. This is a special number, that little, teeny 2 tells us that—no, no, no—there's only two digits in the world, 0 and 1, and we're writing this in base 2, so positions are powers of 2. If we look to the far right, we see there's a 1 there, that's the unit spot, and so we see 1×2^0. Then right next to that we see another 1 to the left and that's sitting in the 2 spot, so we have 1×2. Then we add to that 0 in the 4 spot. That's 2×2. So, we have no 4s. Then that final 1 way off on the left, what spot's that in? Well, it's in the 8s spot, which is the 2^3. So, we see this is actually: $(1 \times 2^3) + (0 \times 2^2) + (1 \times 2^1) + (1)$, which equals $8 + 2 + 1$, which is the number 11 in base 10.

So notice how in this system, 11, which is easy to write in base 10, requires a long string in base 2. That actually happens: the smaller the base, the more we have to write in order capture a number. Now fractional positions are analogous to those in base 10, except now, instead of having a 10th and 100th and $1,000^{th}$, we have a 1/2, a 1/4, and a 1/8—again, powers of 2 in the denominator. So for example, if we take a look at the decimal, the binary number, 101.11_2—and this, again, is expressed in base 2—well, what do we see?

Let's do the whole number part first. I see a 1 immediately to the left of the point, the radix, so that's a 1 in the unit spot. I see 0 in the 2 spot, and I see a 1 in the 4 spot. So I see $4 + 1$, so I see 5. So the whole number part is 5. That's the 101. Now, what about to the right of the radix? Well, I see a 1 in the first spot, which is the one-halves, right? It used to be the one-tenths, but now, in base 2, it's the one-halves spot. I also see a 1 in the one-fourth spot. What I see here is $1/2 + 1/4$, which is 3/4. So, in putting this all together, this number is equivalent to 5 and 3/4 written in base 2.

We could, of course, consider another base. What would a base-3 system even look like? Well, we would again argue by analogy. We're finding a pattern. We would have exactly three digits and those digits we'll call 0, 1, and 2. The positional values would be dictated by powers of 3. So this base-3 system is actually called "ternary"; it's the ternary number system. So, for example, in base 3, the number 2102_3 in base 3—notice that we have a little subscript 3, which indicates that this is base 3.

We start again at the far right, in the unit spot. I see two units. That means I have two 1s. I see zero 3s. That 1 is actually located in the 3^2, or the 9s, spot, and that 2 is located in the 3^3, or the 27s, spot. So this number is actually $(2 \times 27) + (1 \times 9) + (0 \times 3) + (2 \times 1)$, which equals $54 + 9 + 2$, or 65, written in base 10. Again—fractional positions are once again analogous to those for base 10. So the base-3 number 12.202_3—let's see if we can figure this out by looking at the whole number part. Right to the left of the dot, the radix, we see a 2, so that means we have two units. Then that 1 to the left of that tells me I have one 3. So I have $3 + 2$, which gives me 5. So the whole number part is 5. Now, what about the 202? Well, the 2 is in the one-third spot, so I see two-thirds. I have zero ninths, so I have $+ 0$ ninth and then $2/27$, and when I add them together, I see 5 and $20/27$.

So again, just arguing formally by analogy, we actually see these numbers and translate them into more familiar numbers that we're used to in base 10. Are there any other bases left? Well, of course, the answer is yes. In fact, any counting number, any natural number greater than 1—we can't include 1—so 2, 3, 4, 5, 6, 7, and so forth can actually be used as a base. Given any such base, every number can be expressed in terms of that base. So, we can express any number in terms of any base that we wish.

Now, we recall that the Babylonians used base 60 for their number system, which means that they have a unit spot, then a 60s, then a 60^2, and so forth. Some computer languages today use base 16, which is sometimes referred to as "hexadecimal," but of course even this notion of writing numbers in positional form is never without controversy. From the 13th century to the 16th century, Europeans actually disagreed about the advantages of a positional system versus an additive system, like the Roman numerals, for example.

An additive system allowed very speedy calculations using an abacus, because you can add really fast, and didn't require memorizing multiplication tables, which we now subject children to because we have a positional system. Even back then, the prospect of facing large multiplication tables seemed daunting. A little bit more seriously, some worried about that the positional system was more vulnerable to fraud, as someone could radically change the value of a number by simply inserting a single zero to the right of that number. It would make it much bigger. In fact, this possible fraud remains very real, and a very real concern today. That's why on the checks we write we express the amount both in numerals and in words. That's why: because the positional system really is subject to fraud by just adding a zero at the end. Well, after much debate and brouhaha, the virtues of the positional system were finally realized and since its widespread adoption, it has proven to be enormously valuable, not just in its efficiency in writing numbers, but as we'll see in Lecture Nine, in the advancement of the theory of numbers and mathematics in general.

Given the great value of the positional system, it is perhaps not surprising that we find harbingers in ancient times of modern positional systems, as well as holdovers of some very old practices today in our everyday lives. I want to share a few of those with you right now. In terms of the binary expansions where we just had zeros and ones, the Chinese *I Ching* from 2800 B.C.E. contained patterns of solid and broken lines, called "trigrams" and "hexagrams," that corresponded to binary numbers, but they were not commonly used for computation.

So for example, what we'd see here is, for example, here we see a list of eight symbols, these trigrams, of these broken and solid lines. In fact, if we associate each broken line with the number 0, and each solid line with the digit 1, then in fact we can represent these eight symbols as numbers. For example, three broken lines, if we view them in binary, would actually be 000 and this is the number, of course, 0. But the next number over, where we see a broken line, broken line, solid line—well, that solid line would be in the unit spot, and then those two broken lines represent 0, so we'd have the number 1. If we move to the very last number there on the list, we see 111. If we think of that as being a binary number, then I have 1 in the unit spot. I have one 2 and I have one 4. If I add them together, notice I get 7. So that last number is 7.

Well, each one of these trigrams actually represented an aspect of nature. The first one was ether, forest, fire, thunder, wind, water, mountain, and earth. So, the eight elements were represented here in binary, which is a wonderful illustration of the power of just the 0 and 1. One of the most famous icons consisting of two elements is the yin-yang symbol. In fact, the yin-yang symbol corresponds to the solid and broken lines of the *I Ching*. The solid line represents yang, which stands for creativity, and the broken line represents yin, which stands for openness (as in "open-mindedness"). Of course, in the previous lecture, we also saw reflections of the binary system in the Egyptian multiplication technique that we saw known as "doubling," where we kept doubling each time. That was a binary type of operation.

In India, in the 4th century B.C.E., poet and musician Pingala used a binary system to notate musical meters. The modern binary system using 0s and 1s was established by the great German mathematician Gottfried Leibnitz in the 1660s. One hundred years later, the French mathematician and astronomer Pierre-Simon Laplace, who revolutionized mathematical physics, wrote about Leibnitz and Leibnitz's interpretation for giving great grandeur to the binary system. In fact, Leibnitz really gave a religious significance to the 0s and 1s, which Laplace didn't appreciate as we'll hear in this quote. He writes:

> Leibnitz saw in his binary arithmetic the image of Creation....He imagined that Unity represented God, and Zero the Void; that the Supreme Being drew all beings from the void, just as unity and zero express all numbers in his system of numeration....I mention this merely to show how the prejudices of childhood may cloud the vision even of the greatest men.

So you can see he really wasn't accepting of this religious overlay but definitely acknowledges this was a great man who made great contributions. Neither Leibnitz nor Laplace could have ever anticipated the dominant role that binary numbers would play in the computer age, which is built on bits. The word "bits," actually, comes from "binary digits." We slam those two together, and we get "bits." So the binary digits of 0 and 1 gave rise to the bits that we see in our computer. In fact, that is why when we see computers and we see bits, we know that that's the 0 and 1 at work. In fact, the first computer using binary addition was built by mathematician George Stibits, who was a scientist at Bell Labs, and he did this in 1937. He built this in his kitchen, and thus he named it the Model K, which is sort of a cute way of remembering that he built this binary machine in his kitchen.

In this country, the base-3, ternary, system becomes extremely popular during the days of summer, for the ternary system is used in baseball for giving the innings pitched. For example, in this context, if we see a 4.1, that means 4 and 1/3 innings pitched. Notice the 1/3: that's in fact ternary. We note that not all number systems today actually use the positional system that we're describing.

Ironically, in describing our physical position, we don't actually use a positional number system. We designate location on the earth's surface using latitude and longitude in degrees, minutes, and seconds, complete with separate units, and markings, and so forth. So for example, if you wanted to locate Williamstown, Massachusetts, you'd say $42° 71' 22'' $ N and $73° 20' 13''$ W.

However, our division of time shows the influence of the Babylonian base-60 system. Notice that we have 60 minutes in an hour and 60 seconds in a minute—again, potential reflections from the Babylonian's base 60. Here's a fun fact: The stock prices traditionally are measured in 1/8s; maybe you've noticed that. This base may be due to the use of hand signals used in the trading pit. It turned out that in the trading pit you would use your thumbs to indicate whether the prices were up or down, which only leaves 8 digits left to give the price. That's why we have things in eighths rather than in tenths on the stock market, which is interesting.

Now, I just wanted to close with some final thoughts on the power of this new positional system. In fact, I want to have reflections by two great mathematicians, the first one being, again, Pierre-Simon Laplace, who we just heard from a moment ago. Here is his take on this positional system:

> It is India that gave us the ingenious method of expressing all numbers by means of ten symbols, each symbol receiving a value of position as well as an absolute value, a profound and important idea which appears so simple to us now that we ignore its true merit. But

its very simplicity and the great ease with which it has lent to all our computations put our arithmetic in the first rank of useful inventions; and we shall appreciate the grandeur of this achievement the more when we remember that it escaped the genius of Archimedes and Appellonius, two of the greatest men produced by antiquity.

Carl Friedrich Gauss, the 19th-century German mathematician, who, in fact, was so great that people today refer to him as the prince of mathematics—he actually was a great admirer of Archimedes, but wrote this about the positional system with respect to Archimedes:

> How could he have missed the discovery of our present positional system of writing numerals? To what heights science would have risen by now, if only he had made that discovery!

What we see here is that both Laplace and Gauss show their great respect for the positional system—and also for Archimedes, by the way, who was considered to be one of the greatest scientists from antiquity, and yet his contributions really did not move the notion of number forward. But the concept of a positional system allowed us to break the barrier and enter the modern world of numbers.

The Magical and Spiritual Allure of Numbers
Lecture 5

How did numbers capture the imagination of our ancestors? Perhaps even the most primitive thinkers recognized that numbers hold the key to enormous power.

Numbers appear in many rituals and beliefs. In Babylonia, 60 was the number of Anu, the god of heaven, and 30 was the number of Sin, the lunar god. Many modern religions specify the number of prayers to be recited. In Islam, the number 5 is a good omen. The number 4 is avoided in Japan because the word for 4, *shi*, sounds similar to the Japanese word for death. In many cultures today, a well-known superstition surrounds the number 13.

In the 6th century B.C.E., Pythagoras—perhaps inspired by the Jains—ushered in the dawn of a new era in how people viewed and studied numbers and mathematics. The Pythagoreans elevated the study of numbers to the highest intellectual level. They distinguished arithmetic (the study of numbers) from logistic (the practical use of numbers, or calculation). They focused on the four disciplines (arithmetic, geometry, astronomy, and music); these four subjects formed the quadrivium, the basis of the liberal arts. They studied numbers as abstract objects, exploring their intrinsic properties.

Pythagoras founded a community of scholars he called the Brotherhood. Women were included as scholars in the Brotherhood, which was unusual for

> **The Pythagoreans elevated the study of numbers to the highest intellectual level.**

the time. The Brotherhood believed that natural numbers were basic to all qualities of matter and living things. They kept their studies secret, passing down beliefs and results in the oral tradition. Our knowledge of Pythagoras's work is based on the writings of scholars in later generations, including Euclid and Aristotle. The community had many strict, peculiar rules.

The Pythagoreans believed in number characteristics. The number 1, *monad* (unity), was the generator of all numbers and the number of reason. The number 2, *dyad* (diversity, opinion) was the first female number. The number 3, *triad* (harmony = diversity + unity) was the first male number. The number 4 stood for justice or retribution, as in the "squaring of accounts" or "let's get this issue squared away." The number 5 stood for marriage (2 + 3 = female + male). The number 6 stood for creation (perhaps because 6 = 2 + 3 + 1). The number 10, *tetractys*, was the holiest number and represented the four elements of the universe: fire, water, earth, and air. Geometrically, *tetractys* was represented by 10 dots arranged in an equilateral triangle; arithmetically, 10 = 1 + 2 + 3 + 4 (an example of a triangular number).

The Pythagoreans revered the relationship between certain ratios and musical harmonies, as well as aesthetic proportions. They did not consider ratios—that is, fractions—to be actual numbers. In their study of music, they observed that strings with lengths in a ratio of 1:2 vibrated in octaves, and strings with lengths in a ratio of 2:3 vibrated in a perfect fifth. They were among the first to study the golden ratio. Two numbers are in the golden ratio if the ratio of their sum to the largest is equal to the ratio of the larger to the smaller. Using the quadratic formula, it can be shown that this golden ratio is equal to the number $\frac{1+\sqrt{5}}{2} = 1.6180339\ldots$. Pythagoras's wife, Theano, was a mathematician whose best work is said to have been on the golden ratio.

Mathematicians have explored figurate numbers, perfect numbers, and amicable numbers. Figurate numbers are those that can be visualized in particular geometric ways, including triangles, squares, pentagons, and pyramids. A perfect number is one that equals the sum of its proper divisors. The number 6 is perfect because 6 = 1 + 2 + 3 and 28 is perfect because 28 = 1 + 2 + 4 + 7 + 14. The number 10, however, is not perfect because $10 \neq 1 + 2 + 5$. Many open questions remain about perfect numbers: Is there an infinite number of perfect numbers? Are there any odd perfect numbers?

A pair of numbers is amicable if the proper divisors of the first sum to the second and vice versa. The numbers 220 and 284 are amicable because the proper divisors of 220 are 1, 2, 4, 5, 10, 11, 20, 22, 44, 55, and 110, which sum to 284, and the proper divisors of 284 are 1, 2, 4, 71, and 142, which sum to 220. Many open questions remain about amicable numbers:

Is there an infinite number of pairs of amicable numbers? Are there any odd–even pairs of amicable numbers? Perfect and amicable numbers are the subjects of some of the oldest unanswered questions in all of mathematics.

Perhaps dice have dots because when they were first made, in 2000 B.C.E., symbols for numbers didn't exist yet.

Humans have played with numbers since at least 2000 B.C.E. Dice made of fired clay dating from 3000 B.C.E. have been found in northern Iraq. The ancient players did not yet have symbols for numbers, which is perhaps why they marked the sides of the dice with dots. We use the same configuration on our modern dice today. Ancient Egyptians played a finger counting game, *morra*, as early as 2000 B.C.E. Each of two players extends his or her hand with any number of fingers showing. Each player simultaneously calls out a number from 1 to 10. If the total number of fingers equals the number called, that player wins a point.

Magic squares have been a source of fascination, entertainment, and superstition for more than 3,000 years. A Chinese book, *Lo Shu* (*The Book of the River Lo*), from 1000 B.C.E. tells the story of a turtle emerging from the river with a pattern of dots carved into its back. When written as digits, the pattern forms a 3 × 3 magic square:

4	9	2
3	5	7
8	1	6

A 3 × 3 magic square is a configuration in which the numbers 1 through 9 are each put in one cell so that the sum of numbers in each row, each column, and the two diagonals equals 15. Analogously, an $n \times n$ magic square is an arrangement of the numbers 1 through n^2 into a square grid (each row containing n numbers) so that the sums of the numbers in each row, column, and main diagonal are equal.

The German artist Albrecht Dürer included a 4 × 4 magic square in his engraving *Melancholia*. Magic squares were thought to protect against the plague. Ben Franklin enjoyed the challenge of constructing magic squares while he was a clerk for the Pennsylvania Assembly. Modern mathematicians have proved many results about magic squares, including existence theorems and results about related structures, such as Latin squares and orthogonal Latin squares. Sudoku might appear to resemble magic squares, but there are no sums involved in Sudoku, just logic. ■

Questions to Consider

1. Verify that 28 is a perfect number. Verify that 1,184 and 1,210 are amicable. (*Hint*: 37 is a factor of 1,184, and 11 is a factor of 1,210, twice!)

	14		11
15			5
12		13	
			16

2. Complete the following magic square. (*Hint*: The sum is 34.)

The Magical and Spiritual Allure of Numbers
Lecture 5—Transcript

Welcome to the magical and spiritual allure of number. Numbers were not only important tools in early civilizations. They were also a source of mystery, entertainment, and even spirituality. Once humankind named numbers, there grew a curiosity to study them in their own right. How did numbers capture the imagination of our ancestors? Perhaps even the most primitive thinkers recognized that numbers hold the key to enormous power. Numbers were used to name the gods in Mesopotamia. In ancient Greece, the Pythagoreans, perhaps inspired by the religious sect the Jains from India, were the first to view numbers as abstract objects and elemental in all matter and living things. They also discovered that the most fundamental harmonies in music arose from ratios of natural numbers. Numbers that were connected with triangles and squares were thought to have special powers. In this lecture, we'll look at some of the ideas of the Pythagoreans and other practitioners of number mysticism.

So-called "perfect numbers" and "amicable numbers" are notions that originated in ancient Greece and continue to capture the imagination of individuals around the world today. Games involving numbers date back as early as 2000 B.C.E. One of the most popular and persistent number challenges is the so-called "magic square." Over 3,000 years old, magic squares have been used to ward off the plague, appease the gods, and entertain the curious. Here we'll explore some of the fascinating appearances of magic squares throughout history, including in the work of Benjamin Franklin. While entertaining and thought provoking, number recreation draws humanity in to explore many more serious and subtle properties of number. Well, let's begin our journey into number mysticism with a look at some rituals and beliefs involving numbers.

The Babylonians associated a sacred number to each of their gods. Sixty was the number associated with Anu, their all-important god of heaven. Now recall that the Babylonians had used a base-60 number system. 60 was their most important number, and perhaps that's why they associated it with their most important god. Thirty was the number associated with Sin, the lunar god, because the Babylonian month was 30 days long. Now, as we saw earlier, many modern religions specify the number of prayers to be recited, and hence the use of the rosary beads that we saw in a previous lecture. In Islam, the number 5 is a good omen. There are 5 elements of the Muslim profession of faith, 5 daily prayers, 5 keys to the mystery in the Koran, 5 senses, and many more (perhaps 5 more).

The number 4 is avoided in Japan because the word for 4, *shi*, sounds very similar to the Japanese word for death. Of course, in many cultures today, there's a well-known superstition about the number 13. Many people dread Friday the 13th. Many people, in fact, don't like staying in hotels that have a 13th floor. In fact, some hotels don't even have a 13th floor. In fact, Scottish airports have no gate 13, instead what they have is gate 12B. So there you have it.

We now turn our attention to the famous cult of Pythagoras. In the 6th century B.C.E., Pythagoras, perhaps—as we noted in Lecture Three—inspired by the religious sect known as the Jains in India, ushered in the dawn of a new era in how people viewed and studied numbers and mathematics in general. The Pythagoreans elevated the study of numbers to the highest intellectual level. Now, they actually had a distinction between what they called "arithmetic," which was the abstract study of numbers that we're partaking in right now—and they separated that from what they called "logistic," which to them was the practice of using numbers for calculation. So, they really focused on the more abstract notion of arithmetic and left the calculation issues aside.

Now, in their studies overall, they actually focused on four disciplines: arithmetic, geometry, astronomy, and music. Let me just say a word about geometry in this context. Here, the ancient Greek sense of geometry also included the measurement of lines and computing the areas of objects. So, embedded within geometry, in fact, was number theory; so number theory was part of geometry. So when we say "geometry" in the Greek sense, that was also number theory. So we see: arithmetic, geometry, astronomy, music—two of those actually contained number theory. Now those four subjects formed what was known and continues to be known as the quadrivium, and later on this evolved into what we think of as the liberal arts. Here we see that, in fact, the foundation of the liberal arts (really, half of the foundation) sits on number theory, and we see how central it is in our understanding of things.

Now, the Pythagoreans studied numbers as abstract objects, exploring their intrinsic properties. Now, why did they put so much energy in understanding the abstract notion of number? This is an important question for us to think about. Well, one of the Pythagoreans' core beliefs was that the divine ideas, which created and maintain the universe, are those of number. Thus, the way to attain perfection and become closer to the gods was to study arithmetic. In fact, Pythagoras himself is said to have said:

> Number is the first principle, a thing which is undefined, incomprehensible and having in itself all numbers.

Here we can see how Pythagoras viewed number in the same incomprehensible and undefined light that he actually might have viewed the gods. So we see this connection.

Now, to place their number contemplation in context, I'd like to say a few words about the group now known as the Pythagoreans. Pythagoras himself founded a community of scholars that he called the "Brotherhood." Now, in fact, the Brotherhood was not a collection of men, but really a community of families, and in fact some scholars conjecture it was the inspiration for Plato's *Republic*. Again, we see evidence where the emphasis on number really generated something significant, such as Plato's *Republic*, which seems to be outside the realm of mathematics. Again, numbers influence so much of our history and our culture. Women were included in this Brotherhood as scholars, which was very unusual and extremely progressive for the time. In fact, after Pythagoras's death, his wife, Theano, and his daughters led the Brotherhood. So we see this evidence of women being equals in this endeavor.

The Brotherhood believed that the numbers—now remember, from the Pythagorean point of view, numbers were the natural numbers (one, two, three, four, and so on)—were basic to all qualities of matter and living things. They kept their studies secret, passing down beliefs and results in the oral tradition. Now in fact—let me just say a word about how the studies would take place. When a new member of the Brotherhood would enter, they would take a vow of silence, and so they would sit quietly and listen to the more senior scholars speak and pontificate, and then after some unit of time, the junior members would be allowed to speak. This might sound strange, but really it mirrors traditional education as we know it, if you think about it for a moment. Because traditionally, students come into a classroom and they say nothing; they're very quiet, and the instructor pontificates on the subject, and then once the students graduate, they're allowed to speak. But you can see a reflection of that. I think that's what was going on during this period. Well, since they really adopted an oral tradition, our knowledge of Pythagoras's work is based on the writings of scholars in later generations, including Euclid and Aristotle.

Now, in many ways, they were a peculiar group. The community had many strict and unusual rules, including, for example, a prohibition on eating beans. That was not allowed. You couldn't go hunting, and you were not allowed to wear wool. Now, they dressed in white, and they slept on white linen bedding, and they were vegetarians. Members who betrayed the Brotherhood's secrets were severely punished, and we'll talk about that in a future lecture, in fact. The Pythagoreans believed that numbers were the secret principle of reality, and as such, numbers had individual traits and characteristics, and they actually took on almost a personal, human form, and I thought I would share with you some of the numbers and characteristics that the Pythagoreans held.

The most important number to the Pythagoreans was the number 1, which they called the *monad*, which, to them, meant "unity," and, in fact, we still use the word "unity" to represent 1 quite often. This was the generator of all numbers, which makes sense because if you take 1 and you add it to 1, you get 2, and if you add it to 1 again, you get 3. This was their notion of number, was the counting numbers. So, 1 gave birth to all the numbers, so it was the most important and also was given the assignment of being the number of reason. So 1 was the number of reason. The number 2 they called the *dyad*, which represented diversity, and opinion, and even the diversity of opinion. This was also the first female number, so this is the first number that actually has a sex associated with it. In fact, the idea of having the female number hooked up with this notion of diversity of opinion might, in fact, be the inspiration of the old adage that it's a woman's prerogative to change her mind. It really might have come out of this Pythagorean sense of the number 2.

The number 3 was referred to as the *triad*, and it represented harmony; and I really like this because it represents harmony because 3 is actually 2 + 1, and 2 meant diversity, and 1 meant unity, so if you take diversity and add unity, you do get harmony—so, a really beautiful way to see the number 3. Also, 3 was considered the first male number—so much for the progressive nature of the Pythagoreans! They believed that if you took woman, which was 2, and added reason, which was 1, you got male. So, this is not my opinion, I'm just reporting the news here—so much for the progressive nature.

The number 4 was associated with a square since you can use four dots to actually produce the corners of a square, and 4 also stood for justice, or retribution. So that was the quality that was assigned to 4. We actually see reflections of this today in our expressions the "squaring of accounts" or "let's get the issue squared away." You see, again, we see reflections where numbers actually ended up being the key ingredients to even phrases that we use in our everyday lives.

The number 5 stood for marriage, which makes sense if you think about female being 2 and male being 3. If you take 2 and 3 and add them, you get 5—so that was marriage. The number 6 stood for creation, quite possibly because if you take a female and a male—that's a 2 and a 3—and then you add a child, an offspring, that sum would be 6. Now the number 10, which they called the *tetractys*, was the most important and most holy number to them. It represented the four elements of the universe in their eyes: fire, water, earth, and air. Geometrically, the *tetractys* was represented by 10 dots arranged in an equilateral triangle. We have one dot on top, and then 2, and 3, and then 4. Notice that if you add up those dots (1 + 2 + 3 + 4), it actually yields 10. The 10 lines that form this triangle, or the numbers 1 + 2 + 3 + 4, those 4 numbers, are what represent the fire, water, earth, and air. In fact, the Pythagoreans had a chant that they would actually use, a prayer if you will, that actually was to the number 10 and to this *tetractys*. I wanted to share with you a translation of that prayer that the Pythagoreans would use:

O holy *tetractys*, thou that containest the root and the source of the eternally flowing creation! For the Divine Number begins with the profound, pure Unity until it comes to the hole Four; then it begets…the never tiring holy Ten, the keyholder of all!

So, you could see that they really liked 10; 10 was a biggie.

Well, moving beyond the numbers of the gods in the Pythagoreans' minds, which were the natural numbers 1, 2, 3, 4, and so forth, the Pythagoreans actually admired the relations between certain ratios of natural numbers and musical harmonies, as well as aesthetic proportions. They didn't consider ratios (in other words, fractions) to be actual, God-given numbers. Fractions had no real existence to them beyond ratios of natural numbers. Let me just say a word about that. It's a little bit peculiar because we think of fractions as numbers, just like we think of 7 as a number. But remember, the Pythagoreans lived in an entirely different world. They had a different view of the cosmos. So, while here it seems peculiar that a number like 1/2 wouldn't be considered a number, we have to put ourselves back in a time where, in fact, that would make sense, where, in fact, the numbers were clearly defined as 1, 2, 3, 4, 5, and so on, and 1/2 is merely a combination of 1 and 2 in an interesting way, looking at that ratio. So they saw them as ratios and not as God-given numbers.

In their study of music, they observed that strings with lengths in a ratio of 1 to 2 vibrated in octaves and strings with lengths in a ratio of 2 to 3 vibrated in a perfect fifth. So they discovered several such fundamental musical ratios, and you could see the mathematics actually incorporated in the sound, in the music that they would produce. The Pythagoreans were among the first to study the "golden ratio." Two numbers are in the golden ratio if the ratio of their sum to the largest is equal to the ratio of the larger to the smaller. So let me say that again. If we have two numbers, we have a large one and a small. If we take a look at the ratio of the sum of these two to the larger one, then that ratio should be equal to the ratio of the larger one to the smaller one. When we have two numbers that enjoy that equality, then we say they're in a "golden ratio." Now, we could use the quadratic formula to solve for what that golden ratio is. While we won't do it here—it's not important—let me just report that the golden ratio numerically works out to be $1 + \sqrt{5}$

, all divided by 2. If you were to work this out on a calculator and look at the decimal expansion, you'd see 1.618033, and so on. It continues forever. Well, Pythagoras's wife, Theano, was a mathematician whose best work is said to have been in the golden ratio, although none of her writings have been uncovered, maybe because she didn't write since this was more of an oral tradition. We have no evidence of it, but this is the conjecture.

The Pythagoreans studied special numbers that had certain geometric or arithmetic aesthetic appeal. One such collection is the "figure-it numbers." The figure-it numbers are those that can be visualized in particular geometric ways, including triangles, squares, pentagons, and pyramids. So for example, a triangular numbers are numbers for which, if you took that many dots, you could form perfect equilateral triangles. So for example, the smallest triangular number would be 3, because three points, of course, can produce a perfect equilateral triangle. Then the next triangular number would be 6, because we take these three and then add three more numbers here to get a larger equilateral triangle, and you can build larger and larger such numbers.

The perfect squares, the square numbers—4 is the smallest one because 4 points will generate a square. Then the next one would be 9, which would be 1, 2, 3, 1, 2, 3, 1, 2, 3, forming a 3-by-3 square. So the next square number is 9, and so forth. Those are the perfect squares that we're actually familiar with. Then the pentagonal numbers would be 5 at first, because 5 points will form a perfect pentagon. Then the next pentagonal number is 12, as you can see, because you can put them together to make a larger pentagon. Well, a perfect number is a number that's actually equal to the sum of its proper divisors. Now remember, a proper divisor is a number that divides evenly into a number, and "proper" means that it's not the number itself. So, a proper divisor is a number that's smaller than the given number that divides into it evenly. Let's take a look at an example to really solidify this idea in our minds.

Let's consider 6. Well, 6 is actually considered a perfect number, because let's first of all find all the divisors or all the numbers that divide in evenly into 6. Well, there's 1 (divides evenly into everything), 2 (because 6 is even), and then 3 (because 3 divides evenly into 6.) So the only divisors of 6 are 1, 2, and 3. Notice something really amazing. What if you add up those numbers? $1 + 2 + 3 = 6$, and that makes it a perfect number. Let's take a look at another example. Let's take a look at 28. I claim that 28 is another example of a perfect number. Well, let's find all the numbers that divide evenly into 28. We always have 1. Then there's 2. Four divides evenly into it, as do 7 and 14. So we have a bunch. But what happens if we just add those numbers? What if we add $1 + 2 + 4 + 7 + 14$? Well, we actually get 28, an amazing coincidence! Twenty-eight, therefore, is a perfect number.

You might be saying, Gee, it sounds like perfect numbers are all over the place. Well, they're quite scarce indeed. Let's take a look at an example that's not a perfect number. Let's consider 10 and wonder if, in fact, this is perfect or not. Well, what are the numbers that divide evenly into 10, the proper divisors? Well, there's 1, there's 2, there's 5. Those are the only numbers that divide in evenly into 10; those are the proper divisors. What happens when you add up $1 + 2 + 5$? You get 8, and so you actually see that you don't get 10—so, not so good. Now, perfect numbers have actually captured the imagination of individuals throughout history and for centuries. The 5th-century church father Saint Augustine of Hippo was so moved by the perfect number 6 that we just saw that he actually declared—and I want to read this to you because this is an amazing quote showing how important this religious man took the perfect number 6 to be:

> Six is a number perfect in itself, and not because God created the world in six days; rather the contrary is true. God created the world in six days because this number is perfect, and it would remain perfect, even if the work of the six days did not exist.

A really dramatic sentence for a religious individual, saying that 6 being perfect almost trumps the number of days required to create Earth in this religious context. Let me say that there are many open questions about perfect numbers that are still unanswered. Are there infinitely many perfect numbers? We only have finitely many, and we don't know if infinitely many perfect numbers exist. No one knows an example of an odd perfect number, which is quite mysterious. So, do they exist or not? No one knows.

Well, numbers that also attracted the attention of the Pythagoreans were what are called "amicable pairs of numbers." Now, a pair of numbers—if we take two numbers together, we call them amicable if a very peculiar thing happens. If, in some sense, we happen to look at the proper divisors of one and add them up, we actually get the second number. Conversely, if we take the second number and look at its proper divisors, when we add those up, we get the first number. So we kind of see a little reflection of the perfect number, but not quite. We need two numbers here. Let's look at an example. Let's consider 220 and 284. I claim these are amicable numbers, and let's verify that—so we have to do a little calculation.

First, let's find all the proper divisors of 284. It turns out those divisors are 1, 2, 4, 71, and 142, which means that each of those numbers divides in evenly into 284. Well, what happens when you add up those numbers? Well, when you add up those numbers, you get the first number, 220. So you can see that 220 is sort of linked up with 284. But it has to work the other way as well. So let's consider the proper divisors of 220. Well they turn out to be—there are a lot of them, by the way—there's 1, there's 2, there's 4, 5, 10, 11, 20, 22, 44, 55, and 110. All those numbers divide in evenly into 220. What happens when you add all those up? You get the second number, 284. So you can really see these two numbers come together, and that's why they're called amicable. They're very friendly to each other.

Now, again, there are many open questions about amicable numbers. Are there infinitely many pairs of amicable numbers? To this day, we don't know. We only know finitely many examples. Are there any odd–even pairs of amicable numbers? That is, are there any pairs where one is an odd number and one is an even number and yet they're amicable? No one knows if that's possible or impossible, and no one has an example. Well, questions involving perfect and amicable numbers remain some of the oldest unanswered questions in all of mathematics. So, we're really seeing things that still capture the imagination, even though they come from antiquity. Well, let's leave these open questions of the cult of Pythagoras for the moment, although we'll return to the Pythagoreans and their view of number in Lecture Eight. But now, let's enjoy some recreation with numbers and just have some fun.

Now, to generate a random number for games and things, we have been known to toss dice. In fact, this is a really ancient tradition. Dice made of fired clay dating from 3000 B.C.E. have actually been found in northern Iraq. At the time, of course, the ancient players didn't have symbols for numbers.

The dice were actually before the numbers, and this is perhaps why they marked the sides of their dice with dots to denote the quantities. Amazingly, we use the same configuration for our modern dice today. Five thousand years later, what we see, in fact, are the same exact configurations that they used so many millennia ago. Absolutely amazing. That's where this comes from. So the reason why we have dots on dice is because we didn't have numbers. The dice came first. Well, some number games only required the digits on our hands. Here's an ancient one that the Egyptians played with their fingers. It was called *morra*, and it was played as early as 2000 B.C.E.

So, you have two players. I'll be both players. But they'd each, at the same time, extend one hand and some number of fingers, and simultaneously each player would yell out a number. If that number matched with the sum of the number of fingers shown by both players, then that person gets a point. For example, if the person would say "six," then they would lose. If the person would have said "three," they would have gotten a point. Not a very exciting game, but remember, this was before cable TV, so who knows?

Well, much more interesting are the grids of numbers known as magic squares. Magic squares have been a source of fascination, entertainment, and superstition for over 3,000 years. A Chinese book, *Lo Shu* (*The Book of the River Lo*), from 1000 B.C.E., tells the story of a turtle emerging from the river with a pattern of dots carved into its back. As the legend goes, the 21st-century B.C.E. Emperor Yu found a divine tortoise while he was bathing. The markings on the tortoise's shell formed a 3-by-3 grid of numbers (in the form of dots, of course, since we had no numbers at the time). That was considered to be a message from the Heavens instructing the emperor how to govern. So, I don't think this method is used anymore, but back then this is how Heaven would relate to the emperor.

Well, if we write the grid of numbers down (from the turtle's back)—now notice, in fact, if you look at it—for example, in the upper left corner, you see 4 dots in the form of a square. That's number 4. Right next to it, to the right of it, you see a line of dots. You count them, there's 9. Next to it, you see 2 dots. That's 2, and so forth. The middle one, for example, has a cross-type thing with 5 dots. That's number 5. So we see the numbers 4, 9, 2, 3, 5, 7, 8, 1, 6. Well, by the 5th century B.C.E., it was noticed that the 9 numbers in the grid exhibited some amazing arithmetic properties. First, each number, each digit, from 1 to 9 appears in the grid exactly once. Take a look and see for yourself—1, 2, 3, 4, 5, 6, 7, 8, 9. They're all there, and they all appear once. Secondly, and perhaps more profoundly, the sum of any row, or column, or any of the two diagonals—if you add up those sums, it always equals the same value, 15. In fact, check it out. If you look at the first column, you see 4 + 3 + 8, and you see 15, and that holds for all the columns, all the rows, and the two diagonals.

When they discovered this, this amazing, incredible structure, it was viewed with magical, mathematical, and spiritual interest. Today, as a result, we call these grids "magic squares." More generally, any time we take a natural number n and consider the square, an n-by-n square grid, it's called a magic square if we can arrange the numbers from one out to n^2 in such a way that each column and each row contains exactly n numbers, and if you sum any row, any column, or the two diagonals, then the answer you get is always the same. You get the same sum. In fact, the great German artist Albreckt Dürer included a 4×4 magic square in his engraving *Melancholia*. What's really neat about this particular magic square is that first of all, the sum of each row, column, and diagonal turns out to equal 34, so that was the sum number that they all agreed upon. But the really neat thing that I like is that he put a coded message in here, and the code was that if you look at the two numbers in the middle on the bottom row, that indicated the year of the engraving—that the engraving was made, 1514.

Well, magic squares were thought to protect against the plague. Even Benjamin Franklin enjoyed the challenge of constructing magic squares while he was a clerk in the Pennsylvania Assembly. Apparently, the Pennsylvania Assembly wasn't as interesting as constructing mathematical objects such as magic squares. He actually produced at least two 8-by-8 magic squares. Now, modern mathematicians have proved many results about magic squares that we won't go into—existence theorems (that they exist at all sizes), results involving related structures such as Latin squares or even orthogonal Latin squares, if you can imagine such things.

Now, the current hot game is the game of Sudoku, which may appear to resemble the magic squares, but there are no sums involved in Sudoku. In

Sudoku, the numbers are just placed in a particular way, and so they're just abstract symbols, really. There's no arithmetic involved. We have to make sure that every row and every column in every row has the numbers from 1 to 9 embedded in it, and every small 3-by-3 square has all the numbers from 1 to 9 inside of them too. So really, it's a game of logic rather than a game of arithmetic, and so, in fact, there's no reason to actually have the symbols, per se; in fact, we could replace the symbols of the digits by any symbol, and it would still be a fine Sudoku puzzle. But here we're using the ordination property, that we're so comfortable counting 1, 2, 3, 4, 5, 6, 7, 8, 9 that we can immediately see which digit is missing—hard to do if we use other symbols. So, again we see how numbers are so innate in everything that we do and how we think. Anyway, the great magic and mystical allure of numbers continues with us until this very day and has a long and interesting human tradition.

Nature's Numbers—Patterns without People
Lecture 6

Numerical structure, beauty, and pattern existed long before our ancestors named the numbers. We'll open the life of numbers with an exploration into numbers within nature by considering the uninviting, rough facade of a pineapple, and simply counting.

We begin our exploration into the nature of numbers by considering numbers in nature. Focusing on an ordinary pineapple, we notice two interlocking collections of spirals on the pineapple's façade and ask the natural question: How many spirals are there in each direction? We count and see 8 and 13 spirals and are surprised that the counts are not equal. We look at a daisy and a coneflower and are immediately drawn to the spirals in their centers. We find 21 and 34 spirals, but we are no longer surprised that these values are different. If we count the spirals on the side of a small pinecone, we find 3 and 5 spirals.

Note the spiral in a daisy's center. This pattern in nature creates a special list of numbers.

We now list the numbers we find in order and move from numbers in nature to the nature of numbers. We see: 3, 5, 8, 13, 21, 34. If we add any two adjacent numbers on our list, the sum is the next number on our list. If we generalize this pattern, then we could extend the list indefinitely: 3, 5, 8, 13, 21, 34, 55, 89, 144, 233 .... We could even extend our pattern and our list of natural numbers in the other direction: 1, 1, 2, 3, 5, 8, 13, 21, 34, 55, 89, 144, 233 .... This abstract pattern can be used to better understand our physical world.

Why do nature's spiral counts so often appear on this special list of numbers? The daisy is an example of a composite flower—each floret is an individual flower that grows from the stem and gradually moves outward toward the circular boundary. Biologists hypothesize that these florets position themselves so that they have as much room around them as possible. If we pack a circular disk with florets in this fashion, it can be proven that the spiral counts are always two consecutive numbers from the list we just discovered.

The history of these special numbers goes back more than two millennia. This famous sequence was studied in India. In the 3rd century B.C.E., Pingala included these numbers in his commentary on Sanskrit grammar. In the 6th century C.E., the Indian mathematician Virahanka studied meter patterns of long and short syllables and showed how some meter counts corresponded to these numbers. The numbers in this famous sequence are known as the Fibonacci numbers, named after Leonardo of Pisa, an Italian mathematician whom many believe was one of the greatest European mathematicians of the Middle Ages.

The Fibonacci numbers are connected to the golden ratio studied by the Pythagoreans.

In 1202, Fibonacci published an influential text titled *Liber Abbaci* (*The Book of Calculating*) that introduced the Hindu-Arabic numerals to the West. He also strove to popularize the system. His text explained how to perform arithmetic using the decimal system that we use today rather than the analogous calculations using the more cumbersome Roman numerals, which were still popular in Italy at that time. In chapter 12 of *Liber Abbaci*, he posed a question aimed at calculating how many rabbits will be in an enclosed space after 12 months, given their reproductive habits. We see that the number of pairs of rabbits present is equal to the Fibonacci numbers. This solution was the first time this list of numbers appeared in the literature of Western mathematics.

The Fibonacci numbers continue to attract the attention of mathematicians and math enthusiasts. Many results have been found about and involving the Fibonacci numbers. Today, there is a research journal named *The Fibonacci Quarterly*.

The Fibonacci numbers are connected to the golden ratio studied by the Pythagoreans. If we consider the ratios of consecutive Fibonacci numbers—$\frac{1}{1}, \frac{2}{1}, \frac{3}{2}, \frac{5}{3}, \frac{8}{5}, \frac{13}{8}, \frac{21}{13}$, and so forth—then these fractions approach the value of the golden ratio. For example, $\frac{13}{8} = 1.625$, $\frac{21}{13} = 1.615$ ...; and further out, $\frac{144}{89} = 1.6179....$

Though there are natural numbers that are not Fibonacci numbers, in 1972, Edouard Zeckendorf proved a beautiful result: If a natural number is not a Fibonacci number, then it can be written uniquely as a sum of nonadjacent Fibonacci numbers. We notice that $4 = 3 + 1$, $6 = 5 + 1$, and $7 = 5 + 2$. This expression of natural numbers into Fibonacci numbers is now known as the Zeckendorf decomposition. To express a natural number N in this manner, we apply a "divide and conquer" strategy. We first find the largest Fibonacci number less than the natural number N, then write N as the sum of that Fibonacci number and a remainder. We then repeat this process with the remainder. For example: $30 = 21 + 9 = 21 + 8 + 1$. This "divide and conquer" strategy can be used to prove this result in general. ∎

Questions to Consider

1. Consider two consecutive Fibonacci numbers. Square each and add the results together. Repeat this calculation with various consecutive Fibonacci numbers and see if you can find a pattern in your results.

2. Look for Fibonacci numbers in the world around you: Count the number of petals in flowers, the spirals in pinecones and pineapples, or the lobes on leaves.

Nature's Numbers—Patterns without People
Lecture 6—Transcript

Once our ancestors identified and named the abstract objects known as numbers, they gave birth to ideas that immediately took on a life of their own, attracting the interest and capturing the imagination of curious minds throughout human history. In this part of our course, we'll explore the early steps of that historical journey to discover and appreciate the structure and beauty of numbers in their own right. We'll start with the realization that numerical structure, beauty, and pattern existed long before our ancestors named the numbers. We'll open the life of numbers with an exploration into numbers within nature by considering the uninviting, rough facade of a pineapple, and simply counting. From there, we'll move to some beautiful flowers. Our studies of fruit and flora will lead us to discover the famous Fibonacci numbers. Some believe that this sequence was discovered in India over 2,000 years ago.

We'll see how Leonardo de Pisa, also known as Fibonacci, came upon this sequence of numbers as an answer to a question he posed in his influential arithmetic text published in 1202, involving the reproductive habits of rabbits. His seminal work actually introduced the Hindu-Arabic number system to the West. We'll return to this beautiful pattern that both nature and Leonardo incorporated in their works and make a striking observation first established by Edouard Zeckendorf in 1972. Every natural number can be expressed as the sum of different Fibonacci numbers. Thus, we'll see that every natural number, in some sense, is either a Fibonacci number or a not-so-distant cousin of several different Fibonacci numbers. This attractive observation demonstrates that not only are the Fibonacci numbers a natural part of nature, but also a natural part of the natural numbers.

Well, it seems fitting to open our exploration into the nature of numbers by considering the numbers in nature. Now, one life lesson we take away from the study of numbers is that the best way to understand our world is to examine it closely, and in this case we'll focus on an ordinary pineapple. So, if you happen to have a pineapple handy, this is an excellent time to grab it. That's what I'll do. So I grab a pineapple right here. Quite often, just looking at things closely reveals structure, and in a numerical context, we look and we see if we see quantities within the side of the pineapple. We quickly, actually, see some—for we see that there are spirals that are going up in this direction, and then if we look really closely, we actually see that we have spirals going up in the other direction, namely this way. Well, one of the most powerful ways to understand our world better through number is to just ask how many. How many is an excellent way to actually make progress in understanding. So in this case, I want to ask how many spirals are going up this way on this façade, and how many are actually going up this way on the pineapple.

Now of course, a pineapple doesn't know right-handed spirals from left-handed spirals. So we'd expect the same number of spirals to occur each way. Well, let's now verify that by just simply counting. So let's count and see what we see. Well, so I'll try this right now. Now, this is like one of those cooking shows where I've done this in advance, because it's pretty hard to count them, but I've made it a lot easier by actually marking them with a ribbon. Isn't that pretty? If you look closely, you can see those spirals in really quite dramatic depth, and you might notice that one of the spirals actually is marked with these yellow dots. That's there because I once tried to count the spirals without those yellow dots, and I counted for about three days. So this allows me to know where I start so that I will know when to stop.

So we'll start right on the yellows, and we'll count together and see how many we see. Well, we see 1, 2, 3, 4, 5, 6, 7, 8, and notice now we're back to where we started. So there's 8 going in this direction. Let's now count the spirals in the other direction and see if we get 8 as well. So if we try that, and voila, here we are. Let's count the spirals in this direction and we'll start with the yellows, so we see 1, 2, 3, 4, 5, 6, 7, 8—uh-oh, more than 8—9, 10, 11, 12, 13. So we're getting 13 in this direction, which is really surprising. Well, of course, as we said at the onset of this course, a surprise is a great moment because it's a moment when, in fact, our intuition runs counter to reality, and it's a great moment of learning, because now, in fact, we can retrain our intuition and actually make new discoveries. So, this is now a great moment.

Well, now, before we talk more about the pineapple, let's turn our attention to the beautiful flowers we see around us. In particular, I want us to look at the daisy and the coneflower. Now, when you look at these, we're now immediately attuned to the spirals in their center. If we focus on the daisy, for example—the yellow spirals we see there—we see two sets of interlocking spirals. Now, of course, what must we do? We can't curb our enthusiasm; we have to count. We have to count those spirals and see. Do we think that we're going to have the same number of spirals going one way versus the other way? Well now, no—because in our experience with the pineapple, our thinking is that in fact our spiral counts may be different. So already you can see by counting we get to have a new world-view, and so our intuition is no. Well let's count. So we'll count in one direction, let's see what we get. Here we go—1, 2, 3, 4, 5, 6, 7, 8, 9, 10, 11, 12, 13, 14, 15, 16, 17, 18, 19, 20, 21. So, we have 21 in this direction, and now let's count the other direction and see what we have. I'm going to count this pretty fast, let's see if you can keep up—1, 2, 3, 4, 5, 6, 7, 8, 9, 10, 11, 12, 13, 14, 15, 16, 17, 18, 19, 20, 21, 22, 23, 24, 25, 26, 27, 28, 29, 30, 31, 32, 33, 34. Okay, finally. So our modified intuition was correct—different number.

If we were to consider a pinecone, we could count the spirals on its side (if we found one, walking in a park), and we'd see, on a small pinecone, 3 going up one way and 5 going up the other way. Well, now, what if we now list our numbers that we just found in order of their size? Let's now move from the numbers in nature and consider the nature of these numbers. So let's just look at these numbers now abstractly. What we see are 3, 5, 8, 13, 21, 34. Now let's look at these numbers and see if we see any pattern here. Well, we see an amazing coincidence: If we add any two adjacent numbers from our list, that sum is the next number appearing on our list. So, 8 + 13 equals the next number, 21, and so forth. Well, this amazing coincidence actually can be generalized. What if this pattern were to continue? Well, then, we could actually extend the list indefinitely. The next number would be 55, because that's 21 + 34. The number after that would be 89, because that would be the sum of 34 + 55, and so forth. We'd see 144, we'd then see 233, and so on. You see the numbers are getting bigger and bigger.

Even more interesting, we could also extend our pattern in the other direction and actually figure out what comes before. Right? What would come before the 3 if the pattern were to continue? Well, then it would have to be some number that, when I add it to 3, I get 5. That would be a 2. Before the 2 I'd see a 1, because that would be the number when I add to 2 generates 3. What would come before that one? It'd be another 1, because 1 + 1 would give me the 2. Well, this pattern of numbers can be used to better understand our physical world, and this is the power of math. Because what we do now is we look at the world, and we see structure. We see number. Then we abstract it, and when we abstract, all of a sudden we see structure that we didn't see before. With that abstract structure, we look back into our everyday world and see that world with a greater clarity.

In this example, if we look at a giant sunflower, we could guess the spiral counts, and we might guess 55 and 89, and for the spirals of a certain size for certain sized sunflowers, we actually see 55 and 89. This image that, in fact, we're seeing, is an image that I took when I was in Greece (a very romantic place), where I was running a course on number theory and the theory of numbers. So, a wonderful place, of course, as you can imagine, to talk about numbers.

Now, why do nature's spirals and their counts often appear on this list special of numbers? Well, let me give you one theory that seems to be plausible. The daisy is an example of what's called a composite flower. That means that each little floret that we see, each one of those little yellow dots, is, in fact, an individual flower, and they grow from the very center of the stem and gradually move outward. So, the image we should have is if this were, in fact, the image of the flower, then the baby florets are pumped in, in some sense, right in the center, and then they slowly move out. So, the florets on the very, very outer part near the white flowers on the perimeter, in fact, are the oldest, and the florets right in the center are the most recently born flowers.

Well, biologists have hypothesized that these florets position themselves to have as much room around them as possible. So, in fact, we have this circular flower, and little florets are being pumped into the center, and then, of course, as more are being pumped in, these current flowers are being pushed away, and they migrate out to the boundary. But they migrate and position themselves in such a way that they jostle—if you will—so they have as much room around them as possible. This actually makes a little bit of sense when you consider that, in fact, they might need photosynthesis, and they have the sun affecting their growth, and so they might want as much space around them as possible to get as much light around them as possible. It's a reasonable theory. Well, if we assume that theory, we can actually prove mathematically that if we now pack the disk so they fill up space in this circular region in this fashion where they keep juggling and jiggling until they totally pack space in that constrained space—that the florets will, in fact, generate two sets of spirals, and those spiral counts will be numbers on our list. That's a mathematical theorem.

Well, it's amazing to see that we might have a description for why the pattern in nature actually occurs. But, let's take a moment now to consider the history of these special numbers and how they fit into our story of number and let the actual nature of it drift into the background. The 3rd-century B.C.E. Indian poet Pingala included these numbers in his commentary on Sanskrit grammar. This was to help the devout understand and attain a better purity in the cadence of certain Indian ritual chants. Later, also in India, in the 6th century C.E., mathematician Virahanka studied meter patterns of long and short syllables and showed how some meter counts actually corresponded to these lists of numbers. Now today, this famous sequence of numbers is known as the "Fibonacci numbers." They're named after Leonardo de Pisa, an Italian mathematician who many believe, in fact, was one of the greatest European mathematicians of the Middle Ages. Leonardo's father was named Bonacci, and Fibonacci is a shortening of the Latin *Filius Bonacci*, which translates to "the son of Bonacci."

In 1202, Fibonacci published an influential text entitled *Liber Abbaci* (*The Book of Calculating*) that actually introduced the Hindu-Arabic numerals to the West. He also strove to popularize this system, which we now use. In fact, his book opens with a very dramatic quote. His book opens with:

> The nine numerals of the Indians are 987654321. With these nine figures and with the sign 0, which in Arabic is called Zeff-ear-um, any desired number can be written.

A very dramatic beginning to this—and as an aside, it's interesting to note that Fibonacci refers to zero here as a sign rather than a number. You can see that even though zero is considered a number here, there is a distinction between zero, which came along so much later in our understanding of number, even here in 1202. Now, his text explained how to perform arithmetic using this decimal system that we use today rather than the analogous calculations using the more cumbersome Roman numerals, which were still popular in Italy at that time. Of course, in fairness, old habits are hard to break, and even Fibonacci, who championed the cause of the base-10 system that we use today, actually returned to the more familiar Babylonian base-60 system of the time to actually solve an equation during a mathematics competition. So even though he was touting this great message, there were moments where he'd actually revert back to the way he learned as a child, which, again, makes some sense, I think.

Well, returning now to the Fibonacci numbers themselves. In Chapter 12 of his book, he posed an intriguing question involving the reproductive habits of rabbits. Now here is the translation of his question. (It's a little bit cryptic. It's a word problem, if you will.)

> A certain man had one pair of rabbits together in a certain enclosed place, and one wishes to know how many are created from the pair in one year when it is the nature of them in a single month to bear another pair, and in the second month those born to bear also.

Well, quite cryptic, in fact—and let me say what's going on here in our own words. Our goal here is to count how many rabbits will be in the enclosed space at the end of 12 months. Now the question, I hate to admit, is not particularly well posed, so we'll make some assumptions as to the question. First, we'll assume that every pair of rabbits born consists of a male and female rabbit, so therefore reproduction is possible. No pair of rabbits will die prematurely before the year is out. We'll also assume that the first pair of rabbits with which the man starts is a pair of babies. Thus, at the end of the first month, we would still have just one pair. Let me try to actually enact this out, but this will be a totally wholesome reenactment.

Well, we begin with the babies. So here we have a baby pair of rabbits at the very beginning of the first month. Now, at the end of the first month, all they do is mature because they're too young to actually reproduce, so I'll put them over here. But at the end of the first month, we actually see that they are all grown up. So, this is now the same pair of rabbits, but now a month later. Well, now that they're matured, they could actually produce offspring, and they decide to do so in the next iteration. So, in the next iteration, of course, they're still around. So here they are. They're still around, they're wonderful. But now they produce some offspring and here are the offspring. Aren't they cute? Little pair—one boy, one girl. Great. Now what happens at the end of the following month? Well, of course, this pair is still here, it will never, ever go away; that pair is always here. Now, this pair actually grows up and now is mature enough for the following month to reproduce. But this pair reproduced as well, because every month they're going to reproduce from now on (now that they're adults). So, they actually reproduce yet again, and they produce another pair of babies.

Well, what happens now in the next iteration? So, at the next iteration, both of these adult pairs live on and reproduce. But also, this baby pair has now come of age, so they actually now are grown up. But don't forget that these two pairs of adults have reproduced yet again, so now we see two pairs of bunnies, which I'll just fit in here. I think we can fit them in here. As you can see, this little pen of space, this confined space, better be large enough for all of them. But let's see the pattern in how many rabbits we have during the months. We start with 1, at the end of the month we still have 1, then we have 2, then we have 3, then we have 5, and you can actually see that at the end of the next month we're going to have 8. Why? Well, because all of the three adults are going to reproduce three new pairs, and all these pairs are going to live. So we have 1, 2, 3, 4, 5, and then we add 3—we see 8.

Following the rules, we see that the number of pairs of rabbits present are always equal to Fibonacci numbers. Now, how many do we have at the end of the year? Well, I could leave this to you to figure out if you want, but I'll just give you a hint. It's 233 pairs in the man's enclosed place. So, I hope the enclosed place is sufficiently large. Now, this solution was the first time that this list of numbers appeared in the literature of Western mathematics, even though nature has held this list much, much earlier. Fibonacci numbers, throughout the ages, have been a source of inspiration and continue to attract the attention of mathematicians and math enthusiasts from around the world. There are many results about Fibonacci numbers. New theorems involving the Fibonacci numbers continue to be discovered by mathematicians today. This is a very, very active area. In fact, there is a mathematics journal named *The Fibonacci Quarterly*, where people publish articles on the Fibonacci numbers.

The Fibonacci numbers are connected to the ancient and aesthetically appealing golden ratio studied by the Pythagoreans. I want to remind you of what the golden ratio is and then try to elucidate that connection. Well, the golden ratio is a number that is equal to $1 + \sqrt{5}$, all divided by 2. If we were to calculate that number, we would see 1.6180339, and the decimals continue. Well, what does this very cryptic, mysterious number have to do with the more natural Fibonacci numbers, all of which are examples of natural numbers? Well, we saw consecutive Fibonacci numbers appearing in the spirals of nature. So we see them as two next to each other—one and then the next one appearing in the spiral counts.

Let's figure out, in fact, how fast they're growing by taking a look at the relative size of two consecutive spiral counts. In other words, let's compare two of the adjacent Fibonacci numbers as we go further and further out in our list, by looking at their relative size, or their ratio. So let's divide. We'll divide—1/1 (the first two Fibonacci numbers), then 2/1, then 3/2, 5/3, 8/5, 13/8, 21/13—to compare how these numbers are growing relative to each other. Now, if we look at the decimal expansion for that, we see some interesting structure in the decimal. For example, notice that 13/8 is equal to 1.625, and 21/13 has a decimal that begins 1.615. If we go a little further out, we see 144/89 = 1.6179, and so forth. Even further out, 610, a large Fibonacci number, divided by 377, the Fibonacci number right before it, equals 1.618037, which, notice, agrees to the golden ratio's decimal expansion to five places. Well, in fact, the further out we go—and you might want to enjoy trying this as an exercise to see that if we take two consecutive Fibonacci numbers, the larger they are when we look at their ratio and divide the larger by the smaller, we get a number that's getting closer and closer to this magical golden ratio.

Two comments about this observation. First of all, the golden ratio we now see, in some sense, is intrinsically involved in nature. Many people believe that the golden ratio appears in works of art and has aesthetic value, and this remains open, although some artists have been known to deliberately incorporate this number, the golden ratio, in their artistic works—even in works of music for example. But for our purpose here, the interesting foreshadowing is that we have a collection of rational numbers, a collection of fractions, which are heading closer, and closer, and closer to this more complicated number. This is an interesting point that will actually be the key to unlocking more complicated numbers—that when a number is exotic and complicated, what we tend to do is, in fact, try to approximate it by simpler numbers, in this case, by fractions. So this theme (that we can approximate a complicated number by a succession of simple numbers) is a very important one that will allow us to move the frontiers of mathematics and number theory forward.

Well, for Fibonacci fans, it might be sad that there are natural numbers that are, alas, not Fibonacci numbers. The smallest example of this would be 4. Notice that 4 is not on our list of Fibonacci numbers, nor is 6 or even 7. So this could be a source of some strife for those who are real fans of the Fibonacci numbers. Well, happily, in 1972 Edouard Zeckendorf proved a beautiful theorem, and his result states that if a natural number is not itself a Fibonacci number, then it can be written, uniquely, as a sum of nonadjacent Fibonacci numbers. Now, this is a mouthful, but it's basically saying that, if you're a natural number, you're either going to be a Fibonacci number, or I can write you as a sum of Fibonacci numbers in a particular way.

So let's take a look at an example to have this make a little bit more sense and to be real for all of us. Let's pick 4, which, of course, we already saw was not a Fibonacci number. Notice, very easily, that there are two numbers that are Fibonacci numbers that sum to 4—in particular, 3 and 1. 3 + 1 = 4, and you'll notice that 3 and 1 are the only two ways that we can add nonadjacent, different Fibonacci numbers to produce 4. Notice that they are nonadjacent in the sense that there's a Fibonacci sandwiched between 1 and 3—in particular, 2. What about 6? Well, now we can write 6 as 5 + 1, two Fibonacci numbers. That's the only way to write 6 as distinct Fibonacci numbers that aren't next to each other. Seven also can easily be written as

5 + 2. So, we see what we mean by taking a non-Fibonacci number and expressing it as a sum of Fibonacci numbers. Zeckendorf actually proved that, in fact, there's only one way of doing this if you, in fact, do not allow repeats and do not allow the Fibonacci numbers to be next to each other.

This decomposing natural numbers and expressing them into Fibonacci numbers is now known as the "Zeckendorf Decomposition." I want to explain to you why this is true and how you can actually express any natural number you wish as the sum of Fibonacci numbers in this way. It's actually a neat process that illustrates an interesting idea. So, to express a natural number in this manner, we apply a "divide and conquer" strategy, which will be very important in our next lecture. So, the rule is that we first find the largest Fibonacci number that's less than the natural number we wish to express. So think of it this way. We have this number that we wish to write as the sum of Fibonacci numbers. We start going backwards in our list of natural numbers until we hit the very first Fibonacci number. So, that's the largest Fibonacci number smaller than our number. Then we take our number and write it as this Fibonacci number plus whatever we need to. So add a remainder, if you will, to make it equal to the original number.

Now, we've actually made progress, and this is the "divide and conquer" idea. For example, let's take a look at 30, which is a fairly large integer, a natural number. The first thing I do is I go backwards from 30 until I hit my very first Fibonacci number, and we'd see 21. Then I ask, "Okay, that's a Fibonacci number. What do I have to add to 21 to make it equal 30?" Well, the answer is 9. So, I've made some progress, I wrote 30 as 21 + 9. Now, alas, 9 is not a Fibonacci number, but notice two things. First of all, I have a Fibonacci number 21 in the picture, and 9 is a number smaller than my original 30. This is why I divided and conquered. All of a sudden, I've reduced the complexity of my question to an easier one, Now what do I do? Well, what I do is I look at 9, and I say, "How can I express 9 as the sum of Fibonacci numbers?" I use the same process. So, I divide and conquer again. I start at 9, and I go backwards in my list of natural numbers until I hit my very first, or the largest, Fibonacci number that's less than 9. In this case, it's easy to see that number's 5.

What do I have to add to 5 to make it equal to 9? Well, the answer is 4. So now I see that 30 = 21 + 5 + 4. Now I have 21 and 5, which are both

Fibonacci numbers, but 4, unfortunately, is not. But we see what to do—we divide and conquer again. We see, as we saw before, that the number 4 is equal to $3 + 1$, so I see that 30 can be expressed as $21 + 5 + 3 + 1$. There I've written 30 as the sum of Fibonacci numbers, using this method of finding the biggest Fibonnaci number that's smaller than my number and continually dividing and conquering. Zeckendorf proved as a theorem that, in fact, this method always works. There's only one way to represent numbers in this fashion, and in fact, if you force that the numbers are always different from each other, and that the Fibonacci numbers that we use are never next to each other, then this way of writing the numbers is, in fact, unique. This result can be proved in general, and that's what Zeckendorf did, although you can see how the method can be used to work in any way.

This notion of expressing natural numbers in terms of smaller, special natural numbers really foreshadows one of the most important ideas from number theory, an idea whose genesis can be found in the ancient work of Euclid. Euclid's historic observations are really at the heart of the next lecture. But for now, we should appreciate that, all around us, there's not only beautiful nature, but within that nature there are beautiful patterns, and those patterns involve number.

Numbers of Prime Importance
Lecture 7

The formal study of prime numbers dates back to the ancient Greeks, when Euclid first established the fact that there are infinitely many primes. In fact, Euclid's proof that there are infinitely many prime numbers is considered by many to be one of the most elegant arguments in all of mathematics.

What is a prime number? A prime number is a natural number greater than 1 that cannot be written as a product of two smaller natural numbers. The first few prime numbers are 2, 3, 5, 7, 11, 13, 17, 19, and 23. Why is 1 not a prime? The primes that evenly divide a number reveal intrinsic features of that number (e.g., if the prime 2 evenly divides a particular number, that tells us that the number must be even). The trivial fact that any number is itself multiplied by 1 (e.g., $6 = 1 \times 6$) leads to no new insights into that number. Therefore, 1 should not be viewed as a prime number. The more theoretical explanation is that 1 is the *only* natural number whose reciprocal (namely, $\frac{1}{1}$) is also a natural number. Because of this special property, 1 is called *unity*. In advanced mathematics, unity is never considered a prime.

Around 300 B.C.E., Euclid was the first to prove theorems involving prime numbers. He was the author of 13 books known as the *Elements of Geometry*. Many believe these to be some of the most important treatises ever written. Euclid pulled many ideas together for the first time into a unified whole and established the notion of rigorous proof that remains with mathematics to this day.

Books VII, VIII, and IX of the *Elements* contained more than 100 theorems from number theory. Euclid's Proposition 14 stated that every natural number greater than 1 can be expressed uniquely as a product of prime numbers (except for the rearrangement of the factors). For example: $12 = 2 \times 2 \times 3 = 2 \times 3 \times 2$. This important result is now known as the "fundamental theorem of arithmetic." We note that the uniqueness aspect of this result would no longer hold if we considered 1 a prime number ($6 = 2 \times 3 = 1 \times 2 \times 3$).

Proposition 20 of Book IX states that prime numbers are more than any assigned multitude of prime numbers. Today, we would rephrase Euclid's result by stating that there are infinitely many primes. In order to inspire the beautiful idea in Euclid's proof, we explore how to use the primes 2 and 3 to find a third prime. We wish to create a number that is not a multiple of 2 or 3. We first multiply the numbers 2 and 3 together, then add 1: $2 \times 3 + 1$. We note that $2 \times 3 + 1 = 7$, a number that has neither 2 nor 3 as a factor; thus, there must exist a third prime (in this case, 7 itself happens to be a prime).

Using the previous idea, we prove Euclid's theorem in general. Suppose we have a finite list of primes (2, 3, 5, 7, ..., p, with p denoting the last prime number on our list), and we wish to show that there exists a prime number that is not on our list. We consider the number $(2 \times 3 \times 5 \times 7 \times ... \times p)$ + 1 and notice that none of the primes on our list can divide evenly into this number. By the fundamental theorem of arithmetic, there must exist a prime that divides this number that was not on our original list. Because this argument can be applied to any finite list of primes, we conclude that there are infinitely many primes.

Our understanding of the primes has advanced since Euclid's time. How many primes are there up to a certain point? We know by Euclid's theorem that there are infinitely many primes. Is there a way, however, to know how many primes there are up to any particular natural number n? We write $P(n)$ to represent the number of prime numbers less than or equal to n. For example, $P(5) = 3$ because there are three primes less than or equal to 5 (2, 3, 5). Similarly, $P(20) = 8$ (the eight primes are 2, 3, 5, 7, 11, 13, 17, 19). Is there a formula for $P(n)$ in general? This question remains unanswered to this day. Many mathematicians throughout the ages contemplated this question, including Carl Friedrich Gauss (1777–1855), whom many consider the "Prince of Mathematics."

The prime number theorem can be explained as follows. Many individuals (including Gauss) noticed that the number $P(n)$ was closely approximated by $\frac{n}{\ln(n)}$; $\ln(n)$ is known as the natural logarithm, and its value can be found on any scientific calculator. This result was finally proven to hold by Jacques Hadamard and Charles de la Vallée Poussin independently in 1896. The prime number theorem implies that as n gets larger and larger, $P(n)$ gets closer and closer to $\frac{n}{\ln(n)}$. More precisely, as n gets larger and larger, $\frac{P(n)}{\frac{n}{\ln(n)}}$ approaches 1. Because the ratio above *approaches* 1, there is always some ever-shrinking gap between that ratio and 1. That gap can be viewed as an "error." Exploring how the "error" in the previous result is shrinking is a very active field of study (an important part of analytic number theory) and is connected to the famous open question known as the Riemann Hypothesis.

Finding the largest prime known to date is an ongoing quest that has captured the imagination of thousands of people and computers. The largest prime known today was found in September 2006 using a number of supercomputers. It is $2^{32582657} - 1$ and has 9,808,358 digits; printing it out in 12-point font would require more than 2,500 pages. This large prime is an example of a Mersenne prime (a prime of the form $2^n - 1$); this is the typical form of the large primes that are found today.

Finding the largest prime known to date is an ongoing quest that has captured the imagination of thousands of people and computers.

Beyond the prime 2, the closest two adjacent primes can be to each other is two numbers. Two primes whose difference is 2 are called twin primes. For example, (3, 5), (5, 7), (11, 13) are twin primes, while (7, 11) are not twin primes. The Twin Prime Conjecture states that there are infinitely many twin primes. Although many mathematicians believe that this statement is true, no one has yet produced a complete proof of it.

Can every even natural number greater than 4 be expressed as the sum of two odd prime numbers? This is known as the Goldbach conjecture. Notice that $6 = 3 + 3$, $8 = 3 + 5$, $24 = 5 + 19$, and even $1,000 = 3 + 997$. This question, first stated in a letter by the Prussian mathematician Christian Goldbach to the Swiss mathematician Leonhard Euler on June 7, 1742, is one of the oldest unanswered questions in mathematics today. It is known that the conjecture holds for all even numbers up to 3×10^{17}. There is a \$1 million prize for the first correct and complete proof of this conjecture. ■

1. Apply the strategy Euclid used to prove there are infinitely many primes to provide an argument for why there must be a prime greater than 1,000,000.

2. Prime numbers can be viewed as the building blocks, or "atoms," of the natural numbers. What other structures or systems in our lives have atoms?

Numbers of Prime Importance
Lecture 7—Transcript

In this lecture, we'll introduce the concept of prime numbers—those natural numbers greater than 1 that cannot be written as a product of two smaller natural numbers. The formal study of prime numbers dates back to the ancient Greeks, when Euclid first established the fact that there are infinitely many primes. In fact, Euclid's proof that there are infinitely many prime numbers is considered by many to be one of the most elegant arguments in all of mathematics. Here, we'll also discover that the prime numbers are the fundamental multiplicative building blocks for all natural numbers that exceed 1. In other words, every natural number larger than 1 is either a prime number or can be expressed uniquely as a product of prime numbers. Well, there appears no particular pattern to the prime numbers. Great mathematical minds, including Carl Friedrich Gauss, wondered how many primes there were between 1 and any given number. The answer, now known as the prime number theorem, was given independently in 1896 by Jacques Hadamard and Charles de la Vallée Poussin. The study of primes remains a very active area of research today, and this area is known as analytic number theory. We'll close this lecture by describing several questions involving the prime numbers that remain unanswered to this day.

Let's begin by carefully describing these special numbers. In the last lecture, we saw how we could divide and conquer the natural numbers using the Fibonacci numbers and addition. Here, we'll continue this "divide and conquer" theme but leave the Fibonacci numbers behind and now focus on multiplication rather than addition. We define a prime number as any natural number larger than 1 that cannot be written as the product of two smaller natural numbers. So, for example, 5 is a prime number because we can't express it as the product of two smaller natural numbers. Six, on the other hand, is, in fact, not a prime number because 6 can be expressed as 2×3, two natural numbers (each smaller than 6) whose product yields 6. Five is prime; six is not. The first few prime numbers are 2, 3, 5, 7. Now, 9 is not a prime because 9 is 3×3, two smaller natural numbers that give 9. So we don't include 9. But 11, and 13, 17, 19, 23—these numbers cannot be divided and conquered, and in some sense they're the atoms of the natural number. They're the smallest pieces.

Now, why isn't 1 a prime? How come we always talk about these numbers that are greater than 1? Well, the primes that evenly divide a number reveal intrinsic features of that number. For example, if the prime 2 evenly divides a particular number, that tells us that the number must be even. Thus, the primes will allow us to discover, in some sense, the personality of natural numbers, or the makeup of the numbers (an arithmetic personality, if you will, rather than the more human personalities that the Pythagoreans associated with numbers). The trivial fact that any number can be viewed as itself multiplied by 1 (for example, $6 = 1 \times 6$) leads to no new insights into that number. Therefore, 1 shouldn't be viewed as a prime number. The more theoretical explanation, which isn't so important for us here, is that 1 is the only natural number whose reciprocal, whose inverse (namely 1/1, which is 1), is also a natural number. Because of this special property, today we refer to 1 as a "unit," which comes from *unity*, from the Pythagoreans. In advanced mathematics, units are never considered as prime numbers.

We now celebrate the work of the great Greek mathematician Euclid. Euclid, around 350 B.C.E., was the first to prove theorems involving the prime numbers. Now, he was the author of 13 books known as the *Elements of Geometry*, and many believe these to be some of the most important treatises ever written. Euclid pulled many ideas together, for the first time, into a unified whole and established the notion of rigorous proof that actually remains with mathematics to this very day. Euclid's Books Seven, Eight, and Nine of the *Elements* contained over 100 theorems from number theory. Again, remember that the Greek sense of geometry included number theory because they were concerned with measuring lengths. That's how they viewed number—as lengths of things. Now, his Proposition 14 stated that every natural number greater than 1 can be expressed uniquely as a product of prime numbers, except for possibly rearrangement of the factors. Let's look at an example. Take 12. Well, we could write 12 as $2 \times 2 \times 3$. Those are all prime numbers. Now we could also write it as $2 \times 3 \times 2$, or even, $3 \times 2 \times 2$, but notice that, in all those cases, all we did was rearrange those factors. We always had two copies of the prime 2 and one copy of the prime 3. There's only one way to write 12 as a product of primes if we ignore the order that we write the factors in.

This important result is now known as the "fundamental theorem of arithmetic," and in mathematics, "fundamental" doesn't mean easy or basic; it means important. So this is, in some sense, one of the most important theorems in arithmetic. By the way, we note that the uniqueness aspect of this result would no longer hold if we considered 1 as a prime number, because then we could write 6, on the one hand, as 2×3 and, on the other hand, as $1 \times 2 \times 3$. So, if 1 were a prime, we'd no longer have unique factorization. Well, we now arrive at the age-old question: How many primes are there? Well, Proposition 20 of Book Nine states the following: "Prime numbers are more than any assigned magnitude of prime numbers." Well, I want to pause here to see exactly what this means. "Prime numbers are more than any assigned magnitude of prime numbers." Suppose we think of a collection of prime numbers. Let's say 100, so we look at 100 prime numbers. Euclid's theorem states that the prime numbers are more than that, which means there are more than 100 primes.

Let's consider a different number, maybe 1 million. So, the assigned magnitude will now be 1 million. Well, Euclid's theorem says that the prime numbers are more than 1 million prime numbers. So, what he's saying here, in today's language, is that there are infinitely many prime numbers. Now, Euclid wasn't thinking in the language of infinitely many, because infinity really was not a notion that was being considered at the time, but this is how he phrased it. Any magnitude you think of—the primes will be even greater than that. Today, we'd say there are infinitely many prime numbers. Euclid's proof, which we'll see for ourselves in just a moment, is considered by most mathematicians today to be one of the most elegant proofs in all of mathematics. Perhaps it's a little surprising upon first blush that there should be a notion of aesthetics or beauty in mathematics and mathematical proof, but indeed there is; and we'll see for ourselves several theorems and their proofs that are considered aesthetically appealing to mathematicians. I'll try to articulate what the aesthetic appeal is so that we can hone our own sense of aesthetics within a mathematical and a numerical realm.

Let's now consider the beautiful idea of Euclid that allowed him to rigorously prove that there are, in fact, infinitely many prime numbers. Now, one strategy would be to say, "Well, let's just list them all." Well, of course, that's not a good argument, because if there really were infinitely many, as we'll show, then that list would take us forever to write. A mathematical argument, in fact, has to be a finite number of steps of logic, reasoning, and previous theorems. The proof has to end. So we're up against an interesting challenge. We have to prove infinitely many things exist in only finitely many steps. So, we have to be really clever. Happily, Euclid was very clever for us. Instead of actually tackling this harder question, let's warm up to it to build some intuition into Euclid's wonderful idea by trying to do something very, very simple. Remember, when we're faced with a hard challenge, let's not do it. Instead, let's retreat and do something easy. So, we'll attempt to simply find a prime larger than 3, which seems a little frivolous, because you could immediately say, "Well how about 5." Well, yes, but by that method, you're using the fact that 3 is a small prime, and we know facts about small primes.

What I'd like to do is just use the primes 2 and 3 and try to find another prime, but never using the fact that 2 and 3 are small and familiar to us. For if we could craft an idea that never used the fact that 2 and 3 were small, and we could build another prime, then we could maybe extend the idea to prove this for any number of primes. Euclid's idea was to create a number that is neither a multiple of 2 nor 3. Well, that natural number, by the fundamental theorem of arithmetic, has to be able to be written as a product of prime numbers. But if 2 and 3 are not factors in this number, then 2 will not appear in the list of primes, 3 will not appear in the list of primes, and yet we know that there is this list of primes. So that list of primes must be primes that are not 2 or 3. We've just found a prime that's not 2 or 3. That's his idea—to build a number that definitely will avoid 2 and 3 as factors. Well, here's what he did. He first multiplied the two primes 2 and 3 together. Well, 2 and 3—if you take that product, of course, that is a multiple of both 2 and 3, so it seems like this is not a good choice. But then he does something truly ingenious: He adds 1. So, he looks at 2 multiplied by 3 and then adds 1.

Now, let's just first note that that number is 7 and notice that 7 is indeed a prime number that's neither 2 nor 3. So, that's exactly what we were trying to do. But I want us to think about this abstractly and not actually compute the fact that these numbers are so small that we can see 7. Let's look at the representation of 7 as $(2 \times 3) + 1$. How could I look at just that number and see that 2 is not a factor of it? Well, because we could see that the $+ 1$ means that 2 won't divide evenly into $(2 \times 3) + 1$ because I have a remainder of 1. Similarly, because that 3 is being multiplied by the 2 in front there, I see that 3 can't be a factor of this number because there's a remainder of 1 again when I divide by 3. So therefore, without even computing the number 7, I see that neither 2 nor 3 can be a factor of $(2 \times 3) + 1$, because the $+1$ messes up the works. Therefore, there must exist a third prime. Now in this case, the number 7 happens to be prime itself, but that just happens to be a happenstance of the small number.

Using this idea, we can now prove Euclid's theorem in general. So, let's consider this. We're going to take the exact same idea. This is now the complete proof that Euclid came up with. Suppose we have a finite list of prime numbers—for example, 2, 3, 5, 7, and all the way out to a last prime number. I don't know what that last prime number will be called. I'll call it p. So I'll use the letter p just to distinguish the last prime number on this finite list. Maybe there are 20 of them, maybe there are a million of them; but it's a finite list of prime numbers and p will denote the last one. Well, we now wish to show that there exists a prime number that's not on this list, and what do we do? Well, we apply Euclid's idea. We consider the number we get by taking 2, and multiplying it by 3, and multiplying it by 5, by 7—all these primes—up to p. We look at that product, and then what do we do? We add 1. So we add 1 to this number, and we notice that none of the primes on our original list can divide evenly into this number because there's always this remainder of $+1$ at the end. For example, if we divide this big number by 5, because 5 occurs in this product, that $+1$ is going to be the remainder. So, it doesn't go in evenly. Five doesn't go evenly into this number, nor does the last prime, p, that we have, because, again, we see a remainder of 1 when we divide this product $+1$ by p.

Now, by the fundamental theorem of arithmetic, we conclude that there has to be a prime dividing this number, because every natural number can be expressed as a product of primes. So, there's a prime that's a factor of this number and divides in evenly into this number, and yet it can't be any of the primes from our original list from 1 to p. This implies that there must exist at least one other prime that's not on our list. Because this argument can be applied to any finite list of primes (not just up to p, but any number you want), we conclude that, in fact, the number of primes can't be bounded by any number. Therefore, there are infinitely many. If we think there are only finitely many, we take their product and add 1, and there must be a prime dividing that enormous number; and that prime is not on our list, so we missed that. We can do this for any finite collection. Therefore, there are infinitely many primes.

Now, this theorem is truly beautiful and elegant. Why? Well, because we're proving a dramatically profound mathematical result, a result that really allows us to understand numbers in a richer way. Yet the argument itself—while complicated and requiring some serious contemplation to take it all in—the idea is very beautiful in that it's so basic. We just take a product of numbers and add 1. Now you might look at this later and say, Gee, that seems straightforward, I guess. But the genius was to come up with the idea of multiplying all those numbers together and adding 1. That's by no means obvious. Once a mathematician sees that idea, she can certainly go through and complete the proof, but finding that idea really requires some genius. So this is considered one of the most beautiful arguments, theorems, in all of mathematics.

I want to offer a few words about some modern advances in our understanding of the prime numbers. Well, how many primes are there up to a certain point? We know by Euclid's theorem that we just talked about that there are infinitely many primes, but is there a way to know how many primes there are up to any particular number? For example, how many primes are there up to 5? Let's call that number of primes P(5) (we read that as "P of 5"). How many primes are there up to 20? Let's call that number of primes up to 20, P(20). How many primes are there up to 1,000, from 1 to 1,000? How many primes are in that list of natural numbers? Let's call that P(1,000), so whatever that number is, we'll denote it as P(1,000), and more generally, if I'm given any natural number n, let me call P(n) the representation of the number of primes that are less than or equal to n.

For example, P(5) would equal 3 because there are three primes that are less than or equal to 5—namely 2, 3, and 5. There are three primes. Therefore, P(5), the number of primes that are less than or equal to 5, would equal 3. Similarly, we have that P(20), which represents the number of primes less than or equal to 20, would actually equal 8, because there are 8 primes that are less than or equal to 20, and they are 2, 3, 5, 7, 11, 13, 17, 19. So, we see, in fact, that many.

P(1,000), which I'm not going to recite for you, is equal to 168. That means that there are 168 prime numbers between 1 and 1,000. So, that's what this function means. Now the question is, is there a formula for P(n), in general? That is, is there a formula that tells us how many primes there are up to any given point? Well this question remains unanswered to this very day. Many mathematicians throughout the ages contemplated this question, including the great 19th-century mathematician Carl Friedrich Gauss. Gauss is considered by most mathematicians today to be one of the greatest mathematicians of all time and, really, the father of modern number theory as we know it. Gauss once wrote that "Number theory is the queen of all mathematics." Given this royal theme along with the stature and respect that the mathematics community has for him, Gauss is often referred to as "the prince of mathematics." So, you get a sense of the royalty that the math community has bestowed upon this great mind.

Well, let's now return back to this P(n), the number of primes there are up to a number n. Many individuals, including Gauss himself, noticed that the number P(n) was closely approximated by something very interesting, and that number is $n \div ln(n)$, the natural logarithm of n. Now, what is the natural logarithm of n? For our purposes, it's not that important, but let me just say a word about it. First of all, if you want to compute the natural logarithm, you can just do it on a calculator. You can get the value fairly easily on any scientific calculator. But if you want to think about what it means—basically, the natural log of a number is roughly how many digits that number has. More precisely, it's approximately 2 times the number of digits. In fact, even more precise—$ln(n)$ is equal to around 2.3-something times the number of digits of n. For our purpose here, let's just think of $ln(n)$ as, roughly speaking, the number of digits that's required to write the number n out. So, for example, the number 362 would have a value around 3, because there are three digits. For our purposes, that's fine. Well, the conjecture, the observation, that people made was that the number of primes up to the point n, P(n), is approximately equal to $n/ln(n)$, or roughly the number of digits that n has.

Now, this statement was finally declared to be a theorem in 1896 when two mathematicians, working independently, offered a proof of the result. Now today, this important result is known as the prime number theorem, and it was proved by Jacques Hadamard and Charles de la Vallée Poussin. Now, the interesting thing here is that they proved this totally independently in the same year. Amazing, right? Well, remember my copper orb. When our understanding of mathematics grows to a certain point, then these big questions naturally are ready to fall, and before that point we need a dramatic, big, new idea before we can make progress. So here, mathematics and our mathematical understanding, humankind's understanding of math, got to a point where this prime number theorem was ready to fall—to be proven. It turned out that two individuals, at two different places at the same time, saw that insight and proved the theorem. This happens quite a bit in mathematics when the orb goes out.

Well, the prime number theorem gives us that as n gets larger and larger, as the number gets larger and larger, the number of primes between 1 and n gets closer and closer to this funny number $n/ln(n)$. Now, how can we make that more precise? Let me try to actually state the theorem in some precision. How will we know if two numbers are the same? We know that two numbers are the same—one way is if we divide one into the other and get the answer of 1. For example, one way, a silly way, to see that 7 is the same as 7 is to divide 7 by 7, and we get 1. If two numbers are identical, their quotient will yield 1.

Well, in our context, the theorem actually says that if we look at the ratio of the number of primes up to a certain point and look at the ratio between that and this other thing, $n/ln(n)$, that ratio will get closer and closer to 1. It will be coming closer and closer together as n gets larger and larger. So in fact, the prime number theorem actually can be stated as—this function $P(n)$, the number of primes up to $n \div n/ln(n)$—if we take those two quantities and compare it, as n gets larger and larger, this ratio *approaches* 1. So these two values are getting closer and closer to each other. That's the actual statement of the theorem, in fact.

Because this ratio only approaches 1, for any particular fixed value of n that you think of, there's always going to be a little gap between the actual value of primes up to that point and this peculiar number $n/\ln(n)$. This gap is often viewed as the "error" or an "error term," and exploring how the error in this result is shrinking as the end gets bigger is a very active field of study, an important part of analytic number theory, and is in fact connected to one of the most famous and important open questions in all of mathematics today, known as the Riemann Hypothesis (named after Bernhard Riemann, the great 19th-century mathematician who first conjectured an answer). Well, it's interesting that even after thousands of years of serious study, many questions remain unanswered about the prime numbers.

Even though there are infinitely many prime numbers, you might wonder, What is the largest prime number that we know is prime today? Finding the largest prime number known to date is an ongoing quest that has captured the imagination of thousands of people and computers. Because the methods used to establish that these numbers are indeed prime—the numbers typically have a very special form. So, these large numbers that are established to be prime look the same way. The largest prime known today was found in September of 2006 using a number of supercomputers all collaborating and working together. Now, I'll tell you what the number is, let me just describe the form first. It's going to be 2 raised to a really big exponent, and then I'm going to subtract 1. So, the number is: $2^{32,582,657} - 1$. So take 2 to that really big power, take your answer, and subtract 1. That number has 9,808,000 digits. If you decide to print out that number in a 12-point font, it would require over 2,500 pages to have that printed out. That number is prime, and that's the largest prime that we know of today. But of course, we also know that there are infinitely many primes, so there are many more out there. In fact, most of them are still out there waiting for us to find them.

This large prime is an example of what's called a Mersenne prime. That's a prime that's of the form $2^n - 1$. Any prime of that form is called a Mersenne prime, named after the 17th-century French monk Marin Mersenne, who studied primes of this form. Again, this is a typical form of the large primes that we tend to generate in today's age. Well, in fact, after the 23rd Mersenne prime was found in the late 1960s, the members of the Department of Mathematics at the University of Illinois at Urbana were so excited that they actually put that number on their postage stamp. That number, by the way, was $2^{11,213}$, and again, minus 1. By the way, the largest prime known today (that prime that has over 9,800,000 digits) is only the 44th Mersenne prime known. So, this large one that we just found was the 44th Mersenne prime. Now, while we know that there are infinitely many primes, no one knows if there are infinitely many Mersenne primes. That is, no one knows if there are infinitely many primes of the form 2 to a power, minus 1. Again, we see this is a refined question, because we know that there are infinitely many prime numbers, but the question is, are there infinitely many prime numbers that have a certain look?

Another refinement comes from considering pairs of prime numbers. Now notice that beyond the prime 2, the closest two adjacent, or consecutive, primes can be from each other is 2, because between every prime number we have to have at least one even number. After 2, all the primes are odd numbers. We couldn't have a big, even prime number because then 2 would divide into it evenly, and we could break it apart into smaller pieces. So all the primes are odd after 2, and therefore the closest they could be apart would be 2 apart if there's one number in between them. Two primes whose difference is 2 are called "twin primes." So, for example, 3 and 5 are twin primes, because they're both primes, and they're 2 apart from each other. Five and 7 are also an example of twin primes, as are 11 and 13, because they differ by 2. These are all twin primes, but 7 and 11, even though they are consecutive primes, they're not twin primes because their distance from each other is not 2.

One question that is still open to this day is, are there infinitely many twin primes? The Twin Prime Conjecture states that indeed there are infinitely many twin primes. While many mathematicians believe this statement to be true, no one has yet to produce a complete proof, and remember—in mathematics, just because we think something is true is not enough; we have to prove it for certain. Another famous unsolved question is known as the Goldbach conjecture. This question asks, can every even natural number greater than 4 be expressed as the sum of two odd prime numbers? Notice, for example, that $6 = 3 + 3$, two odd primes; $8 = 3 + 5$, again, two odd primes; $24 = 5 + 19$, again, two odd primes. Even 1,000 can be written as $3 + 997$, and 997 is a prime, as is 3.

So the question is, can every even number bigger than 4 be expressed as the sum of two odd primes? It's a tricky question, because usually prime numbers are multiplied together (as in the fundamental theorem), and here we're doing something peculiar: We're trying to add them. Well, this question, first stated in a letter by Prussian mathematician Christian Goldbach to the great Swiss mathematician Leonhard Euler on June 7, 1742, remains one of the oldest unanswered questions in mathematics today. This conjecture, which is known as the Goldbach conjecture, is known to actually hold for every single even number up to 3×10^{17}. That means that if you imagine the number 3 and put 17 zeros after it, all the even numbers before that number can be expressed as the sum of two odd primes. But even this enormous number doesn't prove that it's true always; maybe out there in the stratosphere of natural numbers there's a counterexample—an even number that can't be expressed in this manner. There's actually a bounty on this question. There's a $1 million prize to the first correct and complete proof of the conjecture. But I urge you to consider other investment opportunities before investing your time in working on this notoriously hard question, unless you enjoy the quest of thinking about prime numbers, in which case it would then be a totally worthy endeavor—but don't do it for the money.

The prime numbers are a wonderful collection of numbers that form the atoms of our natural numbers. They've been studied since Euclid and remain a fascination to all who think about them.

Challenging the Rationality of Numbers
Lecture 8

> Thousands of years after the Pythagoreans were faced with irrational lengths, irrational numbers continue to disturb, intrigue, and bother mathematicians.

W hat are rational numbers, and when are they recognized as such? A rational number is a ratio of two integers $\frac{m}{n}$, for which n is positive. The number m is called the numerator and n is called the denominator. Rational numbers are also known as fractions. Although we do not know the first example of rational numbers, there is evidence of fractions in Babylonian and Egyptian writings from around 2000 B.C.E.

A unit fraction is defined to be any fraction having its numerator equal to 1 (e.g., $\frac{1}{2}, \frac{1}{3}$, and $\frac{1}{4}$); the Rhind Papyrus contains sums of unit fractions. Ahmes, the scribe who wrote the Rhind Papyrus, denoted unit fractions by a dot over the denominator. For example, he would write a dot over the number 5 for the unit fraction $\frac{1}{5}$. This dot notation evolved from the hieroglyph for an open mouth. This symbol offers evidence that fractions were used to divide rations of food and drink. Many questions from the Rhind Papyrus concerned dividing loaves of bread and jugs of ale. In addition, the Rhind Papyrus contains a table that shows how, for any odd natural number n from 5 to 101, fractions of the form $\frac{2}{n}$ can be written as the sum of different unit fractions (e.g., $\frac{2}{5}$ can be written as $\frac{2}{5} = \frac{1}{3} + \frac{1}{15}$). An Egyptian fraction is the sum of distinct unit fractions. It is not clear why the Egyptians chose this method of expressing fractions. Fibonacci showed that every positive rational number can be expressed as an Egyptian fraction.

The ancient Greeks—in particular, the Pythagoreans—were one of the first cultures to explore rational numbers as objects of independent, abstract interest. The ancient Greeks had a complicated view of number. They believed that the natural numbers were the true numbers because they were God-given. The rational numbers were really only ratios of natural numbers—they were not as real as the natural numbers. In the eyes of the Greeks, however, the rational numbers formed a gap-free list that could correspond to a line of numbers; the Pythagoreans viewed these numbers as a "flow of quantity."

Euclid made the Greek notion of rational numbers much more rigorous. In *Elements*, he declared that two line segments A and B are commensurable if there exists a third line segment C such that some natural number copies of C, when laid end-to-end, produce the segment A (and similarly for the segment B). If we call the lengths of these line segments a, b, and c, respectively, then we see that there exist natural numbers m and n such that $a = cm$ and $b = cn$. We observe that the ratio of a to b is as follows: $\frac{a}{b} = \frac{cm}{cn} = \frac{m}{n}$. We say that two numbers a and b are commensurable if $\frac{a}{b}$ is a rational number. The Greeks believed that every two lengths (numbers) were commensurable; thus, the Greeks believed that all lengths were either natural numbers or ratios of natural numbers—what we today call rational numbers.

A certain problem revealed the fallacy in this thinking. Suppose we consider a square having a side length of 1 unit. If we draw a diagonal in that square,

The ancient Greeks—in particular, the Pythagoreans—were one of the first cultures to explore rational numbers as objects of independent, abstract interest.

then we can measure its length. In view of the Pythagoreans' belief that all length numbers are rational, the length of the diagonal of our square must be a rational number.

The Pythagoreans made a surprising discovery. Labeling the lengths of our line segments leads us to their discovery. We let h represent the length of the diagonal of the square. Given the Greeks' belief, h must be a rational number (let us call it $\frac{m}{n}$) for some natural numbers m and n. Thus, $h = \frac{m}{n}$.

The Pythagorean Theorem states that the sum of the squares of the lengths of the legs of any right triangle equals the square of the length of the remaining side (the hypotenuse). In our right triangle, we see that $1^2 + 1^2 = h^2$, or equivalently, $2 = \left(\frac{m}{n}\right)^2$; thus, we have $2 = \frac{m^2}{n^2}$. Solving for $\frac{m}{n}$, we find that $\sqrt{2} = \frac{m}{n}$; that is, $\sqrt{2}$ must be a rational number. Therefore, the natural numbers n and m satisfy the equality: $2n^2 = m^2$. By Euclid's Proposition 14, we know that every natural number greater than 1 can be uniquely written as a product of prime numbers. Also, because we are squaring m and n, every prime factor of m^2 and of n^2 must appear an even number of times. There are an odd number of factors of 2 in the number $2n^2$; thus, there are an odd number of factors of 2 in the number m^2, which is a contradiction.

The only assertion that was made without rigorous proof was the Greek belief that all numbers are rational. We now see that this assumption led us to a contradiction. We have constructed a line segment whose length is not a "number" according to the Greek notion of number as rational. This argument showing that $\sqrt{2}$ is not a rational number—similar to Euclid's original proof—is considered by many to be one of the most elegant proofs in mathematics.

The dawn of irrational numbers came despite the aesthetically appealing ideal of a rational world of numbers. We now know that we can find lengths that are not rational. Intuitively, it is clear that the length of any line segment should be a number. We are, thus, at an impasse. The length of our diagonal, $\sqrt{2}$, is not a number (because it is not rational), yet $\sqrt{2}$ does represent a length and all lengths should be numbers. For this reason, we must expand the Greek notion of what *number* means so that it now includes values that are not rational numbers. Irrational numbers are numbers that are not expressible as a ratio of two integers; that is, numbers that are not fractions. Sometime between 2000 and 1650 B.C.E., the Babylonians computed rational approximations to $\sqrt{2}$, but the Greeks were the first to prove it as irrational.

It is suspected that the Secret Society of Pythagoreans reacted skeptically. The Pythagoreans were perplexed by the irrationality of $\sqrt{2}$. Some scholars believe that they did not accept $\sqrt{2}$ as a number. The Pythagoreans called irrational numbers *alogos*, which translates into "unspeakable" or "inexpressible." There are many legends and theories about how the Pythagoreans interpreted this discovery and how they reacted to it. Although we will probably never know their view of this counterintuitive fact for certain, it is clear that this discovery moved our understanding of number forward. Around 370 B.C.E., the Greek astronomer and mathematician Eudoxus, who was a student of Plato, offered a definition of irrational numbers that beautifully foreshadowed the work of mathematicians Karl Weierstrass, Georg Cantor, and Richard Dedekind in Germany in the 19th century. ∎

Questions to Consider

1. Can you modify Euclid's proof that $\sqrt{2}$ is irrational to show that $\sqrt{3}$ is irrational? *Bonus Challenge*: Work through Euclid's argument with $\sqrt{4}$ and determine at what point the proof fails to yield a contradiction. (This is a good thing, because we recall that $\sqrt{4} = 2$ is, indeed, a rational number.)

2. The term *rational* derives naturally from the word *ratio*, and the term *irrational* derives naturally from *not rational*, yet the word *irrational* has a nonmathematical meaning, as well. What impact do you believe this other meaning may have on how people perceive irrational numbers?

Challenging the Rationality of Numbers
Lecture 8—Transcript

Welcome to the rational and not so rational world of numbers. When we count objects that are themselves comprised of smaller parts, we run into trouble when we try to count 1, 2, 3, and so forth, for we may have a quantity that's more than 2 but less than 3. Thus we'd have 2 and some portion—but not the whole—of another. This counting conundrum was resolved early on in humanity by extending the notion of number to include fractional parts. Both Babylonians and Egyptians used fractions, perhaps as early as 2000 B.C.E. The collection of fractions, known today as the collection of rational numbers, satisfied the ancient Greek aesthetic of commensurability and divisibility, dividing objects into a certain number of equal pieces. Although the Pythagoreans did not consider these ratios to be God-given numbers, they were convinced that all lengths could be measured in terms of natural numbers and their ratios. That is, the Pythagoreans viewed the rational numbers as a gapless list that corresponded to a line.

The Pythagoreans were forced to let go of this aesthetically appealing notion of number when they discovered lengths whose measure could not be a rational number. Applying the Pythagorean Theorem from geometry on a special right triangle, we'll see such a length and discover that $\sqrt{2}$ is not a rational number. In other words, it's not a fraction. This unattractive reality furthered challenged the Pythagorean notion of number. In fact, they did not consider these measures as numbers. Today, we call these very real numbers that are not rational numbers, irrational numbers (literally, "numbers without ratios"). The discovery of such perplexing quantities brought to light the surprising subtlety and delicate structure within the world of number when viewed as lengths and measures. Embracing this expanded view of number is a challenge—in fact, a great challenge—for all who face it for the first time, whether they encounter this notion today or, as in the case of the Pythagoreans, over 2,000 years ago. Well, let's begin this lecture with the history of ratios of natural numbers.

In order to define what we mean by a rational number, first let's recall that the integers, which we saw in a previous lecture, are the natural numbers (1, 2, 3, 4, 5) together with 0 and their negatives

(–1, –2, –3, –4, –5, and so forth). All those collectively are known as the integers. A rational number can now be defined as any ratio m to n (or m/n), for which m is an integer and n is a natural number. The number m is called the numerator and the number n is called the denominator of m/n. For thinking purposes—this might sound a little confusing—just think of rational numbers as fractions (natural number divided by natural number), but you're allowed to put a negative sign in front of it, or 0.

The rational numbers are also, of course, known as fractions, and that's what we're familiar with. While we don't know the first example of rational numbers, we find evidence of fractions in Babylonian and Egyptian writings from around 2000 B.C.E. A unit fraction is defined to be any fraction that has 1 as a numerator. So, 1/2, 1/3, and 1/4 are all examples of unit fractions because they have a 1 over something. The Rhind Papyrus, one of the oldest mathematical documents in existence, dating back from around 1650 B.C.E., contains sums of unit fractions. Ahmes, the scribe who wrote the Rhind Papyrus, denoted unit fractions by a dot over the denominator. For example—if he were to use the Hindu-Arabic numeral that we still use now—if he took a 5 and put a dot on top of the 5, that would signify 1/5. Now, this dot notation derived from the hieroglyph for an open mouth. This is really interesting, because this symbol provides evidence that fractions were used to divide up rations of food and drink. In fact, many questions from the Rhind Papyrus concerned dividing up loaves of bread and jugs of ale, which sounds quite attractive, doesn't it?

In addition, the Rhind Papyrus contains a table that shows how, for any odd natural number (let's call it n) from 5 all the way up to 101, fractions of the form $2/n$ can be written as the sum of different unit fractions. That is, any number of the form $2/n$ (where n is an odd number in that range) can be written as 1 over a number plus 1 over a number. Now, this calculation is by no means trivial. For example, let's just consider the first example, 2/5 (2 over 5). How can we write that as the sum of two fractions, each having a numerator of 1? Well, it's not obvious, but it turns out that 2/5 is the sum of 1/3 and 1/15. That is, $2/5 = 1/3 + 1/15$. You can check that by getting a common denominator. We multiplied the top and the bottom there by 5, and we see $5/15 + 1/15 = 6/15$, and if we simplify, we see, in fact, 2/5—so, amazing and highly nontrivial.

Now, an Egyptian fraction is defined to be the sum of distinct unit fractions. It's not clear why the Egyptians chose this method of expressing fractions. Much later, Fibonacci showed that every positive rational number can be expressed as the sum of distinct unit fractions. So in fact, even Fibonacci, all the way out to 1202, when he wrote his *Liber Abaci,* was returning to the same ideas that the Egyptians were considering.

As we've already seen, the ancient Greeks (in particular, the Pythagoreans) were one of the first cultures to explore rational numbers as objects of independent, abstract interest. Given their interest, I want to say a few words about how the Greeks viewed ratios and their notion of rationality. Well, as we've already seen, the Greeks had a complicated view of number. Now of course, that complicated view is only from our vantage point. Remember, it's a different world, a different culture that we live in today. Back in the cosmos of the Pythagoreans and Euclid, their view of the world was so different that their number sense made sense to them. Today it seems complicated, and it's our job to put ourselves in a mind-set where that mind-set of number makes sense.

Well, they believed that the natural numbers were the true numbers, since they were God-given, and the rational numbers were, really, only ratios of natural numbers. They weren't as real as the natural numbers. However, in their eyes, the rational numbers formed a gap-free list that could correspond to a line of numbers. In fact, the Pythagoreans viewed these numbers as a "flow of quantity," which really gives a beautiful poetic way of expressing their view and their idea of number.

It was Euclid who later made the Greek notion of rational numbers much more rigorous. Now, he describes the rational numbers in an interesting way. It's going to be a very geometrical way, but as you remember, in ancient Greece geometry and number theory really came together. Numbers were considered as lengths. So, not surprisingly, to describe ratios, it's going to be in terms of lengths. Well, in Euclid's *Elements*, his famous collection of books, he declares that two line segments are called "commensurable" if there exists a third line segment such that some number copies of the third, when laid end to end, produces the first segment and similarly for the second segment. Let's take a look at an example to make this make some sense.

Here I see two different line segments. The question is, are these two line segments commensurable? Well, if they are, that means that I could find a third line segment for which I could lay end-to-end and see an exact number of copies to produce exactly this, and similarly, that same measurement (or same ruler, if you will) can be used to measure a natural number of copies of the second. Now in this case, if I use this red segment as my unit of measure, what I see here is one unit, two units. So, here I see that in fact two copies of this (a natural number of copies) produce this line segment. What about here? One, two, three. So, again we see that there are a certain number of copies that will produce both this and this. That's what it means for these two segments to be commensurable.

Let's talk about the numbers behind these segments by considering their lengths. So, let me call the length of this shorter line segment a, and let me call the length of this longer line segment b, and let me call the length of our little red unit c. Well, then we see that there exists a natural number such that c times the natural number gives me a. In this case, the natural number—we can see—is 2. So 2 times this length equals this length. Similarly, we see there's a natural number, in this case 3—1, 2, 3. If I multiply the red length by 3, I produce all of these.

Now, let's think about these numbers together. If we actually look at the ratio, we have that a, first of all, equals c times m. Or in our example, a equals c times 2, and b, the length of the longer, would be c times 3. Now, if we observe the ratio of a to b, what do we see? Well, we see that a/b, or "a to b," is equal to cm/cn. Now, notice that we have this common factor of c, this particular length, which now cancels out from this fraction, and we're just left with m/n, which is a rational number. Even though this number might have a funny length, this length might have a funny, funny number. It turns out that m/n, since everything in sight are natural multiples, we see that that's a number.

So, we're now going to generalize from the line segments to numbers. We say that two numbers a and b are commensurable if a/b, or a to b, is a rational number, is a fraction. In our example, we have that ratio being 2/3 (2 over 3), and that corresponds to the length fact that a is 2/3 the length of b, and you can see that here.

Well, this idea of commensurability is connected to the Pythagoreans' study of harmony in terms of ratios of lengths of strings on musical instruments. This was a very common theme throughout their studies. The Greeks believed that every two lengths—in other words, every two numbers— were commensurable. Thus, the Greeks believed that all lengths were either natural numbers or ratios of natural numbers: what we call today "rational numbers."

Well, all is well with this Greek view of number until we attempt to measure a length of a certain line segment. I want to share with you the story of how this line segment came to be. So, let's suppose we consider a square where each side has 1 unit in length. So, each side is a 1-by-1-by-1-by-1 square. Now, if we draw a diagonal on that square, then we can measure its length, of course. Because it's a line segment—it has a length.

Well, in view of the Pythagoreans' belief that all lengths are rational, the length of the diagonal of this square must equal a rational number. It must be a fraction. The Pythagoreans, however, were in for a very big surprise, because that length is not a rational number. Now, to verify this claim, we follow the Pythagoreans' footsteps as they made this disturbing discovery for themselves. So, we're going to do a little, tiny bit of mathematics here, but stay the course, because we're about to make an amazing discovery. Let's first of all give that length a name; let's call it h. So, let's let h represent the diagonal of the square. We don't know its value yet, but we'll find it soon, I assure you.

Well, given the Greeks' belief, we have that h must be a rational number, so we're going to assume that. We're going to assume that h has a natural length, and let's call it m/n for some natural numbers m and n. So, h is now equal to m/n.

Well, we now recall the famous Pythagorean Theorem that we studied in geometry class. The Pythagorean Theorem states that the sum of the squares of the lengths of the legs of any right triangle equals the square of the length of the remaining side, which is the hypotenuse side. So, for example, if we take a look at this right triangle, we see the sides are 3, 4, and 5. Notice that when we take 3 squared, we get 9, when we take 4 squared, we get

16, and when I add those two numbers together, I get 25, which, notice, is the hypotenuse 5, but squared. So 5 squared is 25. So, you can see the Pythagorean Theorem at work.

Well, in our particular right triangle with the square, we actually see that each side, each leg of the right triangle, has a length 1. So, this will be a very easy calculation, happily. We see that $1^2 + 1^2$ is the hypotenuse squared, which is h^2. Well, 1×1 is 1, so we're just basically adding $1 + 1$, which even I can do. That's 2. What we see here is that $2 = h^2$. But remember, the Pythagoreans believed that h is a rational number, we're calling it m/n. So we see the important equality $2 = (m/n)^2$. Well in other words, 2 equals $m^2 \div n^2$. I want us to remember this equation just for a moment. We're going to come back to it in just a second.

But as an aside, I want to note that we could actually solve for the number m/n. I see that $2 = m^2/n^2$, so if I take the square root of both sides, I discover that m/n is equal to $\sqrt{2}$. In other words, we actually computed the length of the diagonal; it's the number $\sqrt{2}$. The Pythagoreans believed that this number is, in fact, a rational number. It's a fraction, m/n.

Well, now let's return to the equation $2 = m^2/n^2$. That is the equation before we actually took the square root. Well, we can see that the natural numbers must satisfy a new relationship. If I just multiply that equation by n^2, if I multiply both sides by n^2, the equality remains, and I'm left now with $2n^2 = m^2$.

Well, we're now going to apply Euclid's Proposition 14, which states that every natural number greater than 1 can be uniquely written as a product of prime numbers. Now, let's think about this, because this is the really tricky part. Because we are squaring m and n, every prime divisor of m^2, and also n^2, must appear an even number of times. Let's think about it. It's kind of like Noah's Ark in a way: If we have n times n, then every prime factor in n is going to appear twice—one in the n and the other in the times n. So in fact, all the primes appearing in n^2 are going to appear an even number of times, because I have a copy of n and another copy of n. Every prime here is going to actually appear here. So, we have an even number of primes. This is also true when we look at the primes of m^2 and its factorization. So, in particular, what we see here is that we have an even number of the prime 2 appearing in the prime factorization of m^2. So m^2 has an even number of the prime 2 appearing in its factorization. So, there could be no factors of 2, there could be two factors of 2, four factors of 2, six factors of 2, and so on—but an even number of factors of 2.

On the other hand, the number $2n^2$ must have an odd number of factors of 2 in its prime factorization. Why? Because we know that n^2 by itself will have an even number of factors of 2, and there's one more factor of 2 right in front. So, an even number plus 1 produces an odd number. What we see here is that the number of factors of 2 of $2n^2$ must be an odd number. But remember that $2n^2 = m^2$, and we have a contradiction, because we have the same number and, on one hand, we have an even number of factors of 2, and on the other hand, we have an odd number of factors of 2. Euclid tells us that, in fact, that factorization is unique. There's only one way, except for rearranging, and you can't rearrange an even number of factors and all of a sudden make it an odd number. You can't do that. So, we have a contradiction.

Well, this contradiction implies that something is wrong. In mathematics, we don't like contradictions. However, every one of our mathematical steps was, in fact, correct. So, what's wrong? Well, the only assumption that was made without proof was the Greek intuitive belief that all numbers were rational, and we now see that that assumption led to a contradiction—a logical impossibility. Therefore, that assumption must be false. In fact, the length $\sqrt{2}$ is not a rational number. It's not a fraction.

So, we have constructed a line segment whose length is not a number according to the Greek notion of number as rational. This argument, which is very similar to Euclid's original proof, shows that $\sqrt{2}$ is not a rational number, and this argument is considered by many to be one of the most elegant proofs in mathematics. We've actually seen another famously elegant proof, which was the proof of the infinitude of primes that Euclid also proved. It's interesting that Euclid got two of his results and proofs to be in the pantheon of great and elegant results, quite a lucky individual (well no, not really lucky, because he was enormously talented—but quite remarkable).

Now, when we think of this result, what makes it so elegant? Well, if you look back at the argument, you see that, even though it's tricky—it involves the Pythagorean Theorem, we had to take a square root, which is always scary—the truth is the argument just involved us looking at a length of a diagonal, a very basic object. Again, "basic" doesn't mean easy, it just means foundational. By measuring that length and by using this wonderful result about the unique factorization of natural numbers, we run into a logical impasse. Now, that point of view really leads us to question an assumption that was so deeply ingrained in the Pythagoreans and the ancient Greeks that they couldn't even accept it. So this very simple (in a way) argument challenged the Pythagoreans to rethink their entire notion of number. This is why this argument is considered to be so important and so beautiful. Once we understand the reasoning of this mathematical argument—which takes time, by the way—we must face a new world view.

The square root of 2 is not a fraction; it's not a rational length. Despite the aesthetically appealing ideal of a rational world of numbers, we have now seen that there are lengths that are not rational, that are not fractions. Intuitively, it seems clear that the length of any line segment, of course, should be a number. Thus, the Greeks found themselves at an impasse. The length of the diagonal, $\sqrt{2}$, is not a number, since it's not rational. So from their point of view, it wasn't a number, yet $\sqrt{2}$ does represent a length, and all lengths should be numbers. You see the problem. Thus, we must expand the Greek notion of what *number* means so that it now includes values that are not rational numbers.

Thus, we discover irrational numbers: numbers that are not expressible as ratios of integers. We discover that these numbers exist. In fact, the word irrational comes from "without ratio." As an aside, we recall that sometime between 2000 and 1650 B.C.E., the Babylonians computed rational approximations to $\sqrt{2}$, but it was the Greeks who were the first to prove that this value is irrational: not a ratio, not rational.

Well, as we have seen with 0 and the negative numbers, it's always a great challenge for humanity to face the strange-looking and counterintuitive numerical objects and to try to imagine an expanded view of number that includes and embraces these new objects. In this direction, I'd like to share the reaction of the secret society of Pythagoreans to the irrationality of $\sqrt{2}$. The Pythagoreans were genuinely perplexed by the irrationality of $\sqrt{2}$. In fact, some scholars believe that they never accepted $\sqrt{2}$ as a number in their sense.

They called irrational numbers *alogos*, which translates into "unspeakable" or "inexpressible." There are many legends and theories about how the Pythagoreans interpreted this discovery and how they reacted to it. One of the most popular was that the Pythagoreans took an oath. Remember, this was a secret society. They took an oath to never share with anyone outside their Brotherhood the existence of these irrational numbers that they found. Perhaps they thought this would wreak havoc in the street if people realized that not all numbers were natural numbers or ratios thereof. So they were afraid, and maybe they were trying to protect people, or maybe they just didn't want to have this ugly truth revealed. Perhaps this is why they called these quantities *alogos*, "unspeakable." You are not allowed to speak of these numbers outside of the Brotherhood, and anyone who would reveal this secret would be put to death. Now, the legend is that one Pythagorean actually did make public this secret, and the great philosopher Proclus, around the 5th century C.E., gave a brief account of this dark episode in the Pythagorean Brotherhood. I want to share that story with you.

> It is well known that the man who first made public the theory of irrationals perished in a shipwreck in order that the inexpressible and unimaginable should ever remain veiled. And so the guilty man, who fortuitously touched on and revealed this aspect of living things, was taken to the place where he began and there is forever beaten by the waves.

In other words, the Pythagoreans took the guy out on a boat, threw him overboard, and he drowned—pretty serious business. But the writing of Proclus is beautiful, because he says, "And ... the guilty man, who fortuitously touched on and revealed this aspect of living things ..." It's true. It's true that these numbers are not ratios, but the fact that he has confessed this and admitted this in public cost him his life—a wonderful legend.

Well, while we'll probably never know for sure how the Pythagoreans viewed this counterintuitive notion of irrationality for certain, it is clear that this discovery moved our understanding of number a quantum leap forward, and we're greatly indebted to their work and their discovery.

Now some years later, around 370 B.C.E., the Greek astronomer and mathematician Eudoxus, who was actually a student of Plato, offered a definition of irrational numbers that beautifully foreshadowed the work of the 19th-century German mathematicians Carl Weierstrass, Georg Cantor, and Richard Dedekind. We'll see this refined and elegant number development in Lecture Eighteen. But for now, it's interesting to see that, even in ancient times, we see reflections of the rigor that was later required to move the frontiers of mathematics forward even offered at that time.

Today, the collection of all numbers that represent the measures of all possible lengths, along with their negatives, are known as the real numbers. So, the real numbers are any number that represents length on a line, or the negative of it, and of course 0. We'll explore these numbers in greater detail in our next lecture.

Well, even thousands of years after the Pythagoreans were faced with irrational lengths, irrational numbers continue to disturb, intrigue, and bother mathematicians. Again, we see the difficulty in widening our intuition of number to include that which first appears as strange and exotic. But how strange and exotic are these disturbing numbers? The surprising and perhaps even more disturbing answer to this question will be revealed in the next two lectures. The irrational numbers challenge our imagination to take what appears to be abstract and foreign and embrace it as real. This challenge was beautifully articulated by the great 17th-century mathematician Gottfried Leibnitz, who once referred to the notion of irrational numbers as "… a miracle of analysis, a monster of the ideal world, almost an amphibian between being and not being." Enjoy the irrational thoughts of irrational numbers.

Walk the (Number) Line
Lecture 9

Decimal expansions of real numbers will lead us to several surprising and perhaps counterintuitive discoveries. For example, we'll prove that the decimal number 0.9999… forever is actually equal to 1.

How do we represent the real numbers on a line? The term "real numbers" itself is misleading and was probably coined in response to the discovery of yet more abstract imaginary numbers. One view the Pythagoreans held for number was as a "flowing quantity." Even though they were unable to accept irrational numbers, their intuition about a "flowing" list of numbers was correct. We now view this flow in the form of a line.

Flemish mathematician Simon Stevin (1548–1620) may have been the first to consider the number line in a text he authored in 1585, which introduced a systematic approach to the arithmetic of decimal numbers. In 1637, René

The binary mark-up of a number line resembles a ruler, with its unit markings.

Descartes described the "Cartesian coordinate system" that implicitly utilized the number line. Zero takes center stage as we move away from the Greek sense of number. We have a *continuum* of numbers. This concept forms an important connection between points on a line and a notion of number.

Stevin was the first to offer a thorough account of decimal expansions of real numbers. For our purposes here, we will define the real numbers as the collection of all numbers that can be written as an endless decimal expression (e.g., $3 = 3.000…$, $7.5 = 7.5000…$, $\frac{1}{3} = 0.3333…$, $\sqrt{2} = 1.414213…$, $-4.012 = -4.012000…$, and $\pi = 3.1415…$ are all examples of real numbers, each expressed as an endless decimal expansion). Stevin's original notation foreshadowed our current decimal notation. Stevin's system expressed $61\frac{837}{1000}$ as 61 8$^\text{I}$ 3$^\text{II}$ 7$^\text{III}$. By 1653, the notation had evolved so that this number would have been written as 61:837; today, we would write 61.837. We can locate a real number, such as 1.417, on the number line by repeatedly dividing the line into 10 equal-length intervals.

Simon Stevin saw the practical value of the decimal system and argued that the monetary system should be decimal-based rather than based in fractions.

The simplest base is base 2—giving us what are known as binary expansions.

We explore how to express real numbers in other bases. We start with binary expansions. There is nothing particularly special about the decimal expansion (*deci* implies a base of 10). We can consider representing real numbers in other bases. This representation would be equivalent to repeatedly dividing whole-number intervals into another number of equal-length sub-intervals. The simplest base is base 2—giving us what are known as binary expansions. The 17$^\text{th}$-century German mathematician Gottfried Leibniz, who invented calculus independently and concurrently with Isaac Newton, believed that binary expansions were extremely valuable and even held quasi-mystical properties.

The next most basic expansion is the base-3 expansion—also known as the ternary expansion of real numbers. We divide the whole-number intervals of the number line into three equal-length subintervals, then repeat. The 17$^\text{th}$-century French mathematician Blaise Pascal believed that there was nothing innately special about base-10 representations—any other base, such as binary or ternary, was just as "natural."

Long division reveals the decimal expansions. Decimal expansions for rational numbers must eventually become periodic—that is, eventually, they must have an endless "tail" of repeated digits in the decimal expansion. Periodic decimal expansions lead us to other revelations. Revisiting 0.3333… leads us to the converse result: All periodic decimals are rational numbers. We find a surprise hidden in the number 0.9999… as well as in other bases. We can classify the rational numbers as those numbers whose decimal expansions are eventually periodic. This result captures, in spirit, the "finite" nature of rational numbers. ∎

1. (a). Match the decimal numbers 1.5, 0.1, 2.15, and −0.9 to their respective places on the number line below.

   (b). For each point marked with an arrow on the number line below, write the corresponding decimal number it represents.

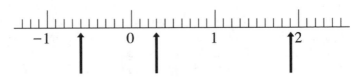

2. (a). Below is a portion of the binary number line, with hash marks to show halves of intervals instead of tenths. Locate the binary numbers 0.1_2, 0.01_2, and 0.101_2 on this number line.

   (b). Below is a portion of the ternary number line, with hash marks to show thirds of intervals instead of halves or tenths. Locate the ternary numbers 0.11_3, 0.22_3, and 0.002_3 on this number line.

Walk the (Number) Line
Lecture 9—Transcript

This lecture is an invitation into the world of so-called "real numbers." In order to build some intuition into this extended notion of number, we'll visualize these values through the use of a relic from our own mathematical past: the number line. While we find evidence of the number line dating back to the 16th century, its first appearance may have come even earlier. This geometric point of view offers us a very useful description of real numbers. Each point on the line corresponds to a real number determined by its distance from a point marked "0" and whether it is to the right of 0 (in which case we say it's positive) or to the left of 0 (in which case we say it's negative). The number line also provides a means of writing real numbers, namely as a decimal expansion. Just as with the natural numbers, here we'll discover that we can express real numbers in bases other than our familiar base-10, or decimal, expansion. As we'll see, these alternate expansions actually can be visualized on the number line itself.

Decimal expansions of real numbers will lead us to several surprising and perhaps counterintuitive discoveries. For example, we'll prove that the decimal number 0.9999... forever is actually equal to 1. This potentially disturbing reality will have a silver lining. We'll use this observation to uncover a very attractive property that the decimal expansions of rational numbers possess. As we'll see, the rational numbers are precisely those numbers whose decimal expansions eventually repeat forever. Thus, we'll find that the number line, together with decimal expansions, offers an elegant and powerful means of expressing and better understanding the real numbers.

The collection of real numbers is a very difficult collection to precisely define, although we can define it and describe the numbers in a more informal manner. The name itself, "real numbers," captures the idea that these numbers represent real lengths of real objects (together with their negative values) and was probably coined in response to the discovery of the more abstract, so-called "imaginary numbers" that do not represent any actual lengths. Now, we'll see these imaginary numbers for ourselves in Lecture Sixteen. In Lecture Eighteen, we'll offer a historical overview of the 19th-century formalization of the real numbers. But for now, we'll offer a geometric means of viewing these real numbers. One view the Pythagoreans held for number was a "flowing quantity." Even though they were unable to accept irrational numbers as numbers, their intuition about a "flowing" list of numbers was correct. We now view this flow in the form of a line, and the collection of real numbers can be viewed as the collection of points on the number line.

It appears that the 16th-century Flemish mathematician Simon Stevin may have been the first to explicitly describe the number line. He discusses a number line in a text he authored in 1585 that introduced a systematic approach to the arithmetic of decimal numbers. In further historical context, I want to note that René Descartes, in 1637, described the famous Cartesian coordinate system from geometry that implicitly utilizes the number line—in fact, two number lines that meet at right angles. As the number line became popular as a means of viewing numbers, we see an interesting shift in our perspective. Let's recall that, to the Greeks, the number 1—*unity*—was the generator or starting point of all numbers because from 1, if you add to 1, you get 2, add another 1 you get 3, and we could build all the numbers that were God-given in the Pythagoreans' eyes. But now, with the number line, we see 0 taking center stage. Everything is centered around 0. Are you to the right of 0 or to the left of 0? Now, we see that 0 has replaced 1 in prominence. This new perspective is a clear movement away from the ancient Greek sense of number. The notion of number had evolved yet again.

Well, with the number line, we have a *continuum* of numbers, which Stevin described poetically as "flowing like a river of magnitudes," which really is a beautiful way of seeing the line without any breaks at all. Now, there's an important connection between points on the line and the decimal expansion of real numbers that I want us to now explore a little bit. This connection will, at last, lead to a precise definition of the real numbers. Stevin was the first to offer a thorough account of decimal expansions of real numbers. For our purposes here, we'll define the real numbers as the collection of all objects that can be written as an endless decimal expression. So as long as you put down a dot and an endless list of digits to the right of that, we'll call that a real number. For example, 3 is a real number because I can express it as 3.000... forever. An endless run of 0s is 3, therefore I'll consider it a real number, because I can express it as an endless decimal expansion; 7 1/2, or 7.5, is a real number because I can write it as 7.5000... forever; 1/3 is a real number because I can write it as 0.3333... forever. The square root of 2, which we studied earlier, is a real number because it has decimal expansion that begins 1.414213, and so forth. Even a random number like –4.012 is a real number because I can write it as –4.012000, and so forth, forever. The famous number π that we'll study in a future lecture is equal to 3.1415926, and so forth, so it's a real number.

Anything that can be expressed with an infinitely long decimal expansion is a real number. These are all examples. Well, as an aside, I want to say a word about the evolution of the notation that we use today for writing decimal numbers. Stevin's original notation foreshadowed our current decimal notation. In particular, in Stevin's system, he would have expressed the number 61 and 837/1000 as the following. He'd write it as 61 8^I 3^{II} 7^{III}. So, we see that he's trying to indicate the positional system of where the 10ths, 100ths and 1,000ths go by these Roman numerals. By 1653, the notation had evolved so that this same number would have been expressed as 61:837, and as an aside, this notation is still used to denote bible chapters and verses. Today we use a radix and simply write 61.837. But you can see how naturally and beautifully that notation evolved from the early, early attempts to express these decimal numbers.

Let's now use the decimal expansion of a real number to locate that number on the number line. For example, suppose we want to find the real number 1.417 and locate that point on the number line. What do we do? Well, I see that the number is between 1 and 2. I look to the left of the decimal point, and that tells me which two consecutive integers this number is going to be in between—in this case, 1 and 2. So, we look between 1 and 2, and now, since this is base 10, or decimal, I take the interval from 1 to 2, and I cut it into ten equally spaced intervals, each having length 1/10. Then I label them 1, 2, 3, 4, 5, 6, 7, 8, 9, and I ask myself, "the next digit, where do I see that next digit coming in?" Well, I see the next digit is 4, and so where do I go? I go between a 4 and a 5, so I know that the number is going to be somewhere in that interval. Okay, that's great. So, again, it's a "divide and conquer" approach. Now, within that tiny interval that has length of 1/10, I repeat the process. So, I take that interval, and I cut it up into ten smaller intervals, each of length 1/100, and I label them 0, 1, 2, 3, 4, 5, 6, 7, 8, 9 for the digits we have in base 10, and I look at the next digit in our number, which in this case is a 1, and that tells me that I'm going to be between the 1 and the 2. So, I look in that tiny interval there, and what do I do? I repeat. I take that small interval, cut it into ten equal pieces, I label the intervals 0 to 9, and then I put my point right on the 7. So, I see 1.417, and I've located that point by whittling down, always dividing into ten equal pieces.

Here again we see the importance of the positional number system as a way of honing in on the number. In fact, Stevin saw the practical value of the decimal system and argued that the monetary system should be decimal based, rather than as fractions, which were used at the time. In fact, he wrote the following:

> In view of the great usefulness of the decimal division, it would be a praiseworthy thing if the people would urge having this put into effect, so that ... the State would declare the decimal division of the large units legitimate to the end that he who wished might use them. It would further this cause also, if all new money should be based on this system of 10ths, 100ths, 1,000ths, etc.

Well, it took over 200 years before governments came to see the wisdom of Stevin's suggestion and adopt a decimal system in their currency.

Of course, there's nothing particularly special about the decimal expansion, although *deci* implies base 10. Just as we saw in Lecture Four with natural numbers, here we'll see that we can write real numbers in bases other than 10 and locate them on the number line. In fact, as we've already seen, the Babylonians found a fractional approximation to $\sqrt{2}$ which they expressed in their own favorite, base 60. So, we've already seen evidence of expansions that are beyond the base 10. Now, visually on the number line, the base determines how many equal pieces we're cutting up each successive interval. With base 10, we saw that we cut each interval up into 10 equal pieces and then repeat, because it was base 10. Well, the simplest base is base 2, giving us what's known as "binary expansions." Now in this case, we have two digits: We have a 0 and a 1, and that's all we're allowed to use. The places that we see the positional units are going to be powers of 2.

In order to see how to interpret the meaning of binary expansions, let's look at a couple of examples, and for each binary expansion, we'll try to locate and pinpoint that value on the number line. So, let's consider the real number 0.101_2. Remember, that little, tiny 2 at the end of that number as a subscript reminds us that this is not a decimal number, but instead is a base-2 number. How would we locate that particular number? Well, we see that that number is between 0 and 1 because it starts with 0 point something. So, we go between the interval from 0 to 1, and since we're in base 2, we'll constantly divide things up into two equal pieces, two equal intervals. We'll cut things in half and then half again. So, here we cut in half, and that halfway point is what we call 0.1 in base 2. Our number is something that starts with 0.1, so we see we're in the right half of this piece. So, we take that right-half interval, and what do we do? We cut it in half again, and we call the first piece 0 and the second piece 1, and we look at our next digit. Our next digit is 0, which means we go to the left, to the 0 interval, which is an interval of length a 1/4. Now we cut that in half again, and we see that, in fact, we land on 1, which is the point right in between there. So what we see now is the exact location of that number, 0.101 in base 2.

And by the way, let's look at that number line. We see that, to get to that point, we first had to travel halfway, so that's 1/2, and then we had to travel an extra 1/8 because we went 1/4 and then half of that, which was 1/8. So, 1/2 and 1/8 actually equals 5/8, which is the value of this number, 0.101. Because the first 1, the left most 1, is located in the 1/2 spot, and then that rightmost 1 is located in the 1/8 spot. So, you can really see how we can visualize these numbers on the actual number line. Notice, by the way, that this binary mark-up of the number line is perhaps even more familiar than our base-10 number line, since it actually resembles a ruler that's marked off in one-half inches, one-quarter inches, one-eighth inches, one-sixteenth inches. In fact, that's the kind of units we use when we actually look at measuring things for real.

Let's look at another example. Let's locate the binary expansion 0.111_2 on the number line. Well, again, how do we proceed? Well, I see this number is between 0 and 1, and so what do I do? Well, I look between 0 and 1, and I cut the interval right in half because it's base 2. Call the first piece 0 and the second piece 1. I look at our number and see that the next digit is a 1, so I'm in the right spot, so I'm in the right interval. I take that interval, cut it in half again, and then what do I see? I see that our next digit called 10 and I call 11. I see that our next digit is a 1, which means again I'm to the right, and then finally, I cut that little interval in half—yet again—and what do I see? Well, I see a 01 and a 1, and I see I'm exactly at 1, so I'm exactly at that hash mark that we put in right there. Again, we could figure out what this number is, because the first 1 starting from the left is in the half spot, so I have 1/2. Then I add another 1, which is located in the 1/4 spot, and then I have another 1, which is located in the 1/8 spot. So I have 1/2 + 1/4 + 1/8. If you add them up, we get 7/8. Sure enough, if you look on this number line, that number is located in the position of 7/8. So there's a way of taking base 2 expansions and seeing how they appear on the number line. Instead of cutting up into ten pieces as in base 10, we cut into two pieces each time, cut everything in half, for base 2.

Now, as we noted in the previous lecture, the important 17th-century German mathematician Gottfried Leibniz, who invented calculus independently and concurrently with Isaac Newton, believed that binary expansions were extremely valuable and even held quasi-mystical and spiritual properties. So, you could see that there was a great import to this, and indeed, this expansion is important. The next most basic expansion is base 3, or base-3 expansion, also known as "ternary expansion." Again, since it's base 3, we now only have three digits, and they would be 0, 1, and 2. So, here we divide our intervals on the number line into three equal subintervals and repeat each time, calling the first interval 0, the second interval 1, and the third interval 2. So, let's solidify this idea by considering a couple of illustrative examples.

First, let's consider the real number 0.121_3 (written in base 3). Now how do we locate that point on the number line? Well, what we do is we first see that that number, again, is between 0 and 1 (because we have a 0). So, we're going to look just between 0 and 1, and then what do we do? Well, the next thing we do is we divide that interval into three equal pieces, divide it into thirds. The first piece we call 0, the next one 1, the next one 2. And we ask, what digit do we see next? Well, we see a 1, which means we're in the middle third of that big interval. Well, we take that interval, and we cut it into three pieces, calling them 0, 1 and 2. And we ask ourselves, "what's the next digit we see?" We see 2, which means that we now move to the rightmost, tiny, third interval that we see there, and then finally we cut that into three pieces, and that last digit of a 1 tells us that we're at the very first new little hash mark that we made. So there we've located the number exactly.

Again, divide and conquer, but the question is, how many pieces do we divide each interval into? That depends upon the base. Let's locate—last example—0.022_3 on the number line. So, what do we do? Well, again I see it's a number between 0 and 1, so I look between 0 and 1. I cut my little interval into three equal pieces, call the first one 0, the second one 1, the last one 2. I look at the next digit, and I see another 0, so that tells me I'm in the first subinterval. What do I do? Cut that up into three equal pieces and look at the next digit, which is a 2, which tells me I'm in the far right of that tiny subinterval. And lastly, I cut that into three little pieces, and that 2 tells me that I'm at that very last, right-most, new hash mark and that I've located the number yet again. So again, it's a divide and conquer, which just takes a little bit of practice. Now once you get used to them, it's not difficult to locate binary or ternary expansions on the number line, although it does take practice. It takes some practice to get used to it.

In fact, the great 17th-century French mathematician Blaise Pascal believed that there's nothing innately special about decimal, our base-10, representations. In fact, any other base would work just as well, such as binary or ternary, and they're just as "natural." In fact, he wrote the following:

> The decimal system has been established, somewhat foolishly to be sure, according to man's customs, not from a natural necessity as most people would think.

Well, man's customs, in this case, might be the fact that we do have ten digits and we've been using these ten digits to count for an awfully long time, and that might be the custom. If we had a creature that had a different number of fingers, the custom might be different. For example, what number system, what base system, might Donald Duck use? Well, we see Donald Duck has, actually, four fingers on each hand, so that would be a total of eight, and so perhaps he would use a base 8 system.

Let's now return to our familiar decimal, base-10, system and look at expansions of real numbers and consider the expansions of rational numbers, the expansions of fractions. Now, we can find the decimal expansion for a rational number or a fraction using long division. For example, if we look at 1/3, we can actually do the division. Divide 3 into 1, and when we do the long division, we see 0.3333, and so forth. So, we see this long division gives us the decimal expansion with which we're familiar. Now, decimal expansions for rational numbers must eventually become periodic. In other words, eventually, they must have an endless tail, if you will, of repeating digits in its decimal expansion, and those repeating digits go on forever.

Now let's see why this assertion holds by looking at another example. Let's look at 11/7 and actually perform the long division. Now, the magic of the classroom—I'll actually perform this long division very quickly. Look, you'll see it right there. See? There we have it all. All right. But the key things I want us to look at are the remainders that we found as went down this long, long division. The remainders we see are 4, and then a 5, and then a 1—and let me just stop for a second. Don't look any further. Let's think about what the remainders are. We remember from long division, a long time ago in our elementary school math classes, that the remainder is always going to be a number—well, in this case we're dividing by 7, so—between 0 and 6, since we're dividing by 7. But what does that mean? It means that if I do this long enough, these numbers are either going to be 0, 1, 2, 3, 4, 5, or 6. Those are the only choices. So, if I do this long enough, at some point we have to either hit a 0, in which case we have a long run of 0s forever, or we're actually going to have something that we've seen before, because we have only six possibilities. So, if we keep taking numbers from six choices, at some point we have to repeat. Now what happens in our particular example here? I see 4, 5, 1, 3, 2, 6, and the next one is a 4. Well, once I hit that 4, I know what the next remainder is going to be. It's going to be 5, and then a 1, and then a 3, and then a 2, and then a 6, and then I hit a 4 again. So, since these remainders repeat, that tells me that the quotient is going to repeat as well. The decimal 1.571428, and that piece will repeat 571428, and so forth. That shows us that, since the remainders come from a finite list of numbers, if we keep repeating, at some point we have to see a repeat of numbers forever or a repeat of just 0s forever. That's perfectly fine.

In fact, this reasoning actually holds for any such long division calculation that we could do, and therefore we actually conclude that all rational numbers, all fractions, have decimal expansions that are eventually periodic—really neat. How about the other way around? Do all periodic decimal representations represent rational numbers? To answer this question, let's revisit the decimal number that we're familiar with, 0.33333 forever, unending. Let's verify that that decimal number is the number 1/3. Now, of course, we know it is, but let's see if we can systematically find a way to verify that. Let's call that mysterious number, which of course we know is 1/3—let's call 0.33333 forever, repeating forever, "?". So, ? = 0.33333. Here's a great trick that I want to share with you. What we do is we multiply both sides of that equation by 10, and by doing so, I see that $10 \times ?$ = (we just move the decimal point one over to the right) 3.33333 forever. Again, an endless list of 3s to the right of the decimal point.

Now, we have these two equalities, and now let me subtract them. Notice that the endless run of 3s that follow that decimal point line up perfectly, like little soldiers, and all cancel away, so we get 0s. What do I see? Well, I have that 3 in front of the decimal, and I subtract 0, so I see 3 on the right. Then, on the left hand side, I have 10?. I take away 1?, I'm left with 9?, and so I see 9? = 3. I can solve this for ? by just dividing both sides by 9, and then see that ? = 3/9, which reduces to 1/3. So, we just proved that 0.33333 forever equals 1/3 through this really clever trick.

Let's try the clever trick one more time and look at the number 0.9999 repeated forever, and we'll be in for a surprise. I love this. Let's follow the same method to see what the number is, what the rational number is. So, we'll call it ?. So, ? is the number 0.99999 forever, unending. Now, what we do is we multiply both sides by 10, because that will move the decimal point over one unit to the right. So, I see $10 \times ?$ = 9.9999 endlessly, and notice that the endless tail of 9s to the right of the decimal point line up perfectly, like little soldiers.

So, when I subtract these two equalities from each other, they all drop out, and I get a run of 0s forever. So 9 − 0 = 9, and if I have 10 ? and take away 1 ?, you know, I've got 9 ?. So, now I see 9 ? = 9. If I divide both sides by 9, I see that ? = 9/9, which is 1. So, we just proved mathematically that 0.9999 forever actually equals the number 1, and this is so annoying and disturbing to people when they first see it, because it sounds as though the number

0.9999 forever should be a number that's a little bit smaller than 1. But in fact, since those 9s go on forever, if we were to try to locate that point on the number line, we'd keep going to the furthest most right interval. So, we'd start here, from 0 to 1, and then we'd go to the biggest 1/10, and then we'd go to the biggest 1/100, and then we'd go to the biggest 1/1000 interval. We'd keep moving closer and closer to 1. Since we're doing this process forever, it never ends. In the limit, we actually are at 1, as we just proved.

Anyway, it's a very counterintuitive fact that 0.9999 forever equals 1. Now, in fact, this fact also holds in other bases. So, for example, let's consider base 3 and look at the number 0.2222 forever in base 3. Let's see what that number equals in base 3. Well, I'll call it ?, so ? = 0.2222_3. Now, to move the decimal point over, remember this is base 3, so we don't multiply by 10, we now multiply by 3. So, I multiply both sides by 3, and I see 3 ? equals—and now that point, that radix, moves one unit to the right—and so I see 2.2222_3 forever, the tail of 2s line up like little soldiers. I subtract, and what do we see? We see on the right we just have the number 2 followed by infinitely many 0s, and then, if I have 3 ?s and take away 1 ?, I've got 2 ?s. So if 2 ? = 2, divide both sides by 2, and ? = 1. So, again we see this is true in all bases, not just in base 10. So, this wonderful, wonderful identity that tells us that 0.999 forever equals 1 carries over to other bases as well, correspondingly.

We're seeing a pattern emerge in these examples. We can actually classify the rational numbers as those numbers whose decimal expansions are eventually periodic. There's a piece that repeats forever in the decimal expansion. This result captures, in spirit, the fact that we know rational numbers precisely. Their decimal expansions are finite in nature in that we only need a finite list of digits which just gets repeated forever. In some sense, the rationals are the simple and understandable inhabitants within the real numbers while the irrational numbers remain subtle and exotic. These character traits only come into focus when we see how these numbers sit within the real number line.

The Commonplace Chaos among Real Numbers
Lecture 10

Within the world of real numbers, our familiar rational numbers are truly the exception, while the at-first exotic irrational numbers are, in fact, the rule.

The rational numbers are precisely those real numbers whose decimal expansions are eventually periodic. The decimal expansion for $\sqrt{2}$ will never become periodic because it is irrational. We can describe "new" irrational numbers via their decimal expansions (e.g., the number 0.101001000100001000001... must be irrational because the decimal expansion never becomes periodic).

Real numbers come in two "flavors"—rational and irrational. Rational numbers are the ones that are the most familiar to us because we use fractions regularly in our daily lives. Irrational numbers are far more exotic; they are difficult to name, and it is challenging to verify their irrationality. Suppose we pick a number at random. The random real number will either be a rational number or an irrational number. What is the probability that a random real number is rational? We can produce a random number by generating its digits through chance. Rolling a 10-sided die will generate random digits. The likelihood that a real number selected at random is rational is 0%, while the likelihood that a real number selected at random is irrational is 100%!

There are, in some sense, "more" irrational numbers than rational ones. Numbers that first appeared to be exotic are more the norm, and numbers that first appeared to be the norm are, in actuality, exotic. Our everyday experiences with numbers do not allow us to gain an accurate sense of how the rational and irrational numbers "fit" together.

Given any two distinct points on the number line (that is, any two unequal real numbers), there is a rational number between them. Our method to establish this fact goes back to the ideas of the great Greek mathematician Archimedes. Because the rational numbers exhibit this property, we say that they are "dense" on the number line (or, equivalently, they are dense in the real numbers).

We look at how to refine our notion of a random real number. If we consider rolling a 10-sided die to generate digits of random real numbers in base 10, how often would we expect the digit 3 to appear in a random real number? We would guess that every digit from 0 to 9 would appear an equal amount of the time (on average); that is, each digit would appear approximately $\frac{1}{10}$ of the time. If we select a real number at random, we expect that $\frac{1}{10}$ of the digits in its decimal expansion will be 0, $\frac{1}{10}$ of the digits will be 1, and so forth to the digit 9.

Normal numbers in base 10 are explained. We might guess that because there are 100 two-digit numbers (from 00 to 99), each two-digit number will appear approximately $\frac{1}{100}$ of the time in the decimal expansion of an average real number. Real numbers for which every length of digits appears the "expected" amount of time in their decimal expansions are called "normal numbers in base 10." In 1909, the French mathematician Émile Borel proved that the likelihood that a random real number is normal in base 10 is 100%.

If the likelihood that a random real number possesses a particular property is 100%, then we say that "almost all" real numbers possess this particular property.

In view of our previous observations, we see yet again through this refined notion that rational numbers, such as $\frac{1}{3}$ and $\frac{1}{2}$, are very strange indeed.

Borel considered an even more general notion of random numbers. A "normal number" is defined to be a number that is normal in every base (that is, normal in base 2, 3, 4, 5 …). Borel then proved an amazing theorem: The likelihood that a random real number is a normal number is 100%. "Almost all" real numbers are normal. If the likelihood that a random real number possesses a particular property is 100%, then we say that "almost all" real numbers possess this particular property. We have seen that "almost all" numbers are irrational and, furthermore, that "almost all" numbers are normal (hence the name). For any random real number, we expect the digits to be such that any particular run of digits will occur its fair share of the time. ∎

1. Assuming that any obvious pattern continues, which of the numbers below are rational and which are irrational?

   0.10110111011110…

   3.7878787878…

   0.123456789101112… (This is known as *Mahler's number.*)

   −81.41783059833167098312… (a number selected at random).

2. When someone says, "Pick a number between 1 and 4," we always choose a whole number, which is certainly rational. Why does this happen? What effect do assumptions about context have on probabilities? (*Fun fact*: When we ask people to pick a number between 1 and 4, most people pick 3—try it!)

The Commonplace Chaos among Real Numbers
Lecture 10—Transcript

Welcome to the commonplace chaos within the real numbers. In the previous lecture, we discovered that the numbers having decimal expansions that eventually repeat are precisely the rational numbers, that is, the fractions. In view of this theorem, we now see that decimal representation of real numbers also offers an alternative means of classifying the enigmatic irrational numbers. This classification offers an alternative and attractive means of producing other irrational numbers. In view of the fact that real numbers are either rational or irrational, a natural question emerges: What proportion of the real numbers is rational and what proportion is irrational? For example, do the rational numbers make up half of the reals? Well, our classification of rational and irrational real numbers by their decimal expansions holds the key to unlocking an intuitive answer to this very question. It also leads to a surprising realization that we'll make mathematically precise, but the realization is that essentially 0% of the real numbers are rational and essentially 100% of the real numbers are irrational. Well, this assertion is totally counterintuitive, because the rational numbers are the ones with which we are most familiar, and the irrational numbers appear as the exotic and rare. Thus, here we discover an important recurring theme: Often the seemingly familiar is the exception, and the exotic is the rule.

While we'll see that the rationals are a very sparse collection within the real numbers, we'll also see that they're spread throughout the number line and are said to be "dense." We'll close this lecture with a description of some incredible facts involving modern, related concepts known as "normal numbers," first studied by the important French mathematician Émile Borel in 1909.

Let's begin by studying the decimal digits of irrational numbers and seeing why we can view these digits, in some sense, as chaotic. First, let's recall one of the main observations that we made in the previous lecture: The rational numbers are precisely those real numbers whose decimal expansions are eventually periodic. Thus, we conclude that the decimal expansion for $\sqrt{2}$ will never become periodic, because we proved—in fact, the Pythagoreans proved even before us—that $\sqrt{2}$ is irrational.

So, think about this. This is absolutely stunning, because you could compute the first few digits for the decimal expansion for $\sqrt{2}$ using a calculator or a computer, but we can't say anything about the entire infinite list. Amazingly, we can say something about this list of digits. We can say that they will never repeat. So, even though we don't know what those digits are, we know a property they possess—because the digits can't repeat. For if they were to eventually repeat forever, that would make it a rational number, a ratio, and we know $\sqrt{2}$ is not. So, this is amazing, because even though we don't know anything about the digits, we actually can say facts that the digits must obey, and this is amazing. This is the power of the decimal expansion.

Now, most of the digits we don't know, of course, and we don't know the list in entirety, but even if we knew the first billion digits, it tells us nothing, because we don't know what happens at the tail. In terms of rationality and irrationality, all we care about is the end game. Well, this insight offers us a new means of describing irrational numbers.

Any number whose decimal expansion never repeats is an irrational number. So for example, we could actually build an irrational number by giving the decimal expansion for it. Here's one, for example. Let's say 0.101001000100001, and so forth. At every stage, I add an extra 0 to buffer between the two 1s, and so I add more and more 0s. Thus, we see this number must be irrational, because its decimal expansion never repeats. The 0s are getting more and more robust as we go down the list. Well, this irrational number seems very contrived, and indeed it was. We had to make sure that we had a system of placing down the digits so that they never became periodic. So, what happens if we pick a real number at random? Well, we know that the real numbers come in two basic "flavors"; they're either rational or irrational. The rational numbers are the ones that are the most familiar to us because we use fractions regularly in our everyday lives. The irrational numbers are the far more exotic ones. They're difficult for us to name, and it's even a challenging task today to verify their irrationality. It's tricky business.

Now suppose that we pick a number at random. Say, for example, we place a pin on the number line while we're blindfolded to select a particular point. Well, the random real number we select will either be a rational or an irrational number. What's the likelihood that the random number will be one of the more familiar rational numbers? Well, if you think about this for a second, a blindfold and a pin are not great tools for producing a random real number. For one thing, we'll almost certainly miss the number line because the number line, of course, is very, very thin, and we'd be blindfolded, and even more dangerously, we might even prick our finger, and that would be awful. So, let's put down the sharp objects and focus on the numbers.

Well, it turns out that we could actually produce a random real number by generating its digits by pure chance. What we could do, actually, is imagine rolling a 10-sided die, because if we were to roll a 10-sided die, which would have all the digits from 0 all the way out to 9, then, in fact—if the die was just as equally likely to come up one side than the other—we would see that we'd be generating random digits and putting it together, stringing it together. We would have a random number. Well, in fact, we can try this.

Now, in terms of rationality and irrationality, all that matters are the digits to the right of the decimal point. So, for our example, let's just suppose we start with 0, and then we'll roll and see what the random is. Now, in fact, you can actually get these dice. Here's a die, and you can actually see the digits. Here we see 0, 8, 2, 6, 4, and if you turn it around, you actually see 1, 7, 3, 5, 9. So, all ten digits are represented here, and we can now generate a genuine, random real number by rolling the die to determine the digits. I'll do this right now, and I'll record the results. So, let's see what number we get. This is going to be an actual random number. We start with an 8. Roll again—a 7. Roll again—a 0. Roll again—a 3. Roll again—a 4. Roll again—a 9. Roll again—a 0. We can keep doing this, but let's stop. Because let me ask a question now. So, we have so far, in our random number, 0.8703490. Now, the question is, is it possible that this now will repeat? For example, let's just take the last three digits, the 490, and ask, is it possible, when I roll this again, and again, and again—this fair 10-sided die—that I'll see a 4, a 9, and a 0, then a 4, a 9, and a 0, then a 4, a 9, and a 0, 490, 490, forever? Well, if I were to do that, then I would have generated a rational number, because it would have a periodic tail, which means it's a fraction.

But let's think about that. What is the likelihood that that's going to happen? Well, if you think about it just qualitatively, it's clear that it's not very likely at all, because if I keep rolling this die forever, can you imagine what an amazing coincidence it would be to see a 4, a 9, a 0, a 490, 490, and never see a 1, or a 6, or a 5 again? That seems remarkable. In fact, the probability is, in fact, 0. Let me try to explain to you how we could see that for ourselves. If I roll this die one more time, what's the probability that I will see another 4? Well, there are ten equally likely possibilities, one of which is the 4. So that's 1/10. So, now it's 1/10 likely that I'm going to have a 4. Once I have the 4, what's the probability that when I roll the next time I'll see the 9? Well, that's going to be 1 out of 10, but I multiply these probabilities together, so that's 1/100. So, 1/100 is the probability—very small, 1/100—that I'll roll a 4 followed by a 9. Now, to roll a 0 after that, that's going to be now 1/1,000. To roll a 4 after the 0 is going to be 1/10,000, and so forth. This product is getting smaller and smaller and, in fact, you could see, is approaching 0.

In fact, if we look at this product for the complete, infinite decimal expansion, mathematicians say this product *is* 0. Therefore, mathematicians declare that the likelihood that a decimal expansion is eventually periodic equals 0%. In other words, we have a sense that the likelihood that a random real number selected genuinely at random is rational, is 0. Now just for fun, let me roll it one more time, and let's see if we get the beginning. Let's see if we get that 4. You ready? And it's a 1. So you see, it was not likely that we're going to repeat 4, 9, 0 forever. So, there you have it.

Well, this result, which is really startling at first, but then, when we think about it, makes sense when you think of the actual physical die being rolled, again, and again, and again, and seeing a finite list of digits repeating forever is highly unlikely. It now becomes a little bit more intuitive that we should believe that finding a rational number at random is extremely unlikely. In fact, mathematically the probability is 0. This result also implies that the likelihood that a real number selected at random, like we were doing here, is going to be an irrational number is 100%. It's 100% likely that it will produce an irrational number, meaning that, as we roll that die, the digits will never eventually repeat and repeat forever.

Well, this counterintuitive and perhaps disturbing discovery offers us a new insight and a wonderful opportunity to retrain our intuition about numbers. Remember, whenever we're surprised in life, that's a moment when learning takes place. Here, this surprise allows us to re-hone our intuition. In some sense, there are "more" irrational numbers than rational numbers. In fact, what we're saying is that there are way, way more irrational numbers.

Now, how can we make this make sense? In fact, this idea of 0 probability of selecting a rational number is still a little disturbing, because of course we know the rational numbers do exist. So, couldn't I just by accident happen to pick one? Well, let me try to give you an analogy to get a sense of why this is really, essentially, impossible, and mathematically isn't possible.

Suppose that we take a piece of paper and just draw a bunch of tiny dots on it, or imagine taking a pin and punching holes in the paper just a few times with the pin to make little, teeny holes—very, very tiny dots or very, very tiny holes—very, very tiny, almost invisible. Well, if you were to mark that, then when you looked at the white piece of paper—suppose you marked it, let's say, with black dots—you would see those black dots. That's all you would see. But why would you see those black dots? It's interesting. The reason why you see the black dots is because there are so few of them that they stand out. That's like the rationals in the real numbers. There are so few of them that they stand out in our minds, and therefore, they're familiar to us. Now, what if we were to close our eyes, and take a pin, and just put it down randomly on that piece of paper that contains just a couple of black dots. What's the likelihood that that pin will land on exactly one of those dots? Well, it's clear that it's extremely unlikely, and really, the probability is 0. Even though the black dots exist, the probability of hitting one at random is 0. That's the same principle that's going on here.

Thus, we see that what first appeared to be exotic are more the norm, and what first appeared to be the norm are, in actuality, the exotic and quite rare. So, the familiar rationals really are the unusual, and these seemingly mysterious irrationals are more the commonplace. It's an interesting realization that, in our everyday experiences and encounters with numbers, it doesn't actually provide us with an accurate sense of how the rationals and irrationals actually "fit" together. So, again, our intuition is off. Now, despite this new insight that on the number line it is nearly impossible to accidentally stumble upon a rational number, we're now going to see paradoxically that the rational numbers are essentially everywhere on the number line. Again, we see these two counter, contradictory ideas coming together. What I'm saying is that the rationals are spread out all over the place. You can't help but be near one, and to see that, let's just look at the interval between 0 and 1 and start running down rationals.

First, we'll put in a half. Now we'll put in 1/3 and 2/3. See how, slowly, those rationals are filling up that interval? If we put in the 1/4, we see a 1/4, a 1/2, and 3/4. We put in the 1/5, and we see 1/5, 2/5, 3/5, 4/5. Put in the 1/6. Put in the 1/7—1/7, 2/7, 3/7, 4/7, 5/7, 6/7, slowly, and keep going. Notice how it's slowly filling up the entire interval with these essentially invisible rational numbers. Well, mathematically speaking, we say that the rational numbers are "dense." That's the word we use. The rational numbers are dense on the real line. Now, let's be a little bit more precise to see exactly what this means.

For the rational numbers to be dense, what we're saying is that they, essentially, are going to be everywhere on the line. To show that the rational numbers are dense, we must show the following: that given any two different points on the number line, we can always find a rational number in between. Let's think about that. Suppose I give you two different points on the number line; I've got to find a rational in between. If the two points were far apart, this is really not a very exciting result. So, from the mathematical point of view, we have to imagine these two points really close together—really close. Even though they're so close together, it turns out there's always a rational in between. Make them closer. We'll still find a rational, as long as the two points differ (so that there's a gap in between), then we could always find a rational. That's what it means to be dense.

Now, I'm actually going to show you why this is a believable fact, even though we see how the rationals fill up the interval from 0 to 1. Let's actually see a method that, in fact, establishes this assertion that goes back to ideas that were originally generated by the great mathematician Archimedes in the 3rd century B.C.E.

So, here's the idea. Let's just take the interval between 0 and 1 and pick two points, A and B, and put them down. There they are. Now, my mission is to try to show that there has to exist a rational in between these two points. So, what do we do? Well, let's just consider the reciprocals of the natural numbers—1/2, 1/3, 1/4, 1/5, 1/6. We notice that as those numbers and the denominators increase, the numbers themselves decrease, because I'm looking at the reciprocal. They're getting closer and closer to 0. They're getting smaller and smaller. Well, I have this fixed interval length from A to B, and I'll just wait. I will just literally wait until the 1 over something is actually smaller than this length. This length is not going anywhere, but those reciprocals are quickly going to 0—1/1000, 1/1001, 1/1002, and so forth.

So, I wait until I have a reciprocal that is smaller than this length. In this example, it looks like maybe 1/13 will fit the bill. What do I do? I start at 0 and start marking off by 1/13—so, 1/13, 2/13, 3/13, 4/13, 5/13, 6/13, 7/13. Look, 7/13 gets right in between that interval, and this method actually will always work. Again, put mathematically, what we just established was that the fact that the rational numbers are dense on the number line. Or equivalently, we could say that the rational numbers are dense in the real numbers. Thus, paradoxically, even though the rational numbers are hard to find by a random game of hide and go seek, they are spread out all over the number line, thus, further demonstrating the depth and complexity of the real numbers.

Let's now consider a more refined notion of a random real number. So, we're going to try to now up the abstraction a little bit here and see what really a random real number should be. Now, if we consider rolling one 10-sided die to generate the digits of a random real number, just as we did before to generate the beginning digits of our random number—well, here's a question. How often would we expect to see the digit 3 appear in that random real number? Let's think about that for a second. So, I'm going to roll this fair 10-sided die forever, and the question is, how often will we see, on average, the digit 3? How often will the die land with a 3 up? Well, we guessed that every digit from 0 to 9 would appear, on average, an equal amount of the time. In other words, each digit would appear about 1/10 of the time, so we'd guess 1/10. In fact, if we select a real number at random, we expect that one 1/10 of the digits in its decimal expansion will be 0, 1/10 of the digits in the decimal expansion will a 1, and so forth down the line, and 1/10 of the digits will be a 9.

To extend this idea further, let's now consider strings of two-digit numbers within the decimal expansion. So, let's look at two digits at a time. How likely would it be to see any particular run of two digits? Now, what does that mean to generate two digits at random? Well, we have to imagine rolling a 100-sided die. (There's something you don't see in Vegas.) It would start with 00—that's the smallest two-digit number—and it would go all the way up to 99, the largest two-digit number. The question is, if I were to generate a number at random, what proportion in that run of digits would I see a particular two-digit number as I look across?

First of all let me show you that these dice actually exist. This is one. In fact, this one actually makes a little bit of noise. It's almost like a maraca, and it's actually a 100-sided die. It almost looks like a golf ball doesn't it, because it has little dimples on it? If you roll it, this would generate two digits at random. For example, if I roll it once—let's see. Hard to read—but that looks like a 10. So, that's a 10. Roll it again, and I see a 43, and so forth.

When I look at a random number, what's the likelihood that I'll see 43, and how often will I see 43? Well, we might guess that, because there are 100 two-digit numbers from 00 to 99, we would see (in an average real number in its decimal expansion) that it would contain the 43 roughly 1/100 times, because there are 100 equally likely possibilities for two-digit numbers, and this is just one of them at random—so, 1/100. So, in general, for any two-digit number, we'd expect to see that that two-digit number appears about 1/100 times throughout this run of the decimal expansion. Now, of course we could consider the exact same issue with runs of three digits. I don't have a 1,000-sided die, but it's the same principle at work. We'd have 1,000 possibilities. The smallest three-digit number would be 000. The largest would be 999. So, the question would be, how often would we expect to see the run of 000 appearing in the decimal expansion for an endless real number? Well, we would expect it would be, on average, 1/1,000, because they're all equally likely.

Real numbers for which every length of digits appears the appropriate or the expected amount of the time in their decimal expansions are called "normal numbers in base 10." What does this mean? It means that a number is a normal number in base 10 if every digit appears, on average, 1/10 of the time, every pair of digits, consecutively, appears every 1/100 of the time, on average, every three-digit number you could think of, on average, appears 1/1,000 times, and so forth (every four-digit number, every five-digit number, and so forth). That's what it means for a number to be a normal number in base 10.

Now, let's return to our randomly generated real number. What's the likelihood that it's a normal number in base 10? So, just rolling that 10-sided die forever, what's the likelihood that, in fact, it will be normal, no matter what the runs of digits we look at? Well, this question remained unanswered until the 20th century, when the influential French mathematician, Émile Borel, provided a complete answer in 1909. He proved that the likelihood that a random real number is normal in base 10 is 100%. That means that, if you pick a number out at random by rolling a 10-sided die forever, every length of runs of digits you could think of will appear the appropriate amount of the time. Absolutely amazing. So basically, in some vague sense, he's proving that the probability that a number is random is essentially 100%. So, if you just look down and randomly grab a number, it's going to be normal in this sense. All the digits will be all messed up and appear in all the different orders that you could imagine, and appear the right amount of the time, the right proportion of the time. By the way, Borel is perhaps the only 20th-century mathematician who actually has a crater on the Moon named after him. There's a little piece of trivia that's amusing.

Well, in view of our previous observations, we see yet again that, through this refined notion, the rational numbers such as 1/3 and 1/2 are indeed very strange. Why is 1/3 so strange? Well, because the decimal expansion for 1/3 is what? It's 0.33333 forever. Imagine rolling a fair, 10-sided die forever and only seeing 3s. It's not likely. One-third is a rare number. What about 1/2? Well, 1/2 is 0.50000 forever. Imagine picking up a fair 10-sided die, rolling it forever, and seeing first a 5 and then 0s forever. It's not going to happen. These rational numbers are rare. So again, we have a different intuitive sense of why these familiar numbers really are the exception.

Now, in fact, just to mention the mathematics again, Borel considered an even more general notion of random number, and this is astounding, but a little bit complicated. So let's see if we can think about this together.

Remember what we just talked about were called "normal numbers in base 10." Borel actually studied numbers that he called "normal numbers," just "normal numbers." What that means—a number is a normal number if the number is normal in every base. So take the number—the decimal number you're thinking of—write it in base 10. It's normal in base 10. Write that same exact number, express it in base 2, and it will be normal in base 2, which means—in base 2 we only have two digits, 0 and 1—we'd see the 0s happening half the time on average, and the 1s appearing half the time. So it's 50/50. We see 0s and 1s, on average, the same amount of the time. If we look at base 3—so, we have 0, 1, and 2—we see each one of those digits appearing how often? A third of the time. Then we look at base 4, base 5, base 6, and so forth. If the number is normal in every single base, then we say it's a normal number. Borel actually proved an even more astounding result: The likelihood that a random real number is a normal number—so, normal in every single base—is 100%. So, if you randomly pick a number, the digits are going to basically be every kind of hodge-podge collection of digits in any kind of order you can imagine, and those orders will be all appearing the right proportion of the time, as if the number were selected by rolling a die.

So, rolling a fair 10-sided die, what we're seeing is an actual, correct way of generating a real number. Now, if the likelihood that a random real number possesses a particular property is 100%, then we say in mathematics that "almost all"—we always put that in quotes, by the way—real numbers possess that property. So, applying this new phrase, we could say "almost all" numbers are irrational, and furthermore, we could say that "almost all" numbers are normal. Hence the name "normal," because "almost all" numbers are normal. Thus, for any random real number we think of, we expect the digits to be such that any particular run of digits will occur the true "fair share" of the time.

Within the world of real numbers, our familiar rational numbers are truly the exception, while the at-first exotic irrational numbers are, in fact, the rule. It's intriguing, actually, to wonder how the Pythagoreans would have reacted to this newfound reality, one in which their God-given natural numbers and even their beloved ratios are totally overrun and outnumbered by the overwhelming dominance of the disturbing irrational numbers. Again, we see the notion of numbers as a constantly evolving object. In the next lecture, we'll take a look at a very exotic collection of numbers sitting nicely and beautifully inside the real number line.

A Beautiful Dusting of Zeroes and Twos
Lecture 11

Cantor's one-to-one correspondence is devilishly clever. It hinges on the fact that every real number can be expressed in a base-2 expansion using just 0 and 1.

Georg Cantor lived in an exciting age for mathematics. He was born in 1845 in St. Petersburg; soon after, his family moved to Germany, where he spent the rest of his life. As a youth, Cantor showed much interest and talent in art, music, and mathematics. In 1866, at the age of 22, Cantor completed his studies in number theory at the University of Berlin after studying under some of the greatest German mathematicians of the time: Ernst Kummer, Leopold Kronecker, and Karl Weierstrass.

In the mid-1800s, there was much interest in subtle and delicate questions involving numbers and functions. A critical question regarding the computation of interest was how to express exotic functions as sums of more familiar trigonometric functions (such as the famous sine and cosine waves we see in visualizations of sound waves). Joseph Fourier, the 19th-century French physicist, developed a famous theory—Fourier analysis—in response to this question. By 1870, it was known that the functions of interest to the mathematicians of the day could indeed be expressed as a (possibly endless) sum of sine and cosine functions. The big question, however, was whether such an expression was unique—that is, is there only one way to write these functions as sums of trigonometric functions? In 1868, Cantor was offered a position at the University of Halle, outside Berlin. Two years later, one of his senior colleagues, Edward Heine, published a paper in which he mentioned this question of uniqueness, noting that many great minds, including Riemann, had been unable to solve this important question. This challenge captured Cantor's imagination, and he set off to tackle this question that the great minds of the day could not answer.

By 1871, Cantor published an answer to the uniqueness question. He proved that if the functions were reasonably "nice" and "well-behaved" (as most functions of the day were), then there is only one way to express them as a sum of trigonometric functions. Cantor soon proved that, in fact, the functions could be "bad" as long as they were "bad" at only *finitely* many numbers; this result was well received. Cantor began to consider functions that were "bad" at *infinitely* many numbers and extended his uniqueness result to such functions as long as these "bad" numbers were spread out; this result turned some heads. He then wondered what would happen if the functions were "bad" at infinitely many numbers (or, equivalently, at infinitely many points on the number line) that were clustered together; it was at this point that Cantor's work became groundbreaking and totally original.

In order to study such delicate infinite clusters of points on the number line, Cantor needed a precise definition of real numbers and a more rigorous treatment of irrational numbers. Cantor started studying exotic collections of real numbers, and soon, this study captured his imagination. His work was highly controversial at first and he suffered greatly; however, in the scope of just five years, his work moved the frontiers of mathematical and numerical understanding out beyond anyone's wildest imagination.

Recall the ternary expansion of real numbers. If we look at the real numbers in base 3 (as Cantor did in 1883), rather than base 10, the collection we will explore will reveal itself much more naturally. We recall that every real number can be expressed in a base-3 expansion. In a base-3 expansion, the only allowable digits are 0, 1, 2; the "positional values" are powers of 3. For example:

$$120_3 = (1 \times 9) + (2 \times 3) + (0 \times 1)$$

$$= (1 \times 3^2) + (2 \times 3^1) + (0 \times 3^0)$$

$$= 15_{10}$$

$$0.1021_3 = (1 \times \frac{1}{3}^1) + (0 \times \frac{1}{3}^2) + (2 \times \frac{1}{3}^3) + (1 \times \frac{1}{3}^4)$$

$$= \frac{34}{81}$$

$$= 0.419753086\ldots_{10}.$$

We can visualize base-3 expansions on the number line by cutting up intervals into three equal pieces.

We will consider all real numbers within the interval between 0 and 1 that can be expressed in a base-3 expansion using only the digits 0 and 2. We will call this collection of numbers the Cantor set or Cantor dust and denote the collection by C. Some examples follow. The numbers 0 and 1 are in this collection (recall that $1 = 0.2222..._3$). The numbers $\frac{1}{3}$ and $\frac{2}{3}$ are in this collection because $\frac{1}{3} = 0.1_3 = 0.022222..._3$ and $\frac{2}{3} = 0.2000000..._3$. Unfamiliar numbers, such as $0.220222000020202022202002222..._3$, are also in this collection. In fact, there are both rational and irrational numbers in the Cantor dust. This collection C appears to have no apparent structure.

We find points in the Cantor set on the number line. We plot the points 0, $\frac{1}{3}$, $\frac{2}{3}$, and 1 on the number line and attempt to pinpoint 0.22022200002020202220 $2002222..._3$. We discover that the numbers in the collection C reside in either the first $\frac{1}{3}$ segment or the third $\frac{1}{3}$ segment of the interval between 0 and 1; the collection C contains no points from the middle $\frac{1}{3}$ of the interval. Within each of these two line segments of length $\frac{1}{3}$, the numbers from C reside in either the first or third $\frac{1}{3}$ of these segments; we essentially remove the middle $\frac{1}{3}$ of each of these smaller intervals. This continual pruning process leads to a geometrical description of this collection of numbers: The Cantor set contains all the numbers (points) that have not been removed. Given this description, the Cantor set is often called the "middle thirds Cantor set."

Cantor extended this notion and defined more general collections of numbers. This geometric description shows that these numbers have some amazing structure—in particular, this collection has a self-similarity property that makes it an example of a fractal collection of numbers. Fractals are images that exhibit a self-similarity property as we zoom in and focus on just a small part of the entire image.

No number in the Cantor set is normal. We recall that normal numbers have base-3 expansions in which each of the digits 0, 1, and 2 occurs one-third of the time. The numbers in the Cantor set do not contain the digit 1 when expressed in base-3 expansions; thus, none of the numbers of the Cantor set are normal. "Almost all" real numbers are normal; thus, we conclude that "almost no" numbers are in the Cantor set. If we were to pick a random real number between 0 and 1, the probability that we happen to select a number from the Cantor set is 0%; thus, we say that "almost no" numbers are in the Cantor set. In some real sense, the Cantor set is a very sparse collection of numbers. This is why we also call this collection Cantor dust—it is essentially invisible on the line segment between 0 and 1.

Even though "almost no" numbers are in the Cantor set, there is a one-to-one pairing of the numbers in the Cantor set with *all* the numbers on the real line between 0 and 1. Cantor proved that even though his set is an essentially invisible collection of numbers inside the interval from 0 to 1 on the real line, that collection of numbers can be placed in a one-to-one correspondence as the entirety of numbers inside the interval from 0 to 1. Cantor challenged Euclid's notion that the whole is greater than the part. The proof for Cantor's claim is elegant. We must find a way of associating each number in the Cantor set with only one number from the line segment between 0 and 1 and, conversely, find a way of associating each number from the line segment between 0 and 1 with one and only one number from the Cantor set. We must remember that every real number can be expressed in a base-2 expansion using the digits 0 and 1; thus, every number from the line segment between 0 and 1 can be expressed as an endless list of 0s and 1s.

To establish that every number from the Cantor set can be paired up with exactly one number between 0 and 1, consider a number from the Cantor set (recall that we express these numbers in base 3):

$$0.22000020202222200020..._3.$$

Change each 2 to a 1, and now view that number in base 2:

$$0.11000010101111100010..._2.$$

This new base-2 number is in the interval between 0 and 1. Pair every number (expressed in base 3) in the Cantor set with a corresponding number (expressed in base 2) in the interval between 0 and 1 by simply changing all the digits that are 2s to 1s and viewing the new list of digits as a base-2 number. Why are there no numbers in the interval between 0 and 1 that have not been paired with any number from the Cantor set in this manner?

Consider the number $0.010110101000101011101\ldots_2$. We can find a number in the Cantor set that is paired with this particular number by simply performing the previous steps in reverse—that is, replacing all the 1s with 2s and viewing the number in base 3. In this case we would see the number $0.020220202000202022202\ldots_3$.

Recall that the Cantor set contains precisely all numbers between 0 and 1 whose base-3 expansions consist only of the digits 0 and 2; thus, the previous number is in the Cantor set and is paired up with the original number we considered from the interval 0 to 1. We now have found a one-to-one pairing between the Cantor set and the entire collection of numbers between 0 and 1, which was what Cantor claimed. The Cantor set is a very counterintuitive collection of numbers and is often used as a counterexample to seemingly reasonable conjectures. ∎

Questions to Consider

1. Which of the following numbers—$\frac{1}{2}, \frac{4}{5}, \frac{1}{3}$, and 0—are in the Cantor set? What about the following numbers written in base 3: 0.202_3 and 0.121_3?

2. The Cantor set can be constructed by repeatedly removing "middle thirds" of intervals starting with the interval between 0 and 1. What kind of set would we get if we started with the interval between 0 and 1 and successively removed middle fifths? Can you draw this collection on a number line?

A Beautiful Dusting of Zeroes and Twos
Lecture 11—Transcript

In this lecture, we'll discover some beautiful and amazing structure within the real numbers that only comes into focus when we think in base 3. In particular, we'll explore the nuances of one of the most famous and vexing collections of real numbers. This collection of real numbers was first described by the tremendously brave, ingenious, but troubled German mathematician Georg Cantor in 1883. His work in general, and his collection of numbers in particular, has led to a better and deeper understanding of many areas of mathematics, especially numbers. Known as the "Cantor set," or "Cantor dust," this collection of numbers is intriguing in its own right, because, as we'll see, the numbers can be described in a very elegant fashion in terms of their base-3 expansions or, equivalently, in terms of their location on the number line. In fact, when viewed on the number line, we'll see this collection is an example of what mathematicians today call a "fractal," a geometric form that has an endless self-similarity property.

Applying our observations from the previous lecture, we'll discover that even though there are infinitely many numbers in the Cantor set, they form a very thinly spread, or "sparse," collection of numbers between 0 and 1. As sparse as this collection is, we'll establish a truly remarkable and counterintuitive theorem. We'll see that, even though this Cantor set is an invisible, dust-like collection, it contains just as many numbers as the entire solid line segment from 0 to 1. This result, and its clever argument that establishes its validity, show that within the jumble of numbers, many surprising phenomena can and do occur. While we'll struggle with these vexing notions in this lecture, it might be comforting to know that the mathematics community struggled with Cantor's revolutionary ideas before seeing the light. In fact, Cantor's set and its remarkable properties were part of his highly controversial, but totally correct, work in cracking the mysteries of infinity. Thus, this lecture foreshadows the original, imaginative, and groundbreaking work of a great mathematician who set the mathematics community on fire with the creative spark of his beautiful ideas.

Before we consider these special numbers, I want to say a few words about the life and mathematics of the man who discovered them, Georg Cantor.

Georg Cantor was born in 1845 in St. Petersburg, soon after which his family moved to Germany, where he spent the rest of his life. As a youth, Cantor showed much interest and promise in art, music, and mathematics, but decided to focus on his mathematical talents. In 1866, at the age of 22, Cantor completed his studies in number theory at the University of Berlin after studying under some of the greatest German mathematicians of the time: Ernst Kummer, Leopold Kronecker, and Karl Weierstrass. We'll say more about these important figures later, in Lecture Eighteen.

Now let's turn to his early mathematical interests. I want to place his work in context, and so I want to first paint for you the mathematical scene of that period. In the mid-1800s, there was much interest in subtle questions involving numbers and functions. Now, for our purposes here, let's just think of a function as a formula. In fact, one of the individuals interested in this study was the very influential mathematician Bernhard Riemann, who we mentioned in Lecture Seven.

The question of interest was how to express certain exotic functions as sums of more familiar trigonometric functions such as the sinusoidal sine and cosine function that we saw sometime back in our school days. So, these functions that we were trying to look at were complicated, wavy functions. You can imagine, for example, visualizing a sound wave (very wiggly functions), and we want to write these wiggly functions in terms of sums of these nice sines and cosines. This theory later became known as "Fourier analysis," named after Joseph Fourier, the great 19th-century French physicist.

Well, by 1870, it was known that the functions that were of interest to the mathematicians of the day, those wiggly functions, could indeed be expressed as a sum, potentially an endless sum, of sine and cosine functions. However, the big question was, is such an expression unique? In other words, is there only one way to write these wiggly functions as sums of trigonometric functions? That was the question. Well, in 1868 Cantor accepted a position at the University of Halle, outside of Berlin. Two years later, one of his senior colleagues there, Edward Heine, published a paper in which he mentioned this question of uniqueness. In particular, he actually noted that many great minds, including Riemann, have been unable to answer this important question.

This challenge captured the imagination of Cantor, and he set off to tackle the question that the great minds of the day could not, themselves, answer. Just within one year, one year later in 1871, Cantor published a solution to this famous uniqueness problem. He actually proved that if the functions were reasonably "nice" and "well-behaved," as most functions of interest of the day were, then there's only one way to express them as a sum of trigonometric functions. Well, Cantor wondered if the result still held for functions that were strange and not so well-behaved. He soon proved that, in fact, functions could be, in some sense, "bad" as long as they were only "bad" at *finitely* many numbers. Well, what do I mean for a function to be "bad" at a number? I mean that if we were to put that number into the formula, it would be like dividing by zero. It's undefined. That's what I mean by "bad." So, his result was extremely well received and actually earned Cantor the respect of his colleagues, and young Cantor was now considered to be a mathematician of the highest caliber.

But still Cantor wondered. He began to consider functions that were "bad" at *infinitely* many numbers, and he extended his uniqueness result to such functions as long as the infinitely many "bad" numbers were spread out on the number line. Well, this result actually turned some heads and raised some eyebrows. People were saying, "Gee, looking at infinitely many bad points, that's a little peculiar." Well, he then further wondered what would happen if the functions were "bad" at infinitely many numbers, or equivalently, infinitely many points on the number line, that were not spread far apart, but were clustered together, and it was at this point that Cantor's work became groundbreaking and totally original.

Well, in order to study such delicate, infinite clusters of points on the number line, Cantor needed a very precise and exacting definition of the real numbers—the work required that—and a much more refined treatment of irrational numbers. We'll highlight this important movement in Lecture Eighteen. Well, Cantor started studying very exotic collections of real numbers, and soon this study captured his imagination, and he actually lost interest in his original goal of writing functions uniquely as sums of sines and cosines functions. He said, "Who cares about that?" These interesting, exotic collections of real numbers were really neat. Well, his totally original work was highly controversial at first, and he actually suffered greatly. However, his work, in the scope of just five years, moved the frontiers of mathematical and numerical understanding out beyond anyone's wildest imagination. This is perhaps the only moment in history in which one individual, in a very short time span of time, moved mathematics a quantum leap forward. For through his work, Cantor realized how to understand infinity.

Now, if you remember, my vision for moving the mathematics frontier forward in our understanding it was building these dimples upon dimples on this copper orb, and what Cantor basically did, in essence, is take a sledgehammer and just put an enormous spike in this orb. It took mathematicians a while just to be able to venture out there, even though the path was already led. That's how innovative the work was. Well, to describe Cantor's collection of numbers, we need to return to the base-3 expansions of real numbers. One of the mathematical morals of this lecture is that sometimes one particular representation of numbers can be much more convenient than another. In this case, we'll look at the real numbers in base 3, as Cantor did in 1883, rather than in the more typical base 10. The collection we're about to explore will reveal itself in a much more natural way when we think of it in base 3.

Let's first recall that every real number can be expressed in a base-3 expansion using just three digits—0, 1, and 2—with the positional locations representing powers of 3. So for example, if we look at the number 120_3, notice that little subscript of 3 tells us this is not the number 120, but it's 1-2-0 in base 3. Well, the unit spot, the 1 spot, has 0. So we have zero 1s. But we have a 2 in the 3s spot, because now, in base 3, the next space over represents 3s. So, we have two 3s. Then lastly, that 1 is in the 3^2, or 9, spot. So, we see that this number is $(1 \times 9) + (2 \times 3) + (0 \times 1)$, which equals 15. So, $120_3 = 15_{10}$.

How about another example? Let's look at one that has a fractional part, 0.102_3. Well, that first 1 we see is in the 1/3 spot. The next 0 to the right is in the 1/9 spot. Remember, now everything is in terms of the powers of 3, and that 2 is located in the 1/27 spot. What we see here is $(1 \times 1/3^1) + (0 \times 1/9$, or $1/3^2) + (2 \times 1/3^3$, or $2 \times 1/27)$. When you put those numbers together, we see 11/27, which is 0.407407407_{10}, and of course it repeats, since it's a rational number.

In Lecture Nine, we saw that we can visualize a base-3 expansion on the number line by cutting up intervals into three equal pieces each time. So as a quick illustration, let's locate the number 0.102_3 on the number line. What do I do? Well, the first thing I do is realize that this number is between 0 and 1, so I look just in that interval, and I cut it up into three equal pieces, each one 1/3. I label the first interval 0, then 1, then 2. I look at the digit I see next, which is a 1, so I know that I'm in the middle third. I'm in the 1 spot, so I'm in that middle third, and what do I do? I take that middle third and cut it up into three equal pieces, and I label them 0, 1 and 2. I see the next digit we have is a 0. That means we're in the left tiny interval. Then I cut that tiny interval into three pieces equal in length, and I go right to the beginning of the third one, the 2, and that's our point. So again, it's a honing-in process that we do, but instead of honing in by tens, we hone in by threes each time.

We're now ready to face Cantor's exotic collection of numbers. Cantor considered all real numbers within the interval between 0 and 1 that can be expressed in a base-3 expansion using only the digits 0 or 2. Okay, let's think about this for a second. So, he's defining a collection of numbers between 0 and 1, and the only condition, the only stipulation, on this collection is that, when you write the number in base 3, you only see the digits 0 or 2. You never see 1; that's how he's defining this collection. This collection of numbers is now known as the "Cantor set," or "Cantor dust." Let me just say word about the word "set." In mathematics, "set" refers to a collection of things. So it's not as though, "ready, set, go." It's not like that. It's "set" as in a "collection." So, the Cantor collection, Cantor set, or they are also referred to as the Cantor dust.

Let's identify some specific numbers that are part the Cantor dust. We begin by noting that both 0 and 1 are in the Cantor set. And why? Well, because 0 is 0.0000 forever. It just contains 0s. Remember, the only rule is that all you can see are 0s and 2s. Now, the number 1 is in there because, remember that we actually established that 1 is equal to 0.2222 forever in base 3. That's a way of expressing 1. In fact, we see that 1 is in the set because it can be written as 0s and 2s and nothing else in base 3—also 1/3 by the similar reason. So 1/3 is in this collection because 1/3 = 0.1 in base 3, which is the same thing as 0.02222 forever in base 3. 2/3 is in this set because 2/3 means I have 2/3, which actually is 0.2000 forever, because the 2 is in the 1/3 spot.

But even unfamiliar numbers that we could just make up are in the set, as long as we make sure we never utter a 1. So for example, the number 0.2202220000202020222_3, and so forth, are also in this collection. In fact, if you think about it, we can see that there will be both rational and irrational numbers in the Cantor dust, because all we have to do is write infinitely long lists of 0s and 2s that never repeat. If it never repeats, it's irrational. If the number does repeat, if the digits do repeat the 0s and 2s, then we know it's rational. So, the Cantor dust contains both rational and irrational numbers and appears to have no apparent structure whatsoever.

Let's now try to visualize the Cantor dust by finding and locating some of dust in the number line. So, first let's just plot the four easy points that we found, 0, 1/3, 2/3, and 1, on the number line. We see that actually spread out quite evenly between 0 and 1. Well, now let's attempt to pinpoint that kind of random number that I generated, and let's just think through the process together. I'm going to be a little bit fast here, but we'll see that there's a groove we're building here.

So the first digit I see in this number is a 2. That tells me that, when I break the interval from 0 to 1 up into three pieces (0, 1, 2), the 2 tells me go to the far right, go to the far right one. Okay, fine. In the far right one I cut that into three pieces. The next digit of 2 tells me to go to the far right in there. Then I take that tiny, tiny, interval cut into three pieces, and the 0 we see tells me to go to the far left of that tiny interval. Cut that into three pieces, the 2 tells me to jump now to the right. So, when we do this, of course we can't do the whole number because it's too many digits. It's hard to pinpoint that point in this image, but we see a process here. We're bouncing from right to left.

So, from the chaos of the digits and all this interval hopping, we begin to see an attractive pattern emerge. The numbers in the Cantor set reside in either the first third of the segment or the last third of the segment of the interval between 0 and 1. So, the Cantor set will either live in the first third or the third third, but not the second third. And why? Because the Cantor set contains no numbers that have the digit 1, and the digit 1 is represented by that middle third. Now, within each one of these two thirds, the first third and the last third, the Cantor set has to reside in either the first or the third third of that and, similarly, the first or the third third of that.

Again, why? Well, because we see that the numbers from the Cantor dust cannot have a second digit of 1, which would put them in the middle spots. Therefore, we see that we're either going to be in the first or the last third, and the middle third will not be used. In fact, those middle thirds are dust free, and this continues. So, imagine repeating this pruning process forever. That is, we successively remove the middle third from each interval that we currently have at each stage of this process, thereby ensuring we never see the digit 1. We'd only see 0s and 2s. So, we keep pruning out the middle thirds, the middle thirds, the middle thirds, and so forth.

Thus, if we do this forever, this process continues forever. We have a geometric description of this collection of numbers. The Cantor set consists of all the points, or all the numbers, that have not been removed from this infinite middle thirds pruning process. Given this geometric construction of successively removing the 1s digit by the removing the middle thirds, the Cantor set is sometimes referred to as the "middle thirds Cantor set," because it's the middle thirds that we keep throwing away. Well, this geometric description shows that this collection of numbers has incredible structure after all, even though it sounded like a jumble of numbers to begin with. In particular, this collection has a wonderful self-similarity property that makes it an example of what's called a fractal collection of numbers. Fractals are images that exhibit a self-similarity property, which means that if we zoom in and just focus on a small portion of it, that little portion looks like the entire original image. So, as you zoom in, you see within the zoom, itself again, and if you zoom closer, you see more. Notice that every time we zoom in to one of those intervals, we actually see another set of itself.

If we start with a triangle now—just for fun—and at each stage we remove the middle third triangle, then we produce a beautiful, two-dimensional version of the Cantor set. So, you can visualize the Cantor set even in two dimensions, and this famous fractal is actually called the Sierpinski triangle. It's an example of a more generalized Cantor set.

I now want us to consider the numerical beauty of this collection rather than just its geometrical beauty. To do this, we're going to pull in some of the ideas from the previous lecture. Well, we saw that normal numbers have base-3 expansions in which each of the digits 0, 1, and 2 occur one-third of the time. Now, since the numbers in the Cantor set do not contain the digit 1 when we expressed it as a base-3 expansion, we conclude that none of the numbers in the Cantor set are normal. Remember, the normal numbers have to see, on average—one-third of the digits have to be 0, one-third of the digits have to be 1, one-third of the digits have to be 2. Here, we never see 1s. Well, that's not normal. Therefore, none of the numbers in the Cantor set are normal.

In the last lecture, we considered Borel's theorem that states that "almost all" real numbers are normal, but we just said that no numbers from the Cantor set are normal. In other words, if we were to randomly select a real number between 0 and 1 by rolling a die again, the probability that we happen to pick a number from the Cantor set is 0%. Thus, we say that "almost no" numbers are in the Cantor set. In some very real sense, the Cantor set is very sparse. It's a very sparse collection of numbers. This is why the collection is also referred to as Cantor dust. These numbers are, in essence, invisible on the line segment between 0 and 1. Of course, this realization is consistent with the geometrical image we saw because it seemed like it was a little dusting of points here and there.

We're now about to face one of Cantor's discoveries that's both truly remarkable and totally counterintuitive. It's one of his results that deeply disturbed his colleagues, in fact. Even though "almost no" numbers are in the Cantor set, Cantor discovered a one-to-one pairing between the numbers in the Cantor set and *all* of the numbers on the real number line between 0 and 1. In other words, Cantor proved that even though his set is sparse and essentially an invisible collection of numbers inside the interval between 0 and 1, that collection of numbers can be placed in a one-to-one correspondence with all the numbers inside the interval from 0 to 1. So, Cantor's saying, If we consider the entirety of all the numbers between 0 and 1, and the Cantor set (that tiny collection of numbers from 0 to 1 that appears to be invisible), then those two collections of numbers are equally numerous. Well, this result is very perplexing and difficult to digest. In fact, one of Euclid's so-called "common notions" that everyone accepted without question was that the whole is greater than the part. As we're now seeing, Cantor, over 2,000 years later, is actually challenging this Greek notion.

We now follow Cantor's steps through the explanation demonstrating that this unbelievable assertion is, in fact, a fact. Now, just as the Sumerian shepherd paired up his sheep with his pebbles in a one-to-one fashion, we must find a systematic way of pairing up each number in the Cantor set with each number from the interval between 0 and 1 in a one-to-one fashion. In other words, we must find a method of associating each number in the Cantor set with one and only one number from the line segment between 0 and 1 and, conversely, find a way of associating each number from the line segment between 0 and 1 with one and only one number from the Cantor set. Now, this task sounds really challenging.

Cantor's one-to-one correspondence is devilishly clever. It hinges on the fact that every real number can be expressed in a base-2 expansion using just 0 and 1. So, 0s and 1s are all you need to write any number in base 2. Well, Cantor's ingenious idea, as we're about to see, is so simple. His one-to-one correspondence involves just switching 2s and 1s and toggling between base-3 and base-2 expansions. Let me say this very, very briefly, informally first, and then I'll explain it a little bit more rigorously. So, I'm going to show a correspondence between all the points on the number line between 0 and 1 and this very invisible Cantor set, and here we go. First of all, let's take a number from the Cantor set. Now, remember, that means that all we see in its base-3 expansion—let's just stop and think about this for a second—are going to be 0s and 2s. Here's Cantor's idea. Wherever you see a 2, replace it by a 1 and think of it as a base-2 number. Then it's going to be some number down here, between 0 and 1, but written in base 2.

So every Cantor point is associated with some number down here by just changing the 2 to a 1 and thinking of it as base 2. Here's the amazing thing: Conversely, if we take any number from 0 to 1 and write it in base 2, all we see are 0s and 1s. Let's replace every 1 by a 2 and view that long, infinite list as a base-3 expansion. Then, in fact, we've got a number in the Cantor set, because it just consists of 0s and 2s. This was his basic idea. Now, let's make this rigorous.

So, first let's pair up every number from the Cantor set with exactly one number from the interval between 0 and 1. So, for example, let's consider the number from the Cantor set that begins 0.2200002020222_3 and so forth.

We're expressing this in base 3. All we do is simply change each 2 to a 1 and now view that number as a base-2 number. So, in this example we would change it to 0.110000101011111 and so forth, but now we're writing the number in base 2. This is some number between 0 and 1. Now using this idea, we could now pair up each number in the Cantor set, expressed as base 3, with its corresponding number, expressed now in base 2, between 0 and 1. In other words, we simply change all the 2s to 1s and view this new list of digits as a base-2 expansion. It's ingenious.

This establishes a way of pairing up each number from the Cantor set with exactly one number from the interval between 0 and 1. Now, here's the big question. With this pairing, are there any numbers between 0 and 1 that are not paired up with any number from the Cantor set? The amazing answer is no. We've paired up every number from 0 to 1. Why? Well, suppose we consider a number between 0 and 1. We first just write it in its base-2 expansion. For example, let's say 0.01011010100010_2 and so forth. To find its partner in this pairing scheme, all we just do is take every appearance of a 1 and just replace it by a 2 and take this new list of digits and view it as a base-3 number. In this case, we'd see the number 0.020220202000_3 and so forth—2, 0, and so on, base 3. Well, since this new number consists of only 0s and 2 in its base-3 expansion, it's in the Cantor set. So every number has one and only one partner. Amazing.

This matchmaking method is the one-to-one correspondence between the Cantor set and the entire collection of numbers between 0 and 1 that Cantor claimed existed. Absolutely tremendous. Well, the Cantor set is a very counterintuitive collection of numbers. In fact, it's quite often used as a counterexample to seemingly very reasonable conjectures. This collection of numbers is at once sparse and thin and yet full and, as we'll see at the end of this course, fat.

Cantor was an extraordinarily original and creative thinker who allowed his mind to wander, and through that journey, made discoveries that at first appeared unbelievable. Unfortunately, as we'll see in the last part of this course, Cantor also had to contend with personal demons that haunted him most of his life. Today, he would most likely have been diagnosed as bipolar, manic-depressive. Sadly, he spent a good portion of the later years in asylums. In the last part of our course, we'll explore this tragic drama, his struggle with relationships—including some ugly mathematical feuds—and of course, his imaginative and profound insights into infinity, which continue to be the foundations for how we view the infinite.

An Intuitive Sojourn into Arithmetic
Lecture 12

Up to this point in our course ... we've been focusing on numbers themselves and discovering how a more sophisticated understanding of nature and of our world leads to a more sophisticated understanding of number. Here, we'll consider various ways of combining numbers in order to widen our view of number further still.

The axioms of combining numbers are included in most of the fundamental rules of arithmetic that we still apply today. In the 3rd century B.C.E., Euclid offered five "common notions"—basic and self-evident truths that did not require proof. Today, we call such statements "axioms." Things that are equal to the same thing are also equal to one another. (Today, we call this property "transitivity.") If equals be added to equals, the wholes are equal. If equals be subtracted from equals, the remainders are equal. Things that coincide with one another are equal to one another. (Today in geometry, we say they are "congruent.") The whole is greater than the part. (Note that this notion is not always true.)

The numbers 0 and 1 have special properties. The number 0 is known as the additive identity because if 0 is added to any given number, the sum is that given number. The number 1 is known as the multiplicative identity because if 1 is multiplied by any given number, the product is that given number.

A review of inverses and operations is useful at this point. Given any number n, there exists a number $-n$ such that its sum with n equals 0. The number $-n$ is called the "additive inverse" of n. Given any nonzero number n, there exists a number $\frac{1}{n}$ such that its product with n equals 1. The number $\frac{1}{n}$ is called the "multiplicative inverse" of n. In an arithmetic sense, addition and subtraction are opposite operations, just as multiplication and division are. This observation was first made by the ancient Egyptians, who viewed the pairs of operations as mirror images of each other. Today, we say they are "inverse operations." Zero is the only number that does not have a multiplicative inverse because when any number n is multiplied by 0, the product equals 0. This statement is not an axiom but a provable theorem. Numbers also satisfy the distributive property: $a \times (b \pm c) = a \times b \pm a \times c$. Thus, for any number n, we see that $n \times 0 = n \times (1-1) = n \times 1 - n \times 1 = n - n = 0$.

Why does a negative multiplied by a negative equal a positive? Multiplying signed numbers follows simple rules. A positive number multiplied by a positive number is positive. A negative number multiplied by a positive number is negative. A negative number multiplied by a negative number is positive.

Why does a negative multiplied by a negative equal a positive?

We can prove that $(-1) \times (-1) = 1$. We first recall that because -1 is the additive inverse of 1, we have: $1 + (-1) = 0$. If we multiply both sides by (-1), we see by the distributive property that: $(-1) \times (1) + (-1) \times (-1) = (-1) \times (0)$. Recall that 0 multiplied by any number is 0. Also, 1 is the multiplicative identity; thus, we conclude: $-1 + (-1) \times (-1) = 0$. Adding 1 to both sides reveals: $(-1) \times (-1) = 1$.

The symbols of arithmetic evolved from numerous sources. Addition and subtraction symbols seem to have derived from Latin. In ancient Latin manuscripts, the word *et* ("and") was used to indicate addition. It appears that the symbol + derives from the *t* in *et*. In a manuscript from 1489, we find the Greek cross used for addition. The Latin cross was also used but printed horizontally. Sometimes, the Maltese cross was used for very formal writings. The minus sign (–) made its first appearance in 1481.

Multiplication and division signs have multiple origins. The multiplication sign (×) first appeared in Fibonacci's *Liber Abbaci*. He employed the notation to indicate what we might consider today as "cross products" of pairs of numbers. Slowly, the notion evolved into the product of two numbers. In 1751, an inverted D was used for division. By 1753, we find the symbol ÷. In the Rhind Papyrus, Ahmes used a dot over a number to indicate a unit fraction. The dot derived from the hieroglyph for an open mouth—indicating that fractions were used to divide rations of food and drink. Writing fractions stacked—using three terraces of type—with a horizontal line between the numbers goes back to Arabic writings. In *Liber Abbaci*, Fibonacci used stacked fractions. The writing of the diagonal line for fractions, such as 2/3, first appears in a text published in Mexico in 1784, with a tilted flourish that slowly evolved into the less ornate /.

The symbols for exponents and powers come from Descartes. Descartes developed the shorthand n^2 for $n \times n$ and n^3 for $n \times n \times n$. This notation allows us to write very large values in a compact fashion (e.g., $1,000,000,000 = 10^9$). Symbols for natural-number exponents also derive from Descartes. In view of Descartes' notation, we have the law of exponents: $(n^a) \times (n^b) = n^{a+b}$. We can combine products, sums, and positive powers to create polynomials; that is, such expressions as: $3n^2 - 2n + 10$ or $120n^{201} + 12.34n^{35} - 111n^7 - 5n + 27$. Polynomials offer an important way to classify numbers in a more refined manner than simply rational or irrational.

The notation for rational exponents goes back to the 14th century. What power of 9 equals $\sqrt{9}$? (We note that $\sqrt{9} = 3$ because $3^2 = 9$). Suppose $9^p = \sqrt{9}$. Then we know that $(9^p) \times (9^p) = (\sqrt{9}) \times (\sqrt{9}) = 9$. By the law of exponents, we conclude that $9^{2p} = 9 = 9^1$. Thus, $2p = 1$ and $p = \frac{1}{2}$. We discover that $9^{1/2} = \sqrt{9}$; that is, $9^{1/2} = 3$. Similarly, $2^{1/2} = \sqrt{2}$. Rational exponents signify roots: $a^{1/b} = \sqrt[b]{a}$ (the bth root of a). Thus, $8^{1/3} = \sqrt[3]{8} = 2$ (because $2^3 = 8$). More generally, $a^{c/b} = (\sqrt[b]{a})^c$; thus, we can deduce a clear meaning for rational exponents—for example, $8^{2/3} = (\sqrt[3]{8})^2 = (2)^2 = 4$.

Nicholas Oresme, one of the most influential philosophers and mathematicians of the 14th century, was the first to write about and make sense of rational exponents. The convention of writing rational exponents became popular after Isaac Newton used the notation in a letter from 1676.

Irrational exponents also date back to the 14th century. Nicholas Oresme was the first to consider such objects. Recall that the rational numbers are dense on the number line; that is, we can find rational numbers as close as we wish to any given real number. We use $2^{\sqrt{2}}$ as an example. We can approach $\sqrt{2}$ (which equals 1.414213562373095048...) using the rational numbers: $\frac{14}{10}, \frac{141}{100}, \frac{1,414}{1,000}, \frac{14,142}{10,000}$, and so forth. We can approach the value of $2^{\sqrt{2}}$ by computing $2^{a/b}$, in which $\frac{a}{b}$ is a rational number approaching $\sqrt{2}$. We compute:

$2^{1.4} = 2.6390158215457885187480003942...$

$2^{1.41} = 2.65737162819302316196830322 5\ldots$

$2^{1.414} = 2.66474965018404354228052965 9\ldots$

$2^{1.4142} = 2.66511908853235146915658056 5\ldots$

$\ldots$

$2^{1.4142135623730950} = 2.6651441426902250984971247 70\ldots$

$2^{1.41421356237309504} = 2.6651441426902251723906107 01\ldots$

and see that these values converge on a particular number. We define $2^{\sqrt{2}}$ to be the number we are approaching in this manner. Specifically, $2^{\sqrt{2}}$ is the real number whose decimal expansion begins $2.6651441426902\ldots$.

Sometimes, our ever-widening vision of number is further enhanced by our understanding of arithmetic and by the machinery of mathematics. ■

Questions to Consider

1. Which of the following expressions are polynomials?

$$5x^3 - 6x + 17, \ \sqrt{4x^2 + 1}, \ \frac{3}{2}x^5 - 0.6x^2 + x - \frac{27}{13}, \ \frac{3x^2 - 8}{x + 1}.$$

2. Simplify $25^{3/2}$ and $8^{2/3}$ as much as possible.

3. The use of variables, exponential notation, and other mathematical shorthand allows for compact expressions and easier manipulation of complicated formulas. To what extent is this shorthand a benefit or a hindrance to learning, working with, and understanding mathematics?

An Intuitive Sojourn into Arithmetic
Lecture 12—Transcript

Welcome to the world of arithmetic. In this lecture, we'll open with a brief historical overview of the familiar arithmetic functions of addition, subtraction, multiplication, division, and exponentiation, and the axioms that accompany them. Many of the basic rules of arithmetic are so obvious and sensible that we accept them without even a second thought. One such "rule" that we learned in our youth actually does not fit this particular mold, and that's the declaration that a negative number multiplied by another negative number produces a positive number. Here, we'll discover that this rule is, in fact, a theorem that can be deduced in an elegant manner from the basic axioms of arithmetic. Thus, we'll finally resolve a conundrum that has mystified many individuals since their early childhood. Before moving on, we'll also take a momentary detour to explore the origins and evolutions of the symbols of arithmetic that have become the mathematical icons of our early school days.

Combining numbers using the basic operations of addition, subtraction, multiplication, and division quickly leads us to some interesting and important numerical questions. For example, what does $2^{\sqrt{2}}$ mean, and is that a number? Here we'll take and pull ideas of Nicolas Oresme, the great 14th-century mathematician and philosopher, and others to make sense of $2^{\sqrt{2}}$ and, moreover, to realize that such a thing is indeed a number.

Up to this point in our course, for the most part, we've been focusing on numbers themselves and discovering how a more sophisticated understanding of nature and of our world leads to a more sophisticated understanding of number. Here, we'll consider various ways of combining numbers in order to widen our view of number further still. So, let's begin with some basic arithmetic rules of the road that we've all seen in grade school and, now, discover their origins.

In fact, a good bit of our basic intuition about arithmetic was foreshadowed by Euclid himself.

Euclid, in the 3rd century B.C.E., offered five (what we call today) "common notions": basic, self-evident truths that did not require proof. In modern mathematics, we call such statements "axioms." The axioms are the statements we just state without proof and upon which we build the edifice of mathematics through theorems and proofs.

So, what were his common notions? The first one was, things that are equal to the same thing are also equal to one another. So, for example, suppose that Deb has the same amount of money as I do and Mike has the same amount of money as I do. Well then, Deb and Mike must have the same amount of money. We call this "transitivity" today.

His second notion was, if equals be added to equals, the wholes are equal. So, in this context, suppose now that I give both Deb and Mike $10 each. Well then, they still have the same amount of money, even though they each have $10 more than they had before. Number three is, if equals be subtracted from equals, the remainders are equal. Well, if I now take back $10 from Deb and Mike, then they still have the same amount of money but might not be too happy with me. Anyway, number four is, things that coincide with one another are equal to one another. Now today in geometry, we call this notion "congruent," and it basically means that two geometric objects are congruent if I could take one and physically slide it on top of the other and they match up perfectly. This would be the notion of congruent. Finally, his last common notion was that the whole is greater than the part. So for example, a whole chocolate cake is greater than any piece of it (and of course, a chocolate cake is greater than anything whatsoever.)

Now by the way, we notice that in the previous lecture, Cantor showed us that this notion is not always true with infinite collections, for he found a collection and a part where they actually had the same number of elements. They were equally numerous because we had a one-to-one correspondence. Well, in the eyes of arithmetic, there are two special numbers, each of which we've already celebrated and that have certainly earned a rightful place of honor in the Pantheon of numbers.

The first number is 0, known in this arithmetic context as the "additive identity," because if 0 is added to any number, the sum of the given number and 0 is, in fact, the given number. For example, $0 + 5 = 5$. So, when we add 0 to 5, we preserve the identity of 5. That's why it's called the additive identity. The second number is the Pythagorean number of reason, unity, the number 1. Here, in this context, we refer to the number 1 as "the multiplicative identity," because if 1 is multiplied by any given number, the product of those two numbers is the given number. For example, $1 \times 5 = 5$, and so, again, we preserve the identity of 5.

Well, since we're here with 0 and 1, I wanted to briefly mention inverse numbers. These are very valuable when we're solving equations. Given any number—say, 5—there exists another number, in this case we call it –5, such that the sum with 5 equals 0. More generally, given a number n, the number $-n$ is called the "additive inverse" of n. In an informal sense, the additive inverse undoes the addition of a number.

In terms of undoing multiplication—given any nonzero number—say, 2, for example—there exists a number whose product with 2 equals 1, in this case namely a 1/2. More generally, every nonzero number n has a reciprocal $1/n$ so that their product equals 1. The number $1/n$ is called the "multiplicative inverse" of n. Arithmetically speaking, addition and subtraction are opposite operations just as multiplication and division are. This observation was first made by ancient Egyptians who viewed each pair of operations as mirror images of one another. Today, we say they're "inverse operations"; less romantic phraseology, but just as descriptive.

Zero is the only number that does not have a multiplicative inverse. Why is that? Well, because any number n, when multiplied by 0, yields 0. Now, this statement is not an axiom. It's a provable theorem, and as we'll see in a moment, in fact, the proof only requires one thing. We need to recall that numbers also satisfy what's called the "distributive property." If I add or subtract two numbers—so take b and add it or subtract it from c, so $b \pm c$—and I multiply that entire quantity by a, that's the same exact number as distributing the a to first the b and then to the c, and look at $ab \pm ac$. So that's the distributive property for any number. Let's take a look at 7 and consider 7×0. Well, I can write 0 as $1 - 1$, so I see $7 \times (1 - 1)$. If I now use the distributive property we just said, I can see that we have $(7 \times 1) - (7 \times 1)$. But 1 is the multiplicative identity, so 7×1 is just 7, so we see $7 - 7$, which equals 0. So, I just proved that $7 \times 0 = 0$ because of the distributive property. Of course, we can apply this idea with any number, not just 7, and thus we've just proven that any number multiplied by 0 equals 0, which is really quite wonderful. We actually get a proof of that.

Now that we're in the mood, let's have a little bit more fun with multiplication and finally figure out why a negative times a negative always equals a positive. Now, of course we know that a positive number multiplied by a positive number is positive, because for example, 2×3 means that I have three copies of 2—so, $2 + 2 + 2$, and it's 6. Similarly, it makes sense that a negative number multiplied by a positive number is going to equal a negative number. For example, $(-5) \times 2$ means I have two copies of –5. That is $(-5) + (-5) = -10$. However, the famous rule that a negative number multiplied by a negative number is a positive number is by no means obvious or even clear. So, let's now see why this rule actually holds.

Now, we'll just consider a specific number to illustrate the idea of the argument, but the argument actually works for all negative numbers. So, now let's prove that $(-1) \times (-1) = 1$. This is going to be a real mathematical proof—so stay with me—but it will be neat to see this finally resolved. Let's first remember what –1 means. That's the first thing we have to do. Well, –1 is the additive inverse of 1. In other words, $1 + (-1) = 0$. That's where we'll start. Now, if we multiply each side of this equation by –1, then we see $(-1) \times (1 + (-1)) = (-1) \times 0$. We now use the distributive property and write this as $((-1) \times 1) + ((-1) \times (-1)) = (-1) \times 0$. Well, we just proved that any number multiplied by 0 is 0, so the right hand side is 0. Remember that 1 is the multiplicative identity, so $(-1) \times 1$ is just –1. We see our equation becomes $(-1) + ((-1) \times (-1)) = 0$, and we're there. All we have to do is add 1 to both sides, and that first –1 becomes a 0 and see $(-1) \times (-1) = 1$. Amazing. By the way, it was Brahmagupta, the 7th-century astronomer who we mentioned back in Lecture Three, who was the first to offer a systematic treatment of negative numbers and 0, including the multiplication of signed numbers. So, a negative times a negative equaling a positive is by no means a new idea, but certainly a wonderful one.

Have you ever wondered where all those cryptic symbols of arithmetic that we know and love come from? Well, if so, this is your lucky day, because

I want to share with you some of the interesting history and evolution of our arithmetic notation. Well, let's begin with addition and our plus sign. In ancient Latin manuscripts, the word *et*, which is Latin for "and," was used to indicate addition, which makes sense, right? "2 and 3 equals 5." So they'd use *e* and then a *t*, and they'd be writing this cursively. Then, after a while, they would just write the *t* and then that *t*, as you can see, slowly evolved into the + sign that we are used to. So, the plus sign that we use actually came from the *t* in *et*, Latin for "and"—really neat, I think.

In a manuscript from 1489, we find the Greek cross "+" used for addition, which is what we use today. The Latin cross was also used but it was printed horizontally. Sometimes the Maltese cross, the very ornate cross, was used for very formal writings. By the way, the minus sign that we use today, a dash, makes its first appearance in the literature in 1481. Well, using the "x"-like symbol as we do today for multiplication came about in an interesting way.

Our multiplication sign first appears in Leonardo de Pisa's, also known as Fibonacci's, important book *Liber Abbaci* from 1202. We mentioned the book earlier. He employed the notation to indicate what we might consider today as "cross products" of pairs of numbers. So for example, he'd make a 2×2 grid of numbers. For example, here you see 2, 3, 4, 6, and then he'd use the cross to indicate which two opposite numbers to multiply. So, he would take the 2×6 and get 12, and then the 3×4 and get 12. He'd record them both, $12 = 12$, and then he'd move on with his calculation. So, this cross that you see to indicate which two were to be multiplied slowly evolved into the product of just two numbers as we use them today.

What about our division sign? Well, in 1751 we see an inverted "D" (a "D" written backwards) used for division. In other words, we'd write 15, and then that inverted "D," 5 would equal 3, because $15 \div 5$ would be 3. Just two years later, in 1753, we find the symbol that we use today with the horizontal bar and the two dots, although originally there were actually two dots on the top and two dots on the bottom, interestingly enough, and somewhere along the line we canceled one of those dots away.

In terms of writing fractions, as we noted in Lecture Eight, Ahmes in the Rhind Papyrus used a dot over a number to indicate a unit fraction. So, although he didn't have our symbol 5, a dot over a 5 in his notation would mean 1 over 5, or 1/5. Now, the dot actually derived from the hieroglyph for an open mouth, indicating how fractions were used to divide up rations of food and drink. As we noted before, many of the questions from the Rhind Papyrus concerned dividing up bread and ale. It was quite a lively bunch of people, at least in the stories of the Rhind Papyrus.

Writing fractions stacked using three terraces of type with a horizontal line between the numbers goes back to Arabic writings. Even Fibonacci, in *Liber Abbaci*, used stacked fractions. The writing of a slash for a fraction all on one line, such as "2/3," first appears in a text published in Mexico in 1784. You can see it's a multicultural contribution from all walks of the world, all points of the globe. In fact, we first see that this is actually almost like a tilted flourish, it was very elegant, and then slowly it evolved—or should I say devolved?—into the less ornate slash that we use today.

Let's now return to the important arithmetic objects known as powers and exponents and leave the notation behind. To start with, let's just consider powers that are natural numbers, the easiest ones. The great 17th-century mathematician and philosopher René Descartes developed a shorthand way of writing powers. He'd write 5^2 to mean 5×5, and 5^3 to mean $5 \times 5 \times 5$. Now, this notation allows us to write very large values in a very compact fashion. For example, instead of writing one billion as 1 followed by 9 zeros, we could just write 10^9 and be done with it. So, it's a very powerful way of notating numbers.

In view of Descartes' notation, let's see that $n^3 \times n^2 =$ something that's just in terms of n. What would it be? Well, I would take n^3, and I would write out $n \times n \times n$, and then I'd multiply that by $n \times n$. So, how many n's am I multiplying? Five, and so I see n^5. So, the rule here is that if we have the same base and different exponents, we just add the exponents together. This is always a tricky rule, and in fact, when I teach my college students this, I actually sing for them a little jingle that I wrote to remind them of the method to do this and it goes like this—and I'm tone deaf—but it goes "when in doubt, write it out," because just by writing out what n^3 ($n \times n \times n$) and n^2 ($n \times n$) mean, we see n^5. We add the exponents.

Now, this song is really an example of a much more general principle known as the law of exponents, which says that if I take $(n^a) \times (n^b) = n^{a+b}$. Now, in algebra, we often combine products, sums, differences, and natural powers to create what are called "polynomials"—and we'll see this throughout our course, this is our first introduction to this idea—that is, expressions such as: $3n^2 - 2n + 10$. We're just adding and multiplying numbers, and we're raising, in this case, n to natural number powers. Or, even more elaborate, $120n^{95} + 12.3n^{35} - 11n^7 - 5n + 27$. Any things like that are known as polynomials. As we'll see in Lecture Fifteen, polynomials can be used to refine our understanding of numbers and thus plays a significant role in the evolution of numbers.

Now it's time to move beyond natural number powers and consider fractions in the exponent. That is, what does it mean to raise a number to a fractional power? Let's answer this riddle by using what we already know and considering a specific example. What power of 9 equals $\sqrt{9}$? Now, recall that $\sqrt{9}$ is 3, because $3^2 = 9$ and the square root undoes the square. So, $\sqrt{9}$ equals 3, so we know that. But the question is, what power of 9 will equal $\sqrt{9}$?

Well, I don't know what that power is, so let's give it a name. It's an unknown. Let's call it p for "power." So, suppose that $9^p = \sqrt{9}$. Well, then, if we just squared both sides, or multiplied the left hand side by 9^p and the right hand side by $\sqrt{9}$, then we see $9^p \times 9^p = \sqrt{9} \times \sqrt{9}$. Well, the $\sqrt{9} \times \sqrt{9}$ is just 9. By the Law of Exponents, we know that $9^p \times 9^p$ is 9^{p+p}, or 9^{2p}. So, we see that $9^{2p} = 9$. So, on the right hand side we see one 9 and on the left hand side we see $2p$ multiples of 9. Well, there have to be the same number of 9s on both sides, so $2p$ has to equal 1. We have one 9 on one side. We have to have one 9 on the other, and so $2p$ equals 1. So we see, if we divide both sides by 2, that p equals 1/2. Therefore, we just discovered an amazing fact that $9^{1/2}$ means $\sqrt{9}$. That is, $9^{1/2} = 3$. Similarly $2^{1/2} = \sqrt{2}$.

These examples, in fact, can be generalized. If we have a fractional number as an exponent (a fraction exponent), then its denominator, the number below, signifies a root. So for example, $8^{1/3} = \sqrt[3]{8}$, which is in turn equal to 2, since $2^3 = 2 \times 2 \times 2 = 8$. In general, we can say that $a^{1/b}$ is the bth root of a. Well, what if our fraction in the exponent has a numerator other than just 1? Well, no problem at all. For example, let's consider $8^{2/3}$. Well in that case, I first take the $\sqrt[3]{8}$, and then I square the quantity. In this case, $\sqrt[3]{8}$ is 2, so I see 2^2, which equals 2×2, or 4. So, $8^{2/3}$ is equal to 4. More generally, we say that $a^{c/b}$ is equal to $(\sqrt[b]{a})^c$.

Nicolas Oresme, one of the most influential philosophers and mathematicians of the 14th century, was the first to write and make sense of fractional exponents. This really was an individual who was a great thinker. In fact, he once noted—and I love this little quote—"Therefore, I indeed know nothing except that I know that I know nothing." I think it's a very brilliant quote, because it really identifies the fact that every time we think we understand something, so much more is left to be understood. (As an aside, some time later, the notion of fractional exponents was rediscovered by our friend Simon Stevin.)

While the symbols for fractional and other exponents were suggested by Oresme and Stevin, the convention is really due to John Wallis, a 17th-century English mathematician, and also to Sir Isaac Newton. The notation of putting a fraction in the exponent became popular around 1676. Now, all notation aside, the important fact is that we can make sense of taking a number like 2 and raising it to a fractional power, like 7/5. It means we must first take the fifth root of 2, which we could do on a calculator, and then multiply that number by itself seven times. By the way—in case you're wondering—that works out to be 2.639015 and so forth. The point is that it's a real number. Natural numbers raised to fractional powers are numbers. Well, all was well with number until Oresme declared and dared to wonder something totally bizarre. What does it mean to raise a number to an irrational exponent? For example, what does $2^{\sqrt{2}}$ mean, and is that thing a number?

The answer is, as we're about to see, yes. That thing does have meaning and it is a number. But how can we see what it means? Since it's way too complicated of an object, we immediately retreat, and we try to inch up to it by constructing objects that we already understand. So, this is an exciting moment, since we're about to discover a new type of number. Now, how can we inch up to $2^{\sqrt{2}}$? Well, in Lecture Ten, we showed that rational numbers, the fractions, are dense in the number line. In other words, we can find fractions as close as we wish to any real number at all. For example, we can approach $\sqrt{2}$, which has a decimal expansion of 1.414213, and so forth, using fractions. 1.4 = 14/10, or 1.41 = 141/100, or 1.414 = 1,414/1,000, and so forth.

In fact, we foreshadowed this idea in Lecture Six, when we saw that the ratios of Fibonacci numbers approached the golden ratio. Well, we'll now use these fractions that approach $\sqrt{2}$ to approach the value $2^{\sqrt{2}}$ by computing $2^{a/b}$, in which a/b is a fraction near $\sqrt{2}$. In this case, we first consider $2^{1.4}$. That equals $2^{14/10}$, which equals $2^{7/5}$, which we just found to be 2.63901, and so forth. If we were to calculate $2^{a/b}$ for each a/b approximating $\sqrt{2}$, then we could produce a list of numbers by just taking the bth root of 2 and then raising it to the 8th power. So for example, we'd see $2^{1.4} = 2.639$ and so forth. If we look at $2^{1.4.1}$, we'd see 2.6573 and so forth. Then we can go down the list and go $2^{1.41421356}$, which turns out to be 2.6651441 and something. Now, look what's happening. We see that these values are converging upon a particular decimal number. The values are looking more and more the same. Thus, we now define what we mean by $2^{\sqrt{2}}$. We declare it to be the number we're approaching in this manner. Specifically, $2^{\sqrt{2}}$ is the real number whose decimal expansion begins 2.66514414 and so forth.

So, in this lecture, we not only saw the origins of arithmetic and those cryptic symbols that we've been taught to use, but we've developed a deeper understanding of what that arithmetic represents and what it actually means, and we've actually proved some theorems involving arithmetic. For example, a negative times a negative being a positive, or any number times 0 equals 0. We see that not only can we make sense of fractions as powers in exponents, but that by inching up to them we can actually make sense of irrational powers. Those strange, new objects are indeed themselves numbers in the truest sense. Thus, we see that arithmetic can be used to enhance our understanding of number and help us widen our ever-growing notion of what number means.

The Story of π

Lecture 13

The term radian first appeared in print in the 1870s, but by that time, great mathematicians, including the great mathematician Leonhard Euler, had been using angles measured in radians for over a hundred years.

In this lecture, we will tell the story of one of the most famous numbers in human history and one of the most important numbers in our universe. We begin with some basics. An interesting observation involving circles brings us to π. Suppose we compare the length of the circumference of a circle to its diameter. We discover that the circumference is slightly greater than three times the diameter.

The number π is defined to equal the ratio of the circumference of any circle to its diameter. This ratio is constant no matter the size of the circle. The first 30 digits in the decimal expansion of π are 3.14159265358979323846264383279. We use the Greek letter π (pi) for this constant because the Greek word for periphery (the precursor to perimeter and circumference) begins with the Greek letter π. The symbol π first appeared in William Jones's 1709 text *A New Introduction to Mathematics*. The symbol was made popular by the great 18th-century Swiss mathematician Leonhard Euler around 1737.

> **The number π is defined to equal the ratio of the circumference of any circle to its diameter.**

Attempts to produce the value of π began early in history. The Babylonians approximated π in base 60 around 1800 B.C.E. They believed that $\pi = \frac{25}{8} = 3.125$. The ancient Egyptian scribe Ahmes offered the approximation $\frac{256}{81} = 3.160493827...$. An implicit value of π is even given in the Bible: In 1 Kings 7:23, a round basin is said to have a 30-cubit circumference and a 10-cubit diameter, implying that $\pi = 3$. In 263 C.E., the Chinese mathematician Liu Hui believed that $\delta = 3.141014$. Approximately 200 years later, the Indian mathematician and astronomer Aryabhata approximated π with the rational $\frac{62,832}{20,000} = 3.1416$. Around 1400, the Persian astronomer Ghyath ad-din Jamshid Kashani computed π correctly to 16 digits.

The number π is everywhere. The number π is used as a measure. Given its connection with circles, we are led to a measure of angle as distance. With the angle measure of degrees, we recall that one complete rotation has a measure of 360 degrees—the approximate number of days in one complete year. We consider a circle with radius 1. Traveling around the circle once would produce a circumference of 2π; thus, every angle corresponds to a distance measure part way (or all the way) around the circle. We call this measure of angles radian measure; thus, 180 degrees = π radians, and 90 degrees = π/2 radians. Radian measure is a much more useful measure of angles for mathematics (including calculus and physics). The term "radian" first appeared in print in the 1870s, but mathematicians, including the great Leonhard Euler, had been measuring angles this way for more than 100 years. The number π appears in countless important formulas and theories, including the Heisenberg uncertainty principle and Einstein's field equation from general relativity.

Using π, we can calculate the area of a circle. Given a circle having radius r, we know its circumference equals $2\pi r$. Now suppose we cut up the circle into tiny, pizza-like slices. If we reposition the slices in an alternating up-and-down fashion, we can approximate the area of the circle by computing the area of the rectangle-like object we have created. We discover the important formula that the area of a circle of radius r equals πr^2, a formula proven by Archimedes.

The 18th-century German mathematician Johann Lambert showed that π is an irrational number in 1761. We now know that the decimal expansion for π will never become periodic. Given that π is not equal to any number of the form $\frac{m}{n}$ for integers m and n, we see that $n\pi - m$ will never equal 0. Therefore, π will never be a solution to a linear polynomial equation $nx - m = 0$ with nonzero integer coefficients m and n. The German mathematician Ferdinand von Lindemann generalized Lambert's work in 1882 and proved that π is not a solution to any polynomial equation with integer coefficients.

Some questions remain open to this day about π. Are the digits of π random; that is, is π a normal number? Is there any pattern in the digits of π? Is 2^δ an irrational number?

The number π has been involved in some amusing antics. Around 1600, German mathematician Ludolph van Ceulen used a method of Archimedes's to compute the first 35 digits of π. He was so proud of his calculation that he requested those 35 digits be carved into his gravestone; his request was honored at the time of his death in 1610. In 1873, British math enthusiast William Shanks computed the first 707 digits of π. It took Shanks more than 20 years to perform the necessary computations. (In 1944, D. F. Ferguson found that Shanks had made a mistake.) In 1897, the Indiana General Assembly passed a bill declaring that π was equal to 3.2 (the state senate postponed the bill indefinitely).

Today, the first trillion digits of π are known, and billions of digits can be generated on a laptop computer with the appropriate software. If we use the first 10 decimal digits of π to compute the circumference of the Earth's equator (assuming we know the exact diameter), we will be off by less than 0.2 mm. The current world record holder for naming digits of π is Akira Haraguchi, who correctly recited the first 100,000 digits on October 3, 2006. In order to remember the first 15 digits of π, remember the following (consider the letter counts in each word): *How I need a drink, alcoholic of course, after the heavy lectures involving ancient constants.* ■

Questions to Consider

1. Convert the following angles from degrees to radians: 45°, 60°, 30°.

2. Why do you think number enthusiasts keep uncovering more and more digits in the decimal expansion of π?

The Story of π
Lecture 13—Transcript

In this lecture, I want to share with you the story of one of the most famous numbers in human history and one of the most important numbers in our universe, the number π. While the origins of π are not known for certain, we know that the Babylonians approximated π in base 60 around 1800 B.C.E. The definition of π centers around circles. In fact, it's the ratio of the circumference of a circle to its diameter. Now, using this definition, we'll see for ourselves that π is a number just a little bit bigger than 3. Here, we'll explore humankind's odyssey to compute, approximate, and understand this enigmatic number, π. These attempts throughout the ages truly transcend cultures.

The constant π helps us understand our universe with greater clarity. In fact, the definition of π inspired a new notion of measurement of angles, a new unit of measurement. This important angle measure is known as "radian measure" and gave rise to many important insights in our physical world. As for π itself, Johann Lambert showed in 1761 that π is an irrational number, and later, in 1882, Ferdinand von Lindemann proved that π is not a solution to any polynomial equation with integers. However, we'll also see that many questions about π remain unanswered. Finally, we'll close this lecture with a number of entertaining stories involving π. These stories include the attempt of the Indiana General Assembly, in 1897, to pass a bill declaring that π equals 3.2; the world record in the number of digits of π memorized and recited by an individual; and even a sneaky method to remember the first 15 of π's decimal digits.

Any discussion of the origins of pi must begin with an interesting experiment involving circles that we can all try. So let's take any circle at all, and I actually have one here. What I want us to do is I want us to compare the length of the circumference, which is the length around, and I want to measure it in terms of the diameter, which is the length across. I actually want to try this right now, and we can actually try this. If you're at a cocktail party or something and there's a coaster, you could actually do this live. Let's go over here and actually try this. So, I'm going to measure, and actually, I have a measuring tool here. This ruler is constructed so it measures the diameter of this particular circle. Now, this experiment will work with any circle at all, so the size of the circle doesn't matter, and you can see, for example, that these units are the diameter. There's one diameter, there's two diameters, there's three diameters, and so forth. You can see the markings correspond perfectly. Well now let's measure and see how we do.

I'm going to measure the circumference, which is a fancy way of saying "the perimeter around." So, I'll start over here—well in fact, I'll start—no, I'll start over here. I go around, and you can see I'm doing this very carefully, trying to be as accurate as possible. I'm coming to the beginning, and I see exactly—here. So, this is the length, which I'll move to this side right here, but no sleight of hand. Watch me—no sleight of hand. That's the length once around.

Now, let's measure it and see what we get. Well, when I put this end here and this end here, what do we see? Well, how many marks do we pass? One, two, three. So, we pass three bits and just a little bit more, and if you look really closely, we can see it's actually a little bit more than 1/10 of the way extra. So, what I see here is three and a little bit more than 1/10. So, just this little experiment here live shows us that that ratio of the circumference to the diameter is going to be a number that's around, or a little bit bigger than, 3.1. Well in fact, this works no matter what size of circle we actually start with. So, no matter what the size of the circle is, the circumference is slightly greater than three times its diameter.

We give this fixed, constant value a name, and we call it π. So, let's say this more precisely. The number π is defined to equal the ratio of the circumference of any circle to its diameter across. As we just noted, this ratio is constant. No matter what size of circle we try this with, that number will be always the same. Now, you can actually list many digits of this number. It begins 3.141592653589, and it keeps going, and so forth.

We'll first take a look at the early history of π and the ancient struggle to pin down its exact value—first, a word about the symbol π. We use the Greek letter π for this number, because the Greek word for "periphery" begins with the Greek letter π. Now, periphery of a circle was the precursor to the perimeter of the circle, which today we call circumference. The symbol π first appears in William Jones's 1709 text *A New Introduction to Mathematics*. The symbol was later made popular by the great 18th-century Swiss mathematician Leonhard Euler around 1737.

Moving from its name to its value, we have evidence that the Babylonians approximated π in base 60 around 1800 B.C.E. In fact, they believed that π = 25/8, or 3.125, an amazing approximation for so early in human history. The ancient Egyptian scribe Ahmes, who is associated with the famous Rhind Papyrus, offered the approximation 256/81, which works out to be 3.16049, and so forth. Again, we see very impressive approximation to this constant. There's even an implicit value of π given in the Bible. In 1 Kings 7:23, a round basin is said to have 30-cubit circumference and 10-cubit diameter. Thus, in the Bible, implicitly it states that π equals 3 (30/10).

Not surprisingly, as humankind's understanding of number evolved, so did its ability to better understand and thus estimate π itself. In the year 263, the Chinese mathematician Liu Hui believed that π = 3.141014. Approximately 200 years later, the Indian mathematician and astronomer Aryabhata approximated π with the fraction 62,832/20,000, which is 3.1416—amazing. Around 1400, the Persian astronomer Kashani computed π correctly to 16 digits—amazing.

Let's break away from this historical hunt for the digits of π for a moment, and consider π as an important number in our universe. Given π's connection with measuring circumferences of circles, scholars were inspired to use it as a measure of angle distance. Now, let's consider a circle having radius 1. Remember, radius is just the measure from the center out to the side. It's half the diameter. The traditional units for measures of angles are, of course, degrees. With degrees, one complete rotation around the circle has a measure of 360 degrees, which, by the way happens to approximately equal the number of days in one complete year and which might explain why we think of once around as 360.

Instead of the arbitrary measure of 360 to mean once around the circle, let's figure out the actual length of traveling around this particular circle, a circle of radius 1, once around. So what's the length? What's the circumference of that? Well, let's see. If we have a radius of 1, then our diameter is twice that, 2, and so we know that the once around will be 2 times π, because the circumference is π times the diameter. What I see here is once around will be 2π. So, one full rotation around, which is an angle of 360 degrees, would be swept out with circumference length of 2π in this particular circle. In fact, what would be half-way around? Well, that would be 180 degrees, and we would sweep out half of the circumference, which, in this case, would be π. Ninety degrees would sweep out a quarter of the circle, and for this particular circle, that would have length π/2, or one-half π.

We're beginning to see that every angle corresponds to a distance measured part-way or all the way around this particular circle of radius 1. In other words, for any angle, we can measure the length of the arc of this circle swept out by that angle. This arc length provides a new way of representing the measure of an angle, and we call this measure of angles "radian measure." So, for example, 360 degrees = 2π radians, those are the units; 180 degrees equals π radians, and 90 degrees would equal π/2 radians. Remember, all these measures are always based on a special circle which has radius 1.

It turns out that this radian measure is much more useful in measuring angles for mathematics and physics than the more familiar degree measure. Now, if you think about it, this fact is not too surprising, since radian measure is naturally connected through the circumference length with the angle, rather than the more arbitrary degree measure that has no mathematical underpinnings, but just represents an approximation through a complete year.

The term radian first appeared in print in the 1870s, but by that time, great mathematicians, including the great mathematician Leonhard Euler, had been using angles measured in radians for over a hundred years. Well, beyond angle measures, π is central in our understanding of our universe. In fact, the number π appears in countless important formulas and theories, including the Heisenberg uncertainty principle and Einstein's field equation from general relativity. So, it's a very, very important formula, a very important number indeed.

There is one famous and ancient formula that involves π that I want us to see together. It's the beautiful formula for the area of a circle. The area of a circle is equal to π, this constant, multiplied by the length of the radius squared—the famous formula πr^2. Now, let's see why this formula actually makes sense. So, we're going to consider a circle that has radius r. Let me move this aside for a second. We're going to actually perform an experiment here, live. Let's consider a circle that has radius r, so the distance from the center of the circle out to the radius is r. That means that that diameter across would be $r + r$, or $2r$. Now, we want to figure out the area of this circle. Now, of course, we can't use this famous formula, because we're trying to derive it.

So, what we'll do here is we'll try to convert this very subtle, complicated shape into a more familiar shape. So again, we take something complicated and try to convert it to something easier (something that we should do in every aspect of our lives). So, what I'm going to do here is I'm going to actually cut the circle up. Now notice, if I were just to cut it up into pieces, it won't change the area at all. So, let's actually try to do this right now, live. So here I go. I just cut it in half. Notice the area won't change at all. I'm going to cut it in half again. I'm going to do this a few times. Now, of course you really want to do this as much as possible. Again, this is a great fun little parlor experiment that you could actually perform at cocktail parties with a coaster, and it really is neat. I really do love this. I'll just cut it up into this size pieces. But of course, the more you do this, the easier it will be to see what we're about to see. Let me just first of all show you that there was no sleight of hand here, that, in fact, these can all be reassembled, and we'd form the complete circle again. It looks like a pizza now, doesn't it? It makes me hungry.

There you can see that the circle is intact. No pieces are missing. We want to find the area of that. The way I'm going to find the area of that is by moving the pieces around just a little bit. By moving the pieces around, we're not going to actually change the amount of material required. What I'm going to do is I'm going to interleaf these pieces, and I'm going to—oh. Before I do that, let's just make one comment about this before I mess this up. The length around this curved circle is what? Well, since the radius is r, the diameter is $2r$, and so the circumference is π times the diameter, which, in this case, would be $2\pi r$. Now, I want us to remember that. This length from here all the way around to here, that total length, is $2\pi r$.

Now, let's make ourselves a new figure, and the new figure I want us to consider has these teeth coming together. I'm going to do this right now live—not as easy as you may think. This is live mathematics unfolding here—very dramatic—and you can see that the pieces fit together quite nicely. Notice that they're inverted up-down, up-down now, but they do fit together quite nicely. If we had smaller pizza pieces, then, in fact, they would fit—the bottom would be less wavy, in fact, is what we see. What I see is something like this. We have to find the area of that. That's what we're trying to figure out. But notice that that now resembles a rectangle. In fact, the smaller you cut the pieces, the finer we have these pizza slices, that in fact the more rectangle-like it looks.

Just to illustrate that, let me just take this last one here and let me just cut this one in half just to illustrate that. Now of course, you'd have to cut them all in half to make this look even better. But just here, you'll see how good it looks just by cutting this and putting this here. So, I'll put this right here, and I'll put this one right here. Look at that. Doesn't that look almost like a rectangle? If you squint your eyes, that looks like a rectangle. Well, the area of a rectangle we actually know. It's the length of the base multiplied by the length of the height. Let's figure out what, roughly speaking, this would be. Well, the height we know. That's just from the center of the circle. It's a pizza slice. So, that's just the radius. So, this height is just r. What about the base? The base is—well, these are wiggly parts that form part of a circle, but what part of the circle? We see that we have half of them down here and half of them up here, so in fact, what we see here is this is half of the entire travel around, or half the circumference. Remember, we figured out the circumference was $2\pi r$, therefore half of it is just πr. So, the length of the base is half the circumference, or πr. So, we have πr multiplied by the height, which is r, and what do we see? We see πr^2. We just gave an excellent heuristic for how we can prove that the area of a circle is indeed πr^2.

Now, this formula was established by Archimedes in the 3rd century B.C.E. However, even before Archimedes, Euclid knew there was a special number that gave the area of a circle once you multiplied it by the radius squared, although he didn't actually explicitly give the value of π. In fact, what Euclid actually did was he gave the earliest proof that the ratio of a circle's area to its diameter squared is always constant. If you want, you can actually work through what that constant ratio would be, which, in this case, would be $\pi/4$, because remember, the diameter is twice the radius. So, you have $2r^2$, and when you square the 2, that's where you get the 4, so $\pi/4$ is what his constant was, which is equivalent to the formula that we have. He actually stated his result in Book 12 of his series *The Elements*, and he stated it as "Circles are to one another as the squares on the diameters." That's how he phrased it, but that's equivalent to what we just proved.

We've seen the importance of π as a number in our natural universe, but what about the numerical personality of π itself? Well, in 1761, the German mathematician Johann Lambert showed that π is an irrational number. It's not a ratio. This important result tells us that the decimal expansion for π will never become periodic. Even though the decimal expansion will never repeat from some point onward, forever, the digits do contain some interesting patterns. It's interesting to think about the fact that even though we don't know all the infinitely long list of digits, we know, by this abstract theorem of Lambert, that, in fact, the digits will never repeat forever. But there are interesting patterns in that screed of digits. For example, starting at the 762nd decimal place, we see a run of six 9s in a row. So, we see 999999, all in a row. But then, of course, it gets jumbled up again.

This place in the decimal expansion of π is sometimes referred to as the Feynman point, named after the Nobel Prize-winning 20th-century physicist Richard Feynman. Why is this run of six 9s starting at the 762nd decimal spot of π named after him? Well, as the story goes, Feynman once said that if he could memorize all the digits of π up to this point, then he could end his long digit soliloquy with: "nine, nine, nine, nine, nine, nine, and so on." Now, we can appreciate Feynman's joke: He would be implying that π ended with an endless run of 9s, which is impossible, because Lambert proved that π is irrational. So, it's a cute little story, I thought.

Now let's get a bit technical, but only for a moment, I promise. Since π is not a fraction—that's what Lambert proved—then it's not equal to any number of the form m/n for any integers m and n, because it's not a fraction. So, π is not equal to m/n. In other words, I could say that $\pi - m/n$ could never equal zero, because they're never the same number, no matter what integers m and n are. If I were to multiply through that not equal by n, I would see $n\pi - m$

will never equal 0. Now, I know this sounds a little bit weird, but basically what it's saying is no integer times π minus another integer will ever equal 0. I'm assuming the integers are not 0, by the way.

Therefore, an equivalent, but more complicated, way of describing what Lambert proved is to say that π will never be a solution to any polynomial equation that looks like some integer x minus an integer equals 0, or $nx - m = 0$. This kind of polynomial equation is called a linear equation. By a "solution," I mean we find a value for x that will actually satisfy that equation. Let's look at an example. How about $5x - 3 = 0$, the solution there would be 3/5, because if we replace the x by 3/5, then we see that 5 times 3/5 is just 3, and $3 - 3 = 0$. So, that would be a solution. Notice that, in this solution, 3/5 is a fraction. So, what we're seeing is that π is never a solution to this kind of thing. So informally, let's just say that Lambert proved that π is not the solution to any very simple and special type of polynomial equation involving integers.

Over 120 years after Lambert's result, in 1882, the German mathematician Ferdinand von Lindemann proved a very important generalization. He proved that π is not a solution to any polynomial equation involving integers. That means you can include x^2s, and x^3s, and all sorts of things, and π will never be the solution to any kind of complicated polynomial of that sort that you can ever think of. While I just wanted to foreshadow this fact here, we'll think about what it means to not be a solution in greater detail in Lecture Fifteen and discover that this result allows us to better understand mysterious numbers such as π.

Although we're concerned with a number of important features about π and facts and formulas involving π, there are many, many, many questions that remain unanswered that involve π. I thought that I wanted to share a few of my favorite open questions, a few of the questions that no one knows the answers to, that involve π. First of all, are the digits of π random? In other words, is π an example of a normal number (which we saw in a previous lecture)? That is, is every digit appearing the right amount of times for the number to be random (like, the digit 1 appears 1/10 of the time, the digit 2 appears 1/10 of the time, and so forth)? No one knows. In fact, is there a pattern to the digits of π? Is there a formula that generates the digits of π? No one knows. We know that the number is irrational, so we know that it will never repeat, but maybe there's some kind of pattern that's more elaborate than just a repetitive pattern. What about a number like 2^π? We looked at $2^{\sqrt{2}}$, but what about 2^π? Is that an irrational number? No one knows.

If you could answer any of these questions, you'd become even more rich and famous than you already are and would certainly gain the respect of generations mathematicians and number fans all over the world, because these are notoriously challenging questions. Well, I want to close this lecture on a light note and offer a slice of life through a slice of π. So, I thought it would be amusing to conclude with some wonderful antics and trivia involving this very famous number.

Around 1600, German mathematician Ludolph Van Ceulen used a method that actually goes back to Archimedes to compute the first 35 digits of π. He was so proud of his calculation that he wanted those 35 digits be carved into his gravestone, and his request was honored at the time of his death in 1610. There's a gravestone out there where you actually see the first 35 digits of π. In 1873, British math enthusiast William Shanks computed the first 707 digits of π. This is in 1873—it's amazing—the first 707 digits of π. It took Shanks over 20 years to perform the necessary computations, 20 years of his life to get the 707 digits of π. But sadly, in 1944, D. F. Ferguson found that Shanks made a mistake in the 528th decimal spot. Whoops! I mean, just amazing—here you dedicate 20 years of your life, and you make a mistake. Oh well.

In 1897, the Indiana General Assembly passed a bill declaring that π was equal to 3.2. There you have it. However, the Indiana Senate postponed the bill indefinitely, and thus, thankfully, it's not against the law in Indiana to say that π equals 3.1415, et cetera. Can you imagine if this were actually a law? That would mean that every mathematician and every math person that would use π correctly would be breaking the law and would be arrested. Can you imagine that? That would be something. By the way, why would Indiana make a law or even propose a law that sounds so silly? Well, there's some sense to it. If you think about it, when you're actually measuring land—and sometimes farm areas are circular in nature—to find the area of that that involves the number π, and for taxation purposes and for deeds and ownership, it's important to know the exact amount of the area. So, computing areas is important (especially of circles). So, if you use a different approximation, the tax on the land might be different, and so they were trying to standardize it. So, they weren't totally kooky, but it wasn't great.

Today, the first trillion digits of π are known, and billions of digits can be generated on a laptop computer with the appropriate software. In fact, π fans celebrate what's called π Day on March 14th at 1:59 pm. Why? Well, because March 14th is 3/14, and 1:59 PM gives us 314159. Now, if you're a real π aficionado then you should be doing this, of course, at 1:59 AM, because really 1:59 PM is 13:59. So really, you need to wake up at 1:59 in the morning on March 14 and start partying then.

How many digits do we need for precision in our calculations? Well, suppose we want to use just the first digits of π to compute the circumference of Earth's equator, assuming, of course, that we know the exact diameter. Well, if you do that calculation, we'd actually be off by less than 2/10 of a millimeter. By the way, the circumference of the Earth's equator is about 24,901.5 miles. So, 1/5 of a millimeter is not bad to be off by. So really, going out to the thousands and billions digits of π is not necessary in a practical sense.

How many digits of π can you name? The current world record holder Akira Haraguchi who, on October 3, 2006, correctly recited the first 100,000 digits of π. By the way, we all know what he recited starting at the 762nd spot. He said, "Nine, nine, nine, nine, nine, nine," but he didn't say, "and so on." He just kept going, and going, and going, like the little bunny.

You want to remember the first 15 digits of π? No problem. Just remember the following sentence, and you have to count the letters of each word. There's actually a famous sentence that's out there, but I changed it. So, this is actually a brand new sentence just for you to help you remember this famous constant π and its beginning decimal expansion. The sentence you have to remember is: *How I need a drink, alcoholic of course, after the heavy lectures involving ancient constants.* If you do a letter count—"how I need," that's 3, 1, 4. If you go through *How I need a drink, alcoholic of course, after the heavy lectures involving ancient constants* you get 3, 1, 4, 1, 5, 9, 2, 6, 5, 3, 5, 8, 9, 6, 9. You can be quite the aficionado for yourself. π has been around forever, and it forms a fundamental constant in our universe and has been a source of inspiration and intrigue for thousands and thousands of years up to today.

The Story of Euler's e
Lecture 14

> The importance and beauty of e has made it a favorite number in many mathematical circles and amongst mathematical fans and groupies all over the world. As a result, we see the digits of e actually appearing in the strangest places. ... For example, in 2004, Google honored this extremely important constant in an unusual way—in its initial public offering, which was $2,718,281,828, precisely the first 10 digits of e.

Compared with the number π, e is new to the number theory scene, but its popularity can only be described as exponential. At the very end of the 17th century, Swiss mathematician Jacob Bernoulli was working through a computation involving compound interest, and he came upon an interesting number. Suppose that we invest $1.00 into a savings account paying 100% interest per year. How much money would we have at the end of the year?

The answer depends on how often the interest is compounded. If the interest is paid once a year, then we end the year with $2.00. If the interest is awarded twice a year, then after the first six months, we would have $1.50 (the interest of $0.50 equals half of 100% of $1.00), and after the second six months, we would earn half of 100% of $1.50, which equals $0.75; we end the year with $2.25. If the interest were compounded quarterly, then we end the year with $2.4414.... If the interest were compounded monthly, then we end the year with $2.61303.... If the interest compounded weekly, then we end the year with $2.6925.... If the interest were compounded daily, then we end the year with $2.71456.... If the interest were compound continuously, then we end the year with $2.718281.... We see that there is a limiting amount to our investment as the compounding becomes more and more frequent. Bernoulli was the first to realize that this process has a limiting value and was the first to compute this special number, which begins: 2.71828182845904523536....

The number known as Euler's e developed during the course of a little more than a century. This special number was first referenced in 1614 in a book by the Scottish mathematician John Napier. In *Miraculous Canon of Logarithms*, Napier describes the theory of logarithms—an idea that he invented. A logarithm is a means of studying exponents. Also in this important book are calculations Napier made that took him 20 years (in which he introduces this special number 2.71828…). The name was first given to this number by Leonhard Euler in 1727, and e is often referred to as Euler's number.

Euler's constant is an extremely important number. It appears in essentially all questions of growth and decay. These are commonly referred to as "exponential growth" and "exponential decay." Often, populations grow at a rate proportional to the size of the population. Modeling this type of growth involves the number e. Computations of the decay of radioactive substances (half-lives) often involve the number e. The number e is one of the most important numbers in the study of calculus.

There are many ways to explicitly and precisely define e. Bernoulli's observation about the limiting value found by compound interest can be stated as the limiting value of $(1+\frac{1}{n})^n$ as the number n (the number of times of compounding) approaches infinity. Another important definition of e arises out of calculus. The number e can be expressed as an infinite sum of ever-shrinking rational numbers: $e = 1+\frac{1}{1}+\frac{1}{(2\times1)}+\frac{1}{(3\times2\times1)}+\frac{1}{(4\times3\times2\times1)}+\dots$.

The product of all natural numbers up to a number n is called "n factorial" and is denoted as $n!$ (e.g., $5! = 5 \times 4 \times 3 \times 2 \times 1 = 120$). Given this shorthand notation, the previous sum that equals e can be written as: $e = 1+\frac{1}{1!}+\frac{1}{2!}+\frac{1}{3!}+\frac{1}{4!}+\frac{1}{5!}+\dots$. This infinite sum can be extended to hold for any power of

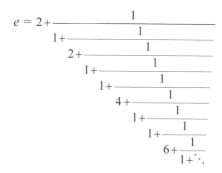

$$e = 2 + \cfrac{1}{1 + \cfrac{1}{2 + \cfrac{1}{1 + \cfrac{1}{1 + \cfrac{1}{4 + \cfrac{1}{1 + \cfrac{1}{1 + \cfrac{1}{6 + \cfrac{1}{1 + \ddots}}}}}}}}}$$

e. This formula is: $e^x = 1 + \frac{x^1}{1!} + \frac{x^2}{2!} + \frac{x^3}{3!} + \frac{x^4}{4!} + \frac{x^5}{5!} + \dots$. Another formula involving *e* is as a continued fraction:

The product of all natural numbers up to a number *n* is called "*n* factorial" and is denoted as *n!* .

Even though this number has been known to us for only about 400 years, it has generated a tremendous amount of interest. In 1815, Joseph Fourier used the infinite sum description of *e* to prove that *e* is an irrational number; we conclude from our earlier discoveries that the decimal expansion for *e* is unending and will never become periodic.

Like π, because *e* is not equal to any number of the form $\frac{m}{n}$ for integers *m* and *n*, we will see that $ne - m$ can never equal 0; thus, *e* will never be a solution to a (nontrivial) linear polynomial equation $nx - m = 0$ with integer coefficients *m* and *n*.

Questions to Consider

In 1873, Charles Hermite generalized this result and showed that *e* (just as with π) is not a solution to any polynomial equation with integer coefficients. Many mysteries remain: Is *e* a normal number? Are the numbers $e + π$, $eπ$, and $π^e$ irrational? ■

1. What do you believe it says about our universe that there exists a constant, *e*, that is abstract and unknowable yet intrinsic to fundamental processes of growth and life?

2. Which number do you believe is more fundamental in our universe: π or *e*? Is it possible that these two extremely important numbers are connected in some natural way?

The Story of Euler's *e*
Lecture 14—Transcript

Compared with the number π, the constant today known as *e* is new to the number theory scene, but its popularity can only be described as exponential. First referenced by John Napier in 1618 and first approximated by Jacob Bernoulli, this number quickly became one of the most important numbers in mathematics. Leonhard Euler, in 1727, was the first to name the number *e*, and although it most likely was not named after Euler himself, today the number is referred to as "Euler's number." With a value approximately 2.7182818 and so forth, this special number is fundamental in our understanding of growth. Thus, in order to help us define this number, in this lecture, we'll consider an excellent $1.00 investment opportunity that offers 100% interest per year. This financial fantasy will inspire the precise definition of *e* as a limiting value of an infinite list of numbers.

Armed with the official definition of *e*, we'll next consider several of its famous features. These properties include a very surprising way of writing *e* as an infinitely long sum of fractions that will be punctuated by actual exclamation points. This important representation allowed Joseph Fourier, in 1815, to devise a very clever proof that *e* is irrational. As we'll see, this punctuated sum can be generalized so as to represent any power of *e*. While the complete decimal expansion for *e* remains unattainable, we'll celebrate a very attractive way of writing *e* as an infinitely nested fraction. Charles Hermite, in 1873, extended Fourier's work and showed that *e*, just like π, is not the solution to any polynomial equation with integers. The number *e* is very important in calculus and has thus captured the admiration of science fans and groupies all over the world. We'll even see a frivolous reflection of this *e* esteem within the IPO for Google, made in 2004.

Well, at the very end of the 17th century, Swiss mathematician Jacob Bernoulli was working through a computation involving compound interest, and he came upon a very interesting number. Now, in order to inspire this real number, I'm going to make you an offer you can't afford to refuse. So, here's the very simple and attractive investment question I want us to consider. Suppose that we invest $1.00 in a savings account paying 100% interest per year. How much money would we have at the end of the year?

Actually, we can't answer this question quite yet, since the answer depends upon how often the interest is compounded. That's why we so often read about APR (annual percentage rate). That's the important number to us as investors. So, let's consider various scenarios in the story.

First, if the interest were paid once a year, then at the end of year, we would have $2.00, because we'd have 100% more at the end of the year. We started with a dollar. That would be $1 + 1$, which is $2.00. If the interest were to be awarded twice a year, then after the first six months, half the year, we would be paid an interest of $0.50, which is half of the 100% of $1.00. Thus, at the halfway point, we'd have a $1.50, and after the second six months, we would earn half of 100% on $1.50, which equals $0.75. So the total amount that we'd have at the end of the year would be $2.25. Now, notice that $0.25 came from the fact that we had a little more money in our account at the midway point, and that money generated some interest. That's the power of compounding.

If the interest were compounded quarterly, then the amount at the end of the year would be $2.44. Now, let me give you the entire decimal, because we'll think of these as numbers, not just dollars and cents. So, the amount we'd actually have precisely would be 2.4414 something dollars.

If the interest were compounded monthly, then we'd have 12 payments, one each month. Then we'd end the year with $2.613 something. Notice the amounts are going up, because we're having more money placed into our account, which, in turn, generates interest. If the interest were compounded weekly, then at the end of the year, we'd see $2.6925 something. If the interest were actually compounded daily, so 365 times we get an injection of a little bit of interest, that will accrue even more, and we'd end the year with $2.71456 and so forth. So, we see about $2.71, not bad.

Finally, what if the interest were compound continuously? That means every instant, every moment, so forth—just a stream of interest being paid. We can figure this out numerically, and it works out to be 2.718281 and so forth. Now by the way, at this point it's not obvious how to compute this amount, but we'll actually come back to that.

Thus, we've discovered that there's a limiting amount to our investment as the compounding becomes more and more frequent. We hit a ceiling, in some sense, at around 2.71 something. Bernoulli was the first to realize that this process of compunding more and more frequently has this limiting value, and he was actually the first person to compute this special number. In fact, this special number that Bernoulli computed begins 2.718281828, and it goes on so forth. This special number was first referenced in 1614 in a book by the remarkable Scottish mathematician John Napier.

Napier was a true Renaissance man, his interests were political and religious in nature, and his recreation was mathematics. Among his long list of credits, he is believed to be the individual who actually popularized the use of the decimal point that we use today. Napier was an extraordinarily brilliant individual, and I want to share with you one of my favorite stories of how clever he actually was. On one occasion, he believed that one of his servants was stealing from him, but he didn't know which one. So, Napier obtained a black rooster and informed his servants that his rooster would be able to tell who the thief was. One by one, he told them, they would be put into a room alone with the rooster, at which time they were to pet the animal.

Without the servants knowing it, Napier covered the black rooster with charcoal dust. The innocent servants, of course, would have no qualms about petting the rooster as they were instructed. However, the guilty parties, being afraid of being caught, wouldn't dare pet the rooster and would simply lie and say that they petted the animal. Napier would then just look at the servants' hands as they came out of the room, and the servants with the clean hands were the guilty parties that he was looking for. Really ingenious, don't you think?

On a more serious note, perhaps his most important mathematical work was his text, entitled *Miraculous Cannon of Logarithms*, in which he describes the theory of logarithms, an idea that he himself invented. So if you ever hated logarithms in school, you now know who you can thank for all that pain and suffering that you may have endured. In fact, a logarithm is really just a means of studying exponents, and that's all I'll say about that.

What's relevant here for us now is that in this important book, the calculations in which took him 20 years to finish, he introduces this special number, 2.71828 and so forth, for the very first time. Today, we denote this special number by a lower case e. The name e was first given to this number by Leonhard Euler in 1727, and often e is referred to as Euler's number. However, given that Euler was known for being very modest, quite a modest mathematician, it's likely that he did not actually name this number after himself.

Moving beyond its name, Euler's constant is an extremely important and a fundamental number in nature. It appears in essentially all questions of growth and decay. These are commonly referred to as exponential growth and exponential decay. Perhaps e comes from "exponential." Often populations grow at a rate proportional to the size of the population. So, in other words, if we have a very large population, then the rate of growth of that population—the people would be reproducing at a much larger rate. Whereas, if we had a relatively smaller population, then that reproduction would be a little bit smaller, and the rate of growth would be smaller. So, this is a standard model that the population growth is proportional to the actual size of the population. Now, the population could be anything. It could be people, it could be bacteria, or it could be money. It doesn't matter. Now, modeling growth of this type involves the number e. All serious financial advisors employ this number, whether they're even aware of it or not.

Computations of the decay of radioactive substances, which are sometimes known as half-lives, often involve the number e. It's called half-life because quite often what we have here is a radioactive substance decaying at such a rate that half of what's there decays at a certain point. Then if you wait that same length of time, half of what remains decays, and so forth. It's called a half life, a very slow way, very gradual way to dissipate quantity. That is also modeled by this number e. In fact, the number e is one of the most important numbers in the study of calculus, which really is the study of how things change. So not surprisingly, if we're talking about things moving, and growing, or shrinking, e plays a central role.

Let's now look at the number e itself and study it in a little more depth. There are many ways to explicitly and precisely define e. Bernoulli's observation about the limiting value found by compound interest rates, that we saw at the opening of this lecture, can be expressed as a formula. Let's now let the number n stand for the number of times we have compounding in one year. Then, after one year, with our $1.00 investment at 100% interest compounded n times, we would end up with the quantity $(1 + 1/n)^n$. That's how many dollars we would have at the end of the year. So once we know how many times we compound, we could actually plug into this formula, and it would tell us how much money we have at the end of the year.

For example, let's take a look at some examples to illustrate the point. If $n = 1$, so there's compounding just once, which is at the end of the year, then this formula gives $(1 + 1/1)^1$. Now, raising a number to the first power is just the same thing as having the number itself, so we don't have to do anything with that exponent of 1. We just see $1 + 1/1$. And $1/1$ is 1, so $1 + 1 = 2$, and of course, that was the answer we arrived at earlier.

If n were 2, which means that we actually have compounding twice, one at six months and then one at the end of the year, then we'd actually see—we'd have the formula gives $(1 + 1/2)^2$, which equals $(3/2)^2$, which equals 9/4. And 9/4 is 2.25, which again we saw for ourselves earlier.

If n equals 4, so we have compounding quarterly, then we look at $(1 + 1/4)^4$, which works out to be 2.4414 and so forth. If we have compounding daily, then n would equal 365, and we'd see the formula giving $(1 + 1/365)^{365}$, which works out to 2.71456 and so forth, and that's how we actually computed those values.

Now, if we compound every second of the year, every second, then n would equal 31,536,000, and we'd have, for the amount of money that we'd have at the end of the year, $(1 + 1/31,536,000)^{31,536,000}$, which equals 2.7182817 and so forth. You can see that we're actually heading toward our number. So we now see that, as Bernoulli saw, namely, that as n gets larger and larger, and actually approaches infinity, the corresponding value of $(1 + 1/n)^n$ is approaching this number e.

Well, another important definition of e arises out of ideas from calculus. The number e can be expressed as an infinitely long sum of ever-shrinking fractions. So, we could actually write e as $1 + 1/1 + 1/(2 \times 1) + 1/(3 \times 2 \times 1) + 1/(4 \times 3 \times 2 \times 1)$, and this pattern continues. Now, the product of all natural numbers from 1 up to any natural number n is called "n factorial," and it's denoted as n followed by an exclamation mark. So, for example, 5! would be the number $5 \times 4 \times 3 \times 2 \times 1$, which in that case, would equal 120. So, given this short hand, we could actually write or speak the previous sum for e in a more compact way. We can see that the formula tells that equals $e = 1 + 1/1! + 1/2! + 1/3! + 1/4! + 1/5!$ and so forth. The pattern continues. Isn't that something? Really quite something. Now of course, 1! is just 1, and 2! is still 2 (because it's 1 times 2), but we keep that ! there so that we can see the attractive pattern throughout.

In fact, this infinite sum can be extended to hold for not just e, but any power of e—for example, e^2, or e^3, or in general, e^x. That generalized formula would be e^x—or e to whatever the power is—$e^x = 1 + x^1/1! + x^2/2! + x^3/3! + x^4/4!$ … Do you see the pattern? We keep having x to the power—for example, the next one is $x^5/5!$ and so forth. So for example, e^2 would be written as $1 + 2/1! + 2^2/2! + 2^3/3!$ and so on. That we could write as $1 + 2/1! + 4/2! + 8/3! + 16/4! + 32/5!$ and so forth. Now, this is an extremely famous and important formula, and we'll apply it ourselves in Lecture Seventeen to make one of the most amazing discoveries in all of mathematics. Now, that's a pretty big build-up, don't you think? But I promise I'm going to deliver on this. The important thing here is that this formula is highly non-trivial, not at all obvious, and really, to derive it, we require the deep ideas of calculus.

As an aside, though, I want to say a word about the evolution of that factorial symbol, the factorial mark, the exclamation mark, and how it came about. In 1774, we find the notation n with an asterisk, or n^*, for this quantity of multiplying all the numbers from 1 up to n together. In 1841, Carl Weierstrass used n', which in mathematics is read "n prime." But by 1816, the notation $n!$ for n factorial was really catching on. The 19th-century French mathematician J. B. Durand once wrote, "There is ground for surprise that a notation so simple and consequently so useful has not yet been universally adopted."

Again, we see an example—by just finding the right notation for some mathematical idea, it actually allows us to understand it better and to make new and further discoveries. The exclamation point was sometimes actually read as "*n* admiration" instead of "*n* factorial." If you were Victor Borge, you'd probably say "*n phhhht*"—so, lots of ways of saying it. Today, in certain computer science circles and in the United Kingdom, sometimes one reads *n*! as "*n* bang." "Bang" is shorter and easier to say than "factorial," so the formula that we saw for *e* would be read as "*e* equals 1 plus 1 over 1 bang, plus 1 over 2 bang, plus 1 over 3 bang." It certainly sounds more hip, doesn't it?

Well, there're many other amazing formulas involving *e*. One such beautiful formula is known as a "continued fraction." It's not a fraction in the truest sense, but it's a nested collection of fractions. In fact, this is actually the area of number theory that houses some of my own research, and so I'm very fond of these expressions. I thought I would share with you the beautiful pattern that occurs when we try to write the continued fraction expansion for *e*. It looks like this. *e* can be written as:

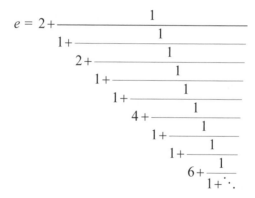

$$e = 2 + \cfrac{1}{1 + \cfrac{1}{2 + \cfrac{1}{1 + \cfrac{1}{1 + \cfrac{1}{4 + \cfrac{1}{1 + \cfrac{1}{1 + \cfrac{1}{6 + \cfrac{1}{1 + \ddots}}}}}}}}}$$

As you look at this, just focus on those first numbers that you see. You see 2, 1, 2, and then 1, 1, 4, 1, 1, 6, 1, 1, 8, 1, 1, 10, and so forth. We see 1, 1, and then the even numbers—absolutely amazing pattern, which is highly unusual, but it shows the deep and rich structure in *e* and how fundamental of a number this number actually is.

I want to close this lecture with a chance for us to celebrate and enjoy the wonders of this magical and all important number, *e*. Now even though this number has been known to humanity for only 400 years, it's generated a tremendous amount of interest and it's had an enormous impact on our understanding of mathematics, science, nature, and our universe.

In 1815, Joseph Fourier used the description of e as that infinitely long sum of fractions with the factorials to actually establish and prove that e is an irrational number. Thus, we can conclude from our earlier discoveries that the decimal expansion for e is never-ending and will never eventually become periodic—again, a wonderful moment where we can say something about this infinitely long list of digits even though we don't know what those numbers are. We can say something about a property that they possess.

Just as we saw with π—because e is not a fraction, in fact, we can say that in a cryptic way. We can say that e is never equal to a number of the form m/n, where m and n are integers. So again, just like we did for π, let's take a moment to reflect on this with e. So e doesn't equal m/n, which is identical to saying that $e - m/n$ can never equal 0, because if they were equal to 0, then they would be the same, and we know that e is not a fraction. If we take this inequality and multiply both sides by n, then we actually see that $n \times e - m$ can never equal 0. Let's just say that in words. What that means is that any integer times e minus any other integer cannot equal 0. Again, I'm going to assume that these integers aren't 0 themselves. This is a cryptic way of describing the irrationality that Fourier just showed us. Thus, we can see that e can never be a solution to a linear polynomial equation of the form integer x minus integer equals 0, or more formally, $nx - m = 0$, where the n and m are integers. For example, let's consider $7x - 5 = 0$. The solution to that would be $x = 5/7$, because if I replace the x by $5/7$, then I see that I can simplify this. The 7s cancel, I'm just left with the 5. And $5 - 5$ is 0, so I see $x = 5/7$ is a solution to that, but notice $5/7$ is a fraction. It's a ratio; it's a rational number. We know that e is irrational, which means that e will never satisfy an equation of that form.

In 1873, Charles Hermite generalized this result and showed that e, just like with π, is not a solution to any polynomial equation with integer coefficients. In other words, you can throw on not just x's, but you can have x^2s, and x^3s, and put really big integers in front of those numbers, multiply by big integers, subtract them, add them, combine them in any way that you want, there's no way to have e be a solution to any such equation, no matter how complicated you try to make it. It's really a stunning result, and these remarks, in fact, about both π and e will be the inspiration for our next lecture, when we study numbers that much deeper.

Just as with π, many mysteries remain with e. First off, we could ask, is e a normal number? Well, that means do the digits of e in the decimal expansion appear the right amount of time if the number were to be selected at random, namely that the digit 0, for example, appears one-tenth of the time, the digit 1 appears one-tenth of the time, all the way out to the digit 9 appearing one-tenth of the time? Well, no one knows for certain. In fact, here's a question: Does the digit 7 appear infinitely often in the decimal expansion for e? No one even knows that. So we know we have this endless decimal expansion that's never repeating for e, but we don't even know if the digit 7 appears infinitely often or if there's a last 7 in this long list of digits. No one knows.

Well, there are many basic questions involving both the numbers π and e together that no one knows the answer to. For example, no one knows if the number $e + \pi$ is irrational. What does that mean? Well, remember that we proved that e is—we didn't prove it but we saw that e is—an irrational number, and we remarked that π also is an irrational number, but what about their sum? Well, there's no obvious reason by just knowing I have two irrational numbers that their sum is, in fact, irrational. Possibly with the sum, there might be tremendous conspiracy in all those digits, and when you actually add that infinitely long list of digits for π to the infinitely long list of digits for e, potentially, all of the sudden there's an amazing conspiracy, and the decimal digits to the sum begins to repeat from some point onward. That could, in theory, happen. We don't know otherwise. That's an open question. In fact, even e multiplied by π, that number—no one knows if it, in fact, is irrational or not. For example, take π^e. No one knows if that number is irrational, although just intuitively it seems obvious to mathematicians and maybe even just to general people thinking about these complicated numbers that they're so complicated that, in fact, it would make sense that all of these quantities, $e + \pi$, $e\pi$, π^e, would all be irrational numbers. So while we believe they're all irrational, remember, in mathematics, we need a complete proof, and complete proofs of these facts (if they're facts) elude us to this day. So, many, many open questions.

The importance and beauty of *e* has made it a favorite number in many mathematical circles and amongst mathematical fans and groupies all over the world. As a result, we see the digits of *e* actually appearing in the strangest places. So, I thought I'd share one with you. For example, in 2004, Google honored this extremely important constant in an unusual way—in its initial public offering, which was $2,718,281,828, precisely the first 10 digits of *e*. So, even Google is gaga over *e*.

By the way, suppose that you actually want to rattle off the first 11 digits of *e*. It's always fun to be able to have a little trivia fact like that to rattle off at parties and in meetings. If you want to actually remember the first 11 digits of *e*, you just have to remember the sentence that I'm about to give you. And again, there are famous sentences like this, and I modified this just for us. So, this is the first time this sentence has ever been uttered, but if you remember it, you'll always remember the digits of *e*, at least the first 11. So here's the sentence to remember: "To express *e*, remember to memorize a sentence in embedded code." Notice that if you take the letter count of each word in the sentence—"To express *e*, remember to memorize a sentence in embedded code"—in the first three words, if you were to count the letters, we'd see 2, 7, and then a 1. If you continue, you would see 2, 7, 1, 8, 2, 8, 1, 8, 2, 8, 4—the first 11 digits of *e*. Well, *e* is indeed a natural constant that's extremely important in our universe. It's intrinsic in nature and in growth. Now, it's a new number to humanity in terms of our discovery of this amazing constant, but it's been with us since the very, very beginning of time.

Transcendental Numbers

Lecture 15

The existence of algebraic numbers goes back at least as far as the Babylonians, who were solving quadratic equations as early as 2000 B.C.E. However, humankind had to wait several millennia before we were able to demonstrate the existence of even one transcendental number.

Algebraic numbers are solutions to certain polynomial equations. We identify such equations as polynomials with integer coefficients. One of the most common practices in mathematics is the solving of equations. The simplest equations we can imagine are of the form: $x - 2 = 0$ or $2x - 10 = 0$. These are examples of linear equations: $nx - m = 0$. If m and n are integers, then the solution $x = \frac{m}{n}$ is a rational number. Linear equations are special cases of the simplest class of equations we can have, known as polynomial equations—sums and differences of numbers multiplied by x and raised to various natural-number exponents, all set equal to 0. Examples of polynomial equations are: $x^2 - 2 = 0$, $5x^3 + 9x^2 - 35x - 7 = 0$, and $-x^{25} + 49x^{17} + x^{10} - 12x = 0$. A value of x that makes the equation valid is a *solution* to the equation. For this lecture, we will consider only polynomials having integer coefficients.

Solutions to polynomials having integer coefficients can be found in special cases through special formulas. For quadratic polynomial equations (equations in which the highest power of x is 2), we have the quadratic formula to solve them; in general, it has been proven impossible to have a formula that will find the solutions to an arbitrary polynomial equation. The desire to solve these polynomial equations goes back to the Babylonians in 2000 B.C.E., who studied the solutions to $x^2 - 2 = 0$. One solution to this equation is $x = \sqrt{2}$; as we have already noted, the Babylonians approximated this value in base 60. Numbers that are solutions to such polynomial equations with integer coefficients are known as algebraic numbers because they are numbers that arise from using algebra to solve the equations. Integers are all algebraic numbers, rational numbers are algebraic numbers, and some irrational numbers are algebraic (e.g., $\sqrt[3]{5}$ is algebraic because it is a solution to the polynomial equation $x^3 - 5 = 0$). Although algebraic numbers have no specific "look" to them, they often involve roots (e.g., $\sqrt{2}, \sqrt[3]{2}, \sqrt{5-7}, \frac{\sqrt{6+\sqrt[4]{22}}}{9}$).

How are transcendental numbers defined? At one time, one of the most important questions in mathematics was: Are all real numbers algebraic numbers? A number that is *not* an algebraic number is called a transcendental number, which led to the question of whether or not transcendental numbers existed. Gottfried Leibniz was probably the first to use the term *transcendental*, but the modern definition of a transcendental number probably did not arise until the next century in the work of Euler.

In 1844, Joseph Liouville produced a particular real number and proved that it could not be an algebraic number; that is, it could not be a solution to any polynomial equation with integer coefficients. His number, now known as Liouville's number, is the first number known to be transcendental. Its value is 1.110001000000000000000001000000000...0001000..., with a decimal expansion of only 0s and 1s (and with the number of 0s between the 1s growing at the rate of $n!$). His number is irrational because its decimal expansion does not become periodic. This result was one of the greatest triumphs in our understanding of number and has since led to some beautiful and important mathematics that continue to be generated by research mathematicians.

Our understanding of transcendental numbers is relatively limited.

Our understanding of transcendental numbers is relatively limited. Establishing the transcendence of a particular number remains a challenging undertaking. Given this reality, Georg Cantor, in the late 1880s, surprised the mathematics community when he showed that "almost all" numbers are transcendental, extending the previous observation that "almost all" numbers are irrational. In 1937, mathematical historian E. T. Bell described this vision in a very romantic manner, comparing the algebraic numbers to stars against a black sky. Once again, we see the theme that the familiar numbers (algebraics) are, in fact, rare, while the more mysterious transcendental numbers are more commonplace.

David Hilbert's seventh question challenged a generation of mathematicians. In 1900, the Congress of Mathematics, held in Paris, invited David Hilbert, one of the greatest mathematicians of the modern age, to deliver an address reflecting on the then-current state of mathematics. He posed 23 open questions that he believed would shape the future of mathematics. Only 10 of his 23 questions have been solved completely.

The seventh question perplexed mathematicians for more than three decades. This question involves a number that we studied earlier: $2^{\sqrt{2}}$. Hilbert asked the following: Let a be an algebraic number not equal to 0 or 1, and let b be an irrational algebraic number. Is a^b a transcendental number? Aleksandr Gelfond in 1934 and Theodor Schneider in 1935 proved that the answer is yes; those numbers are always transcendental. The result is now known as the Gelfond-Schneider Theorem. As a consequence of this powerful theorem, we will discover that $2^{\sqrt{2}}$ is a transcendental number.

Open questions remain. Are any of the following numbers transcendental: $\pi + e$, π^e, $\delta^{\sqrt{2}}$, e^e, π^π, 2^π, and 2^e? We conjecture that all of them are, yet we cannot prove they are irrational. ∎

Questions to Consider

1. Show that $\sqrt[3]{2}$ is algebraic by producing a polynomial equation with integer coefficients for which $\sqrt[3]{2}$ is a solution.

2. Why is it so challenging to produce explicit examples of transcendental numbers?

Transcendental Numbers
Lecture 15—Transcript

In this lecture, using π and e as two famous exemplars, we'll introduce the mysterious world of transcendental numbers. We'll see how numbers fall into one of two different classes. Either they are a solution to some special type of equation, or they're not a solution to any such equation. If a number is a solution, we call it an algebraic number and, otherwise, we call it transcendental, or a transcendental number. The existence of algebraic numbers goes back at least as far as the Babylonians, who were solving quadratic equations as early as 2000 B.C.E. However, humankind had to wait several millennia before we were able to demonstrate the existence of even one transcendental number. Joseph Liouville in 1844 made this wondrous discovery about nature and number by actually constructing one.

As we've seen throughout these lectures, what at first appears exotic and rare are, in fact, commonplace and ordinary. While it took thousands of years to show that transcendental numbers even existed, it took Georg Cantor only 50 years after Liouville's seminal work to prove that "almost all" numbers are transcendental. Even though we know that real numbers are teeming with transcendentals, it remains a very difficult task to produce specific ones even today. In fact, David Hilbert, in his famous address at the 1900 Congress of Mathematics, in which he listed 10 fundamental open questions of mathematics, challenged the mathematical community to prove the transcendence of $2^{\sqrt{2}}$. While many mathematicians, including Hilbert himself, considered this question out of reach, Alexander Gelfond and Theodor Schneider independently gave an affirmative answer to this question in the mid-1930s. We'll close our discussion of transcendental numbers by describing some questions that remain unanswered to this very day.

The main object of interest in this lecture is the notion of a transcendental number. Unfortunately, transcendental numbers are defined indirectly; they're defined by what they are not. They're not the solutions to certain equations, and so I want to open this lecture here with a few minutes about those equations themselves. Now, because these ideas are challenging for all who encounter them for the first time, including the mathematicians themselves, here we have the opportunity to celebrate one of my favorite lessons for understanding difficult notions: First, try to understand simple things deeply. So, let's start at the beginning and see our story slowly unfold.

Well, one of the most common practices in mathematics is the solving of equations. The simplest equations, in my mind, are the equations that look like, for example, $x - 2 = 0$, for which I can see that the solution would $x = 2$. What does it mean to find a solution? It means we have to find a value for x so that when we plug that value in for x in the equation, the equation is valid. Another example would be $2x - 10 = 0$. Again, fairly simple. If we work through and solve that, we'd see the solution is $x = 5$, because if I replace the x by 5, I would see $(2 \times 5) - 10$. By the way, when we write $2x$, that always means that we're multiplying. So, if there's no symbol in between 2 and an x like that, that means 2 times x. So, we see $(2 \times 5) - 10$, which does indeed equal 0. That's a solution.

Well, in fact, these are two examples of what are called linear equations, of the general form something times x minus something equals 0. Now, if m and n are integers in the expression $nx - m = 0$, then the solution $x = m/n$ is a rational number. In other words, it's a fraction. We have a ratio of two integers. So, for example, $3x - 2$—if we were to find the solution to that, we'd see that the answer is that $x = 2/3$, and notice 2/3 is indeed a ratio of integers. It's a rational number.

Linear equations are, in fact, a special case of the simplest class of equations that we have, known as polynomial equations. These are sums and differences of numbers multiplied by x's that are being raised to various natural number powers, and then all set equal to 0.

So, let's now see some examples of polynomial equations that are not linear just to get into the groove. I'm just going to make some up, and these are going to be just random and ugly looking things. I assure you that we're not going to solve them, but I want you to get a sense of what they look like, their general form. For example, $x^2 - 2 = 0$. That's an example of a polynomial equation. And $5x^3 + 9x^2 - 35x - 7 = 0$ is even a more elaborate one, and you can make even really complicated looking ones—for example, $-x^{25} + 49x^{17} + x^{10} - 12x = 0$. Wow, that looks really impressive. But they all have the same form and that's what I want us to see here now. They all have a similar look.

Now, even with these more intimidating polynomial equations, a value of x that makes the equation valid is called a solution to that equation. That is, if we plug in that value for the unknown x, we see the equality holds. Even though it looks complicated and so forth, if we had a value, we could plug in and see if it satisfies the equation or not. So, for example, let's see that $x = 3$ is, in fact, a solution to the polynomial equation $x^2 - 9 = 0$. Well, let's check and see. If we replace the unknown x by a 3, what we see here is $3^2 - 9$. Well, 3^2 is 3×3, which is 9. So we see $9 - 9$, which indeed equals 0. So yes, it's a solution.

Let's see another example and check that $\sqrt{2}$ is a solution to $x^2 - 2 = 0$. So if we look at the polynomial $x^2 - 2 = 0$, and now we substitute $\sqrt{2}$ for the unknown x, then what do we see? Well, we see $\sqrt{2}^2$. Well, remember that when we square a square root, those two functions counteract each other, and we're just left with, in this case, the 2. So, $\sqrt{2}^2 = 2$. But then we subtract 2, which means we yield 0. So in fact, we see that $\sqrt{2}$ is a solution to that polynomial. So, we see solutions.

Now for this lecture, we'll only consider polynomials having integer coefficients. In other words, the only numbers we'll see multiplying all those xs are going to be natural numbers or their negatives, which has been the case in all the examples that we've seen so far. Solving polynomial equations is very, very difficult in general. For quadratic polynomial equations—and by the way, a polynomial equation is a quadratic equation if the highest power of x is 2, so we see x^2s and no x^3s or higher. We can actually employ the so-called "quadratic formula" to solve them, and we actually studied the quadratic formula long ago in our math classes in school. However, in general, it's been proven that it's impossible to have a formula that can find the solutions to an arbitrary polynomial equation. That is, we know for certain that no such big formula exists. This important discovery was made by Evariste Galois, a brilliant young 19th-century French mathematician who, tragically, at the age of 20, was killed in a duel on May 30, 1832. Really, a sad story.

Now, we're not sure today exactly why this duel took place. One romantic theory is that Galois was defending the honor of a young lady. A less romantic, but more likely, explanation is that he was defending his radical political beliefs. What we do know for sure is that the evening before this fateful duel, Galois stayed up all night writing down his mathematical ideas. He was 20 years old, and he was writing and writing, because he realized things might not go so well the next day. And, in fact, the ideas that he wrote were the ideas that resolved this question showing that there is no big solution to all polynomial equations. There's no formula that will work. He proved that that night. In fact, in his last letter to his brother Alfred, he actually wrote, "Don't cry, Alfred. I need all my courage to die at 20." A very tragic story, and an untimely death of an incredible mind.

The desire to solve these polynomial equations goes back to at least the Babylonians in 2000 B.C.E. For example, they studied the solutions to $x^2 - 2 = 0$. One solution to this equation, as we just saw, is $x = \sqrt{2}$. As we have noted several times in the course, the Babylonians approximated this value in base 60. The numbers that are solutions to such polynomial equations with integers are known as algebraic numbers, because they're numbers that arise from using algebra to solve the equations. The integers, of course, are all algebraic numbers, because if you think about it for a moment—take for example the integer 7. If we consider the polynomial $x - 7 = 0$, then 7 is a solution. Well, if you're a solution to such an equation, then we see that you're an algebraic number. So, all the integers are algebraic numbers. Similarly, all the rational numbers, all the fractions, are algebraic numbers, as we saw with the examples we looked at in the opening of this course. Those are two collections of well known algebraic numbers in our minds—the integers, the counting numbers, and the rational numbers.

But of course, we've also seen some irrational numbers that are algebraic. For example, $\sqrt{2}$ is the most famous that we actually proved following the ancient Greeks' proof. But for example, $\sqrt[3]{5}$ is an algebraic number because it's actually a solution to the polynomial equation $x^3 - 5 = 0$. Now to see that, let's actually substitute $\sqrt[3]{5}$ in for the unknown x and see that, indeed, it's a solution. If we take $\sqrt[3]{5}$ and cube it—well, just like with the square root being squared, a cubed root being cubed, those two operations cancel out, so—we just see 5. And $5 - 5 = 0$, so indeed, it's a solution to that, and therefore, $\sqrt[3]{5}$ is awarded the title "an algebraic number."

While algebraic numbers have no specific look to them, they often involve roots. For example, $\sqrt{2}$ is an algebraic number, $\sqrt[3]{2}$ is an algebraic number, as is $\sqrt{5} - 7$, and you can even make more elaborate ones—for example, $\frac{\sqrt{6 + \sqrt[4]{2}}}{9}$.

These are all examples of algebraic numbers, and you can see that these can get very complicated looking very quickly. In fact, this reminds me that Sir Winston Churchill carried the memory of these square roots and cube roots throughout his life. In his autobiography, entitled *My Early Life: A Roving Commission*, he reflects on his bout with mathematics as part of his entrance examination to be admitted into the military academy, and I thought I would read a little excerpt from his book in his own words.

> All my life from time to time I have had to get up disagreeable subjects at short notice, but I consider my triumph, moral and technical, was in learning Mathematics in six months.... Of course what I call Mathematics is only what the Civil Service Commissioners expected you to know to pass a very rudimentary exam. I suppose that to those who enjoy this peculiar gift, Senior Wranglers and the like, the waters in which I swam must seem only a duck-puddle compared to the Atlantic Ocean. Nevertheless, when I plunged in, I was soon out of my depth.

He goes on to say:

> When I look back upon those care-laden months, their prominent features rise from the abyss of memory. Of course I had progressed far beyond Vulgar Fractions and the Decimal System. We were arrived in an "Alice in Wonderland" world at the portals of which stood "A Quadratic Equation."... We turned aside, not indeed to the uplands of the Delectable Mountains, but into a strange corridor of things like anagrams, and acrostics, called Sines, Cosines, and Tangents. Apparently they were very important especially when multiplied by each other, or by themselves!

Finally, he writes:

> There was a question on my third and last Examination about these
> Cosines and Tangents in a highly square-rooted condition which
> must have been decisive upon the whole of my afterlife. It was
> a problem, but luckily I had seen its ugly face only a few days
> before, and I recognized it at first sight.

There you see this vision of mathematics and how these numbers with the
roots can be extremely intimidating. In fact, he reflects on mathematics just
a page or so later, and I wanted to share with you his reflections on his own
discovery of mathematics. He writes:

> I had a feeling once about Mathematics, that I saw it all—Depth
> beyond depth was revealed to me—Byss and the Abyss. I saw, as
> one might see the transient of Venus … a quantity passing through
> infinity and changing its sign from plus to minus. I saw exactly how
> it happened and why the tergiversation was inevitable: and how the
> one step involved all the others. It was like politics. But it was after
> dinner and I let it go!

So there you have a great, great memory from Churchill.

Getting back to our transcendental numbers—once mathematicians found
these algebraic numbers, it led to a very natural question. In fact, it led to one
of the most important questions in mathematics that remained unanswered
all the way through to the 19th century: Are there any real numbers that are
not algebraic? In other words, are all real numbers these algebraic numbers?
So, is every real number a solution to some polynomial equation having
integer coefficients?

A number that is not algebraic is called a transcendental number. So, the big
question was, do transcendental numbers exist? Well, Gottfried Leibniz, in
1682, was probably the first to use the term *transcendental*, but the modern
definition of a transcendental number probably did not arise until the next
century in the work of Euler. The question of the existence of transcendental
numbers remained unanswered until 1844, when Joseph Liouville actually

produced a particular real number and proved that it could not be an algebraic number. In other words, it could not be the solution to any polynomial equation with integer coefficients.

His number, now known as Liouville's number, is the first number actually to be known to be transcendental, and its decimal value is 1.110001, and then a run of 17 0s, then a 1, then an even longer run of 0s, then a 1, and then an even longer run of 0s, followed by a 1, and so forth. That is, his number has a decimal expansion just consisting of 0s and 1s, with the number of 0s between the 1s growing at a very dramatic rate, almost at the rate of $n!$, if you remember the factorial from the previous lecture (which is the product of all natural numbers up to a certain point).

By the way, we can see for ourselves that this number is irrational, namely, it's not a fraction, since its decimal expansion does not become periodic. The runs of zeroes are ever increasing, so we can't have a repeating piece. So, we know it's irrational, but he proved much more; he proved that it's transcendental. It's not the solution to any polynomial equation with integer coefficients.

This result was one of the greatest triumphs in our understanding of number, and continues to inspire beautiful and important mathematical research. This area is known as transcendence, or transcendental number theory, and I must confess that Liouville's result is my personal favorite result in all of mathematics. So, this is a wonderful theorem. It certainly means a lot to me. As we have already noted, in 1873 Charles Hermite proved that e is a transcendental number, I'm just using that language, and nine years later, Ferdinand von Lindemann proved that π is a transcendental number. We are just converting what we said before about not being a solution to polynomial equations with integer coefficients, and are using the phrase "transcendental number" which means the same thing.

Now, the transcendentals form a vast world of mysterious numbers, and our understanding of them is relatively limited, even today. It remains a great challenge and an amazing undertaking to establish the transcendence of a particular number. And why? Well, because in theory, if you want to show that a particular number is transcendental, in some sense you'd have to go through all the infinitely many different polynomial equations with integer coefficients, and check each one, and make sure that this number is not a solution to any of them. Well that, of course, is impossible in finitely many steps, and so people have to use ingenious ideas in order to get around that and prove transcendence indirectly.

Given this reality, Georg Cantor, in the late 1880s, surprised the mathematics community when he showed that "almost all" numbers are transcendental. In other words, the percentage of algebraic numbers is so small that it's effectively 0, and in fact, mathematically it is 0. So, the probability of picking one is, in fact, 0. This extends our previous observation that "almost all" numbers are irrational. Remember, if you pick a real number out at random, it's with certainty that the number will be irrational.

In fact, the famous mathematical historian E. T. Bell, in 1937, described this vision in a very romantic manner. He wrote, "The algebraic numbers are spotted over the plane like stars against a black sky; the dense blackness in the firmament of the transcendentals." A wonderful image. If you look up into the dark sky, you see stars. Of course we see them, because they're the exception, but if you were to just somehow pick one point out in space, the likelihood of selecting one of those stars would be 0. More likely, you would hit the black background, and that dense, black background is the firmament where the transcendentals live. Once again, we see in our course this wonderful theme that the familiar numbers (in this case, the algebraic numbers), like $\sqrt{2}$, are, in fact, the rare numbers, while the more mysterious transcendental numbers are more the commonplace and more the ordinary. Again, we see this wonderful dichotomy between our understanding and intuition versus reality.

Thus, we're presented with another opportunity to retrain our thinking and intuition, although this is by no means easy to do. It's difficult to even grasp the idea of a transcendental number, let alone imagine that, in fact, the real line is teaming with them, and $\sqrt{2}$ and $\sqrt[3]{5}$ that we're familiar with are the exception.

Well, I now want to share the story of a very famous question about transcendental numbers that was posed by the early-20th-century German mathematician David Hilbert. One of the most famous international mathematics conferences is the Congress of Mathematics, held every four years. In 1900, at the dawn of the 20th century, the Congress of Mathematics, held in Paris, invited David Hilbert, one of the greatest mathematicians of the modern age, to deliver an address on August 8. Hilbert took this opportunity to reflect on the then-current state of mathematics and consider where it has been and the next direction in which it should move.

So, in his address, he posed open questions that he believed would shape the future of mathematics. He actually posed 23 questions, although only 10 were presented at the conference. But in the conference proceedings, which he contributed, he listed all 23. Of the 23 offered in 1900, 10 of those questions have been solved completely today, and the rest still remain open. In this famous list of questions, I want us to focus on his seventh question that he describes. The question actually involves a number that we ourselves studied earlier, the number $2^{\sqrt{2}}$. We figured out what that meant in a previous lecture, but he asks here if that number is, in fact, transcendental.

Now, more generally, Hilbert asked the following question. I'm going to state this question in mathematical language, but we'll see that it's similar in structure to $2^{\sqrt{2}}$. He said: Let's let a be an algebraic number that's not equal to 0 or 1, so throw away 0 or 1, but take any other algebraic number you can think of, and let's let b be an irrational algebraic number, for example, $\sqrt{2}$ or $\sqrt[3]{5}$. Then the question is, if I take a^b, is that number a transcendental number? That was his seventh question.

Now, most mathematicians believed that this question would remain unanswered for a long, long time, because it appeared totally out of reach using the current techniques and theorems from number theory of the day. However, in a surprisingly fast turn of events, an answer was provided by Alexander Gelfond in 1934 and then again by Theodor Schneider in 1935. Amazingly, Schneider wasn't even aware of Gelfond's work. Once again we see where our knowledge of mathematics is at a point where a question was about to fall, and two great minds, independently on their own, but simultaneously, saw that avenue and cracked the question open, providing the answer.

They proved that the answer is yes, all those numbers that Hilbert described are indeed transcendental numbers. The result today is now known as the Gelfond-Schneider Theorem to celebrate and honor both men who created the proof. Now, we can actually apply their theorem, and this powerful result, to establish that $2^{\sqrt{2}}$ is indeed transcendental. Let's just do it really quickly to see that the number is transcendental. So, first of all, the number 2 is neither 0 nor 1 (which is one of the stipulations), and notice that it is algebraic, because it's the solution to $x - 2 = 0$. So therefore, it is an algebraic number that's neither 0 nor 1, and so it's an allowable candidate for the base. What about the exponent $\sqrt{2}$? Well, it's an algebraic number, because it's a solution, as we've seen earlier, to $x^2 - 2 = 0$. That's a polynomial, and the Pythagoreans and we ourselves proved that that number is irrational. So, we have an irrational algebraic number; we can use that in the theorem as the exponent. So the hypotheses are fulfilled. We've satisfied the conditions required in order for us to conclude, through the Gelfond-Schneider Theorem, that $2^{\sqrt{2}}$ is indeed a transcendental number.

Many open questions remain, and I want to share some of them with you. In fact, our understanding of transcendental numbers is very, very narrow, very, very limited. For example, $\pi + e$—we don't know if that number is transcendental. And as you might remember, we discussed earlier that we don't even know if that number is even irrational, so we certainly don't know if it's transcendental. And πe, nobody knows; π^e, nobody knows; $p^{\sqrt{2}}$, nobody knows. Notice we can't use the Gelfond-Schneider with $p^{\sqrt{2}}$, even though $\sqrt{2}$ is indeed an irrational algebraic number. Here π is *not* an algebraic number, because we've already discussed that π is an example of a transcendental number, and the Gelfond-Schneider Theorem says nothing about a transcendental raised to an algebraic power, only an algebraic to an algebraic power. Another example is e^e, where we know nothing about it, and π^π, 2^π, and 2^e. You can make up lots of numbers of this sort, and we really don't know anything about them. It's amazing how limited our understanding of numbers genuinely is at the forefront. Now, we conjecture that all of these numbers are transcendental, because they look so complicated, and so forth. But that says nothing; that proves nothing. We have to be able to prove it, and in fact, we don't even know if any of those numbers are even irrational, so there's much work to be done if we really want to understand the transcendental numbers in greater depth.

The point, in this lecture, is to discover that there's different ways of actually approaching and viewing number, and we can classify numbers depending upon whether they're sympathetic to algebraic expressions or not. If they are, then we call them algebraic numbers, and if they're not, we call them transcendentals. We've seen lots and lots of examples of algebraic numbers in our own mathematical histories and certainly in our course, for example, $\sqrt{2}$, and the natural numbers, and the ratios of natural numbers, the fractions. But the transcendental numbers remain enigmatic in our minds. However, in reality, what we're seeing is that the number line is teeming with these transcendental numbers, and they are more the norm than the exception. Quite often—you see this again and again in our course—that which is the most familiar turns out to be the most exotic. So, I hope you will consider thinking about transcendental numbers, an extremely deep idea that continues to intrigue the world of mathematicians and mathematics to this day.

An Algebraic Approach to Numbers
Lecture 16

Cardano was the first to consider square roots of negative numbers. Now, he called these numbers "fictitious" and "meaningless" numbers, which is a wonderful, wonderful way of describing numbers that don't have any sort of physical measurements associated with them or a negative. Today, we call the number i an "imaginary" number.

What is algebra? At the heart of algebra, we find two numerical expressions that are equal to one another. Because some number (or numbers) in one of the expressions is *unknown*, the main mission is to figure out the number that represents that unknown quantity. For example, $2 \times 5 = 10$ is a true equation, but nothing is hidden from sight. The equation $2x = 10$, however, has a mysterious quantity, x. Our mission is to find the number x that makes this equation true; that is, we must find a number with the property that, when multiplied by 2, the product equals 10 (in this case, $x = 5$).

Word problems have challenged people for thousands of years. The book *Greek Anthology* from the 4th century B.C.E. contained many algebraic riddles. For example, six people are to divide a heap of apples. The first person receives $\frac{1}{3}$ of the apples; the second, $\frac{1}{8}$; the third, $\frac{1}{4}$; and the fourth, $\frac{1}{5}$. The fifth person receives 10 apples, and the last person is given only 1. How many apples were there in the original heap?

The ancient Chinese indicated unknowns by their physical position in the equation, while in Hindu works we see the unknown as a dot.

Today, we often refer to the unknown as x. In the Rhind Papyrus, Ahmes used a symbol for "heaps" for unknown quantities. The ancient Chinese indicated unknowns by their physical position in the equation, while in Hindu works we see the unknown as a dot. Brahmagupta and, later, Bhaskara used the names of colors to designate different unknown quantities, while in Arabic works from 900 C.E., different coins were used to represent unknowns and one unknown was referred to as "thing." In 1637, René Descartes adopted the italic letters x, y, and z for unknowns, the notation that is used today. The always-present object in an equation is the equal sign, $=$. Robert Recorde, who published an algebra text in 1557, introduced the symbol $=====$ for equality. This symbol was later shrunk down to the more modest $=$.

We see the need to expand our net of numbers through the desire to find solutions to simple equations. We begin with natural numbers. If we consider linear equations, such as $x + 2 = 5$, we see that these equations have solutions. We run into trouble, however, if we consider $x + 7 = 5$, because there is no number in our pretend universe that satisfies this equation. We now expand our notion of numbers so that these equations have solutions; therefore, we allow negative natural numbers to be considered as numbers.

If we consider the entire world of number to be composed solely of the collection of integers, then we face further troubles when we consider such simple equations as $3x = 5$. There is no integer that satisfies this equation. Therefore, we must allow fractional numbers to be considered as numbers. If we consider the entire world of number to be composed solely of the collection of rational numbers, then we face further troubles when we consider such simple equations as $x^2 = 2$. There is no rational number that satisfies this equation. Therefore, we must allow irrational numbers to be considered as numbers. We call these irrational numbers "algebraic numbers."

Cardano's work in this area was groundbreaking. One of Girolamo Cardano's greatest works was a text he wrote in 1545 entitled *The Great Art*, in which he included a systematic method to solve cubic equations of the general form $x^3 + ax = b$. Cardano also posed the following question: Divide 10 into two parts whose product is 40. The answers are $5 + \sqrt{(-15)}$ and $5 - \sqrt{(-15)}$; however, what does $\sqrt{(-15)}$ mean?

To explore this question, we consider a simpler equation: $x^2 + 1 = 0$. If there is a number x such that it is a solution to $x^2 + 1 = 0$, then $x^2 = -1$. In this case, x cannot equal 0 because $0^2 = 0$ (not -1); therefore, the number x must be a nonzero number. Given that the square of any nonzero real number equals a positive number, there is no real number that satisfies this equation.

For the first time in this course, we find a natural need to expand our notion of number beyond the real numbers of our number line. We call these new numbers complex numbers, and we denote the special number $\sqrt{-1}$ as i. We call the number i an imaginary number. The name i was given by Leonhard Euler two centuries after Cardano's original work. Because i is not real, it cannot be a rational number; hence, it is irrational. The number i is an algebraic number because it is a solution to the polynomial equation $x^2 + 1 = 0$.

Armed with the special number $i = \sqrt{-1}$, we can define all the complex numbers as numbers that can be expressed as x + yi, for real numbers x and y (e.g., $2 + 3i$, $-\sqrt{3} + \pi i$, and $-3i$). We can also combine complex numbers. By definition, i^2 equals -1. We see that $i^3 = -i$, $i^4 = 1$, and $i^5 = i$; that is, we return to the number i. This repeating pattern continues. We can visualize complex numbers by considering a geometric object with an extra dimension. This idea leads us to the complex plane as a means of visualizing complex numbers. The complex number $x + yi$ is the point (x, y) on the Cartesian plane.

Gauss's Ph.D. thesis was an algebraic end of the line for numbers. We recall that our algebraic development of number evolved through our desire to have solutions to polynomial equations with integer coefficients. We have seen that we can even solve such equations as $x^2 + 1 = 0$, although the solutions are no longer real numbers; we just added one new number, the imaginary number i, to build all the complex numbers.

It can be proven that just by including i (and all complex numbers $x + yi$), we have captured all the algebraic numbers. Every polynomial equation with integer coefficients has solutions of the form $x + yi$, in which x and y are real algebraic numbers. In 1799, child mathematics prodigy Carl Friedrich Gauss, in his Ph.D. thesis, proved that every polynomial equation with complex coefficients has a solution of the form $x + yi$, in which x and y are real numbers. This insight shows us that there is no need to create any additional numbers, after we include the imaginary number i, to ensure that all polynomials have solutions. This can be expressed by saying that the complex numbers are "algebraically closed." ∎

1. Do you believe it is appropriate to use the word *imaginary* to describe *i*? Place your response in the context of the history of numbers (recall that Diophantus called negative numbers "absurd").

2. Simplify the following numbers as much as possible: i^2, i^4, i^5, i^{25}, and $i^{1,000,000}$.

An Algebraic Approach to Numbers
Lecture 16—Transcript

We'll open this lecture with a historical look back at algebra—what it means, its symbols, and how it can even distract from star-crossed lovers. Inspired by the study of algebra, in particular, solving polynomial equations, we'll then offer a mathematical story that opens with the natural numbers and naturally leads to the discovery of the integers, the rational numbers, and then the algebraic irrational numbers, such as $\sqrt{2}$. This story brings us to a seemingly innocuous, or perhaps even ridiculous, question. Are all algebraic numbers, in fact, numbers, as we know them to this point? In other words, is every algebraic number a real number on the number line?

This will bring us to Girolamo Cardano, a 16th-century Italian mathematician with an interesting story. We'll soon discover that the answer is no. There are equations that have *no* solutions among the real numbers. Cardano was the first to consider square roots of negative numbers. He called them "fictitious" and "meaningless" numbers. This discovery led to the need to expand our notion of number yet again and thus, in turn, to develop the so-called complex numbers or imaginary numbers. However, in view of our story, these new numbers will not appear to be that imaginary.

Given this new expanded view of number, that is, the collection of all complex numbers, we'll wonder if we've captured every algebraic number in this expanded net of numbers. Carl Friedrich Gauss, in his Ph.D. thesis of 1799, showed that the answer is yes, and surprisingly, much, much more is true. In fact, we'll see that, in the eyes of this algebraic approach to numbers, when we ascend to the complex numbers, we have arrived at the end of the number line, the ultimate notion of number. While this observation is true using this algebraic lens, we'll soon discover there are other mathematical lenses through which we can gaze into the world of number, and through those visions, we actually see and discover different conclusions.

To get started, I want us to return to our heady days of algebra. In fact, what exactly is algebra? At the heart of algebra, we find two numerical expressions that are equal to one another. However, algebra holds a mystery. Some number, or numbers, in one of the expressions are *unknown*. Thus, the main mission in algebra is to figure out the number that represents that unknown quantity. For example, if we consider $2 \times 5 = 10$, well, this is a true equation, but nothing is hidden from sight. However, the equation $2x = 10$—we'll remember $2x$ means 2 times x—has a mysterious quantity, namely that x. Our mission now is to find the number x that makes that equation true. In other words, we must find a number with the property that when it's multiplied by 2, the product equals 10. Of course we see that in this case, our mysterious number is $x = 5$.

Now, when we think of algebra, we might think of those dreaded word problems or story problems. These have challenged some, and tortured others, for thousands of years. The book entitled *Greek Anthology*, from the 4th century B.C.E., contained many algebraic riddles. In fact, here's one from that ancient book: Six people are to divide a heap of apples. The first person receives 1/3 of the apples, the second person receives 1/8 of the apples, the third person receives 1/4 of the apples, and the fourth person receives 1/5. The fifth person receives 10 apples, and the last person is given only 1. How many apples were there in the original heap? That's the kind of question that we've all faced from time to time. Maybe I shouldn't tell you the answer, but just in case you're wondering, the answer is 120, and you can go back and check that 120 satisfies all those conditions. Every person gets the appropriate number of apples that they're allotted.

Today, we often refer to the unknown as x. But Ahmes, in the Rhind Papyrus, represented unknown quantities with a symbol for "heaps," and if you remember, back in this period, numbers were defined by heaps or piles of things, so using a symbol that means "heaps" seems very natural and consistent with that point of view. The ancient Chinese indicated unknowns by their physical position in the equation—how they were placed—while in Hindu works, we see the unknown represented as a dot. Brahmagupta and, later, Bhaskara used names of colors to designate different unknown quantities, while in Arabic works from 900 C.E., different coins were used to represent unknowns and referred to one unknown as 'thing," which is interesting that it was an unknown object, so we called it "thing," even though it represented some number. We just didn't know what that number was at this point.

It was René Descartes, in 1637, who adopted the italic letters *x, y,* and *z*, all lower case, for unknowns. This is the notation that is used today. Now, why the popularity of *x*? Descartes favored *x* since it was the least most common letter in the French language, and so it would be the least confusing. It was the least used letter to start French words. Beyond the unknown, the always present object in an equation is the equal sign, which we denote as two short, little, parallel horizontal lines. Ahmes, in the Rhind Papyrus, used a symbol that meant "it gives" for the equation sign, which again is wonderful, because certainly—you know, 1 + 2, well, it gives 3—so that makes sense.

In texts before 1557, Latin words such as *equales*, abbreviated as *aeq*, were used. So, the letters were actually used to denote the equals. Some actually used a long hyphen for the equal sign. But it was Robert Recorde who published an algebra text in 1557 in which he introduced the symbol of two horizontal parallel lines, although his were actually very, very long at first. This symbol was later shrunken down to the more modest equal sign that is familiar. Recorde's book was also the first English text to introduce our current symbols for addition and subtraction.

Algebra remains an important tool to help answer important questions. However, believe it or not, it can also be viewed as a pastime. The great 12th-century mathematician Bhaskara authored a text entitled *The Gem of Mathematics*. It contains entertaining challenges to keep the interest of the reader, and he wrote these puzzles in verse, and I wanted to offer you an example of one:

> O tender girl, out of the swans in a certain lake, ten times the square root of their number flew away to Manasa Sarovar when the rains came at the monsoon. One-eighth went away to the forest called Sala Padmanee. Three swans remained in the lake engaged in amorous play. How many swans were there in all?

The kind of question that we saw when we were kids and we dreaded, you know—square-root of the number of swans—it's very silly in a way. And legend has it that Bhaskara wrote this book, in fact, for his daughter. And if you notice, it opens with "O tender girl," perhaps referring to his daughter. And the story goes that he actually wrote this book to try to distract his daughter from seeing a certain suitor which Bhaskara didn't particularly like. Although having a question that involves swans engaged in amorous play—I don't know how successful this distraction really was.

Well, let's now witness the growth of numbers within an algebraic garden. In other words, let's exploit our desire to find solutions to simple equations and see how the notion of number evolves and where it takes us. We begin with the numbers that we can really count on, namely, the natural numbers. That is, we will now momentarily pretend that our entire world of number consists solely of the collection of natural numbers. If we consider simple linear equations, such as $x + 2 = 5$, we see that these equations do have solutions. In this case, x would be 3, which is a counting number. It's a natural number.

However, we run into trouble if we consider the equation $x + 7 = 5$. There's no number in our pretend universe that actually satisfies this equation, because we have no number -2, only the natural numbers. Well, as we saw in an earlier lecture, the father of algebra himself, the Greek mathematician Diophantus, around the year 250, wrote about such nonsensical equations. If an equation had a negative number solution, he called the entire equation absurd. Again, note the bias and the prejudice. It's hard for minds, even great minds, to become open to notions and ideas that seem so unnatural and counterintuitive at first. But still, we try to overcome it.

Well, returning back to algebra and our evolving number sense, it seems sensible that simple equations such as $x + 7 = 5$ should have solutions. Thus, the algebra now forces us to expand our notion of numbers so that these equations have solutions. Therefore, we now allow negative natural numbers to be viewed as numbers. We recall from Lecture Three that the great 7th-century Indian astronomer Bhramagupta was the first to offer a systematic treatment of negative numbers. This expanded view leads us to the integers. That is, let's now build the world of the rational numbers. So, we have the integers, which are the numbers 1, 2, 3, 4, and so on, with 0 and with $-1, -2, -3$, but let's see how algebra demands more from us.

If we consider the entire world of number to be comprised solely of the collection of integers, then we face further trouble when we consider simple equations such as $3x = 5$. So, 3 times $x = 5$, and the solution there would be $x = 5/3$. And there's no integer that satisfies this equation, because 5/3 is, in fact, not an integer; it's a ratio of two integers with a denominator of 3. And yet, it seems sensible that simple equations such as $3x = 5$ should have solutions.

Thus, we now expand our notion of number so that these equations do have solutions. In other words, we now introduce fractional numbers into our ever-evolving world of number. This expanded view leads us to rational numbers. Now, as we've seen in previous lectures, fractions go back to the Egyptians and the Greeks. But did the Greeks view rational numbers as numbers or as merely ratios of numbers? Again, this remains open, but in our minds now, algebra dictates that, in fact, we view these rational numbers, or ratios, as numbers. Notice how algebra is beautifully growing our number sense for us just by our desire to find solutions to equations that seem simple and should have solutions within mathematics.

Let's now journey into the universe of algebraic numbers. If we consider the entire world of number to be comprised solely of the rational numbers, in other words, the fractions, then we face further headaches when we consider simple equations such as $x^2 = 2$. Well, we know that one solution to this is, in fact, $x = \sqrt{2}$, because $\sqrt{2}^2$ equals 2. And we've proven that $\sqrt{2}$ is, in fact, an irrational number; it's not a ratio. So, therefore, as the ancient Greeks discovered thousands of years ago and as we discovered in Lecture Eight, there's no solution to this, because there's no rational number that will satisfy this in our current number state. And yet, it seems sensible that simple equations such as $x^2 = 2$ should have solutions. What are we forced to do?

Thus, we expand our notion of numbers so that these equations do have solutions. In this case, we have the new number $\sqrt{2}$ and also $-\sqrt{2}$, which is the other solution. Therefore, we now allow these irrational numbers to be considered as numbers. Now, we call the collection consisting of all these numbers, the irrational and the ones we saw earlier, we call that collection the "algebraic numbers." So now we've arrived at the algebraic numbers. Well, up to this point in our algebraic evolution of number, all the numbers we've seen are numbers that we've already discovered in previous lectures, even including these algebraic numbers, which we saw in the previous lecture. Well, that's about to change, and it's going to change with a character named Cardano.

Girolamo Cardano was a colorful 16th-century Italian physician and mathematician. He was the first to describe typhoid fever; he invented the combination lock, but was always short of cash, and thus was also an accomplished gambler and chess player—quite a character. One of his greatest works was a text he wrote in 1545, entitled *The Great Art*, and it was an algebra text, and in it he included a systematic method for solving certain cubic equations. So, these are equations which have an x^3 in it, but no x^4 or higher powers—just x^3. But even his solution was not without controversy. Niccolo Tartaglia derived this particular solution and shared it with Cardano. Tartaglia claimed later that Cardano promised not to publish it, and thus this led to a very long and bitter plagiarism battle between these two men.

In a different direction, but perhaps even more controversial, Cardano posed the following question in his book *The Great Art*. The question read, "Divide 10 into two parts whose product is 40." In other words, we are asked to find two numbers that sum to equal 10 and whose product equals 40. Sounds like a harmless question. Well, if one were to work it out, the answers turn out to be $5 + \sqrt{-5}$ and $5 - \sqrt{-5}$. Now, it's easy to see that if we take those two numbers ($\sqrt{-5}$ and $-\sqrt{-5}$) and just add them together, they cancel out, and $5 + 5 = 10$. So, the sum being 10 is easy to see, and if one were to very carefully multiply those two numbers together correctly, we'd actually see the product is 40, as it was supposed to be. But the question really is, what does $\sqrt{-5}$ mean?

To face this challenging question, we need to consider a simpler question. So, the question that I want us to think about is, does $x^2 + 1 = 0$ have a solution? Well, let's think about $x^2 + 1 = 0$. If there is a number x that is a solution to $x^2 + 1 = 0$, then if I were to subtract 1 from both sides, I would see that $x^2 = -1$.

We notice that, in this case, x cannot be 0, since $0^2 = 0$, and it's not equal to -1, so $x = 0$ is out. It's not a solution. Therefore, the number x must be a nonzero number. However, let me remind you that, in Lecture Twelve, we proved that the square of any nonzero real number equals a positive number, because a number multiplied by itself is either positive times positive or negative times negative, and we saw that, in both cases, the answer is positive. Thus, since we have a square equaling -1, we see there is no real solution that's going to actually satisfy that. No real number will satisfy this equation.

Yet, it seems sensible that such a simple equation as $x^2 = -1$ should have solutions. Now, this is an interesting moment, because we find a need to expand our notion of numbers yet again. So, let's do just that. Let's now widen our notion of numbers so that these equations do have solutions. Therefore, for the first time in this course, we find a need to expand our notion of number beyond the real numbers of our number line. So, we're finally going to take off from the line.

We call these new numbers complex numbers, and we denote the special number $\sqrt{-1}$ as the lower case, italics i. This special number i is our first number that is not real. It does not correspond to any point on the number line, and as we've noted, Cardano was the first to consider square roots of negative numbers. Now, he called these numbers "fictitious" and "meaningless" numbers, which is a wonderful, wonderful way of describing numbers that don't have any sort of physical measurements associated with them or a negative. Today, we call the number i an "imaginary" number. In fact, Cardano himself wrote about this and said:

> Put aside the mental tortures involved. So progresses arithmetic subtlety the end of which, as is said, is as refined as it is useless.

Implying that the math is correct—we derive this answer—but, in fact, these numbers mean nothing, and they're useless. It turns out that Cardano was, in fact, wrong about the utility of these numbers. These imaginary numbers are used in physics to help us understand and model flows and dynamics of currents and fluids. So in fact, even things that, again, first appeared to be totally abstract and devoid of application and utility turn out to be descriptors of our every day world.

The name i was given by Leonhard Euler two centuries after Cardano's original work. We note that because i is not real, it cannot be a fractional number, because, of course, all fractions live on the real line. Hence, i is another example of an irrational number by it not, in fact, living on the line. However, the number i is an algebraic number, because it is the solution to the polynomial equation $x^2 + 1 = 0$.

Let's check that. If we let $x = i$, remember, i is $\sqrt{-1}$. If I take $\sqrt{-1}$ and square it, remember that the square root and square counteract each other and what we see here is just -1. Then $-1 + 1$ indeed equals 0. So, it is a solution. Therefore, i is an algebraic number.

Now, armed with the special number i, which is $\sqrt{-1}$, we can now define all the complex numbers. So, we can extend this notion yet further. The complex numbers are numbers that we can write as a real number plus real number times i, or, put more mathematically, $x + yi$, where x and y are real numbers from the number line. So, for example, $2 + 3i$ is an example of a complex number, as is $-3i$. These are just two examples; there are, of course, many, many examples of complex numbers, but this is the general form that they take.

Just as with real numbers, we can add, subtract, multiply, and divide complex numbers. So, we can perform arithmetic with this new collection for number. For example, let's compute i^2. Well, we know by its definition that i^2 actually equals -1, but what about higher powers of i? What about i^3? Well, i^3 would just be $i \times i \times i$. Well, $i \times i$ is -1, and $-1 \times i$ is $-i$, so i^3 is $-i$. What is i^4? Well, i^2 is -1, and if we square again to get the fourth power, we see 1. In other words, $i \times i$ is -1, $i \times i$ is -1, and -1×-1, we actually proved ourselves, is 1. What about i^5? Well, that would be $1 \times i$, which is i. So i^6 would be $i \times i$ yet again, which would be -1. In other words, what we're seeing here is that we return to the number i, and then we start to repeat. So, it's a repeating pattern when we take higher and higher powers of i. We see i, and then when we square it, we see -1, then $-i$, then 1, then i, then -1, then $-i$, then 1, then i, then -1, and so forth. It just keeps repeating as we take higher and higher powers of i. We cycle through these four numbers. Well, this repeating pattern actually continues, and we'll actually apply this really neat fact in the next lecture to discover something truly amazing.

How can we visualize these complex numbers? We recall that the real numbers can be viewed as the points on the number line. Well, it turns out that we can generalize this idea and visualize complex numbers by replacing the number line by a flat plane of numbers. So, here we actually see two number lines that meet up at right angles. The horizontal number line is referred to as the real number line, the one that we're used to, and this new vertical number line is referred to as the imaginary axis. And this figure is called the "complex plane," and we can actually visualize complex numbers by plotting their location in the plane. For example, let's consider the complex number $3 + 2i$. How do I find out where that point is in this plane? Well, the 3, the first number, tells me how far to go over in the horizontal direction on the real axis, so I move three units to the right. And then the $2i$ tells me how far up or down I should go. In this case, I go up two units, and that point that we see represents the number $3 + 2i$, and so there's that point represented in the plane. More generally, we can plot the complex number $x + iy$ by moving the number x to the real axis, so that's the horizontal—go over to the value x—and then go vertically up or down the y, and then that dot we see is $x + iy$.

The complex plane really is a generalization of the number line, but we notice that we need to include that extra direction, that up-down direction, for the new number i that we introduced. In fact, this vertical axis, as I said before, is sometimes referred to as the imaginary axis.

Bringing these ideas back full circle to the beginning of this lecture, we recall that our algebraic development of number evolved through our desire to find solutions to polynomial equations with integers. It was the 11th-century Persian mathematician Omar Khayyam who first formalized the notation of algebra by explicitly considering solutions to polynomial equations. He, unlike the ancient Greeks, viewed the irrational numbers as numbers themselves.

We've seen that we can even solve equations such as $x^2 + 1 = 0$, even though our solutions are no longer real numbers and no longer reside on the number line, and these new imaginary numbers are certainly no more imaginary than $\sqrt{2}$. Remember, that was a number which was imaginary to the Pythagoreans, so the phrase "imaginary" really is just one within our minds. But these numbers are just as abstract and just as real as the other numbers (although they're not real numbers as we define them in a mathematical sense).

Incredibly, with just the inclusion of this one new number, the imaginary number i, we were able to build all the complex numbers just using real numbers. And as we've seen, we need the complex numbers, because they are the solutions to certain equations. However, do we need more numbers? Do we need to expand even more? In other words, do we need to expand our notion of number further to insure that all polynomial equations of this form have solutions? In a surprising turn of events, it can be proven that just by including i and all the complex numbers $x + yi$, we've captured all the algebraic numbers. In fact, every polynomial equation has solutions of the form $x + yi$ in which x and y are real numbers. That's it. We have all the algebraics.

This incredible result was established in 1799 by the great child mathematics prodigy Carl Friedrich Gauss in his Ph.D. thesis. Gauss today is known as the prince of mathematics, because he's so great and made so many important contributions. Gauss's amazing and surprising insight shows us that there's no need to create any additional numbers after we include the imaginary number i to insure that all polynomials have solutions. We might note, however, that Gauss still found the number i to be very mysterious. In 1825, he wrote, "The true metaphysics of $\sqrt{-1}$ remains elusive." Even though he was performing work using these numbers, it still wasn't exactly clear what these numbers represented. The incredible fact that, once we include i, we need nothing more in order to describe solutions to polynomial equations can be expressed mathematically by saying that the complex numbers are "algebraically closed," meaning that all algebra questions that we could think of with polynomials can be solved in the world of the complex numbers. There is no further expansion. So, with i in hand, we have come to the end of the number line, or maybe I should say number plane, when we view the world of numbers with this algebraic lens. Of course, the reality is, as we'll soon we see, there are other lenses with which we can look at the world of number, and within those lenses, our view and vision of number can grow further still. Enjoy i.

The Five Most Important Numbers
Lecture 17

Here we focus on the five most important numbers in human history and see how they are, in fact, connected with one another.

The five most important numbers are 0, 1, π, e, and i. The numbers 0 and 1 have special properties. The number 1 is the multiplicative identity ($a \times 1 = a$). The number 0 is the additive identity ($a + 0 = a$). The numbers π and e have enabled us to understand our universe in a deeper way. The number i can be used to generate all the complex numbers. There is one equation that connects the numbers 0, 1, π, e, and i: $e^{\delta i} + 1 = 0$.

With these numbers, we can reach another view of the complex plane of numbers. Every complex number can be expressed in the form $x + yi$. We can visualize these numbers not on the (real) number line but as points on the complex plane. We now look at the complex plane with a different eye and see that points in the complex plane can be expressed in terms of an angle and a length. Every complex number corresponds to a right triangle in the complex plane.

We revisit some basic ideas from trigonometry involving right triangles and their angles. The ratio of various side lengths of a right triangle produces various trigonometric functions. The sine of one of the two non-right angles of a right triangle is the ratio of the length of the opposite side to the length of the hypotenuse. The cosine of such an angle is the ratio of the length of the adjacent side to the length of the hypotenuse.

Suppose we have a complex number $x + yi$ with the property that its associated right triangle in the complex plane has a hypotenuse of length 1 and an angle of A radians. The x value is equal to $\cos A$, and the y value is equal to $\sin A$. Extending this principle to "degenerate" triangles, we see that $-1 = \cos\pi + i \sin \pi$. Note that this formula captures the spirit of Cantor's work in writing functions as sums of sines and cosines. We now state an important result that follows from the ideas of calculus: $\sin A = A - \frac{A^3}{3!} + \frac{A^5}{5!} - \frac{A^7}{7!} + \cdots$, and $\cos A = 1 - \frac{A^2}{2!} + \frac{A^4}{4!} - \frac{A^6}{6!} + \cdots$.

A sum for e^x can be formulated. Given the previous infinite sums, we are reminded of the infinite sum for e^x. Recall that $e^x = 1 + \frac{x^1}{1!} + \frac{x^2}{2!} + \frac{x^3}{3!} + \frac{x^4}{4!} + \frac{x^5}{5!} + \cdots$. We now evaluate this sum for $x = \pi i$. Recall the powers of i: $i^2 = -1$, $i^3 = -i$, $i^4 = 1$, and so forth. We can rewrite our infinite sum into two parts: a real part with no i's and an imaginary part with an i. We notice that these two infinite sums equal trigonometric functions and conclude that $e^{\pi i} = \cos \pi + i \sin \pi$.

Revisiting the number $\cos \pi + i \sin \pi$ brings us back to our five numbers. We now recall that $\cos \pi + i \sin \pi = -1$. We put all these discoveries together and realize that $e^{\pi i} = -1$; equivalently, we have shown that $e^{\pi i} + 1 = 0$. Thus, the five most important numbers come together in one identity. This identity was first recorded in its modern form by Leonhard Euler in the 18th century, though English mathematician Roger Cotes had proved the result 30 years earlier. Euler was also the first to consider exponents with imaginary numbers in 1740, but in 1825 Gauss wrote, "The true metaphysics of $\sqrt{(-1)}$ remains elusive."

We can apply our amazing discovery to prove that e^π is a transcendental number.

We can apply our amazing discovery to prove that e^δ is a transcendental number. Let us assume for the moment that e^δ is an algebraic number (recall that i is an algebraic, irrational number). We conclude from the Gelfond-Schneider Theorem that the number $(e^\pi)^i$ is a transcendental number; however, we have just shown that $(e^\pi)^i = e^{\pi i} = -1$, and -1 is algebraic (not transcendental). We arrive at a contradiction; therefore, our assumption that e^π is an algebraic number must be false. We have just proved that e^π is a transcendental number. ∎

Questions to Consider

1. Locate the following complex numbers as points on the complex plane: $1 + i$, $2 - 3i$, $5i$, and $-0.5 - 2i$.

2. What is implied about mathematics, nature, truth, and the universe that the five most important numbers all come together in one elegant formula?

The Five Most Important Numbers
Lecture 17—Transcript

Before leaving the complex numbers, I'd like us to return to their geometrical universe, the complex plane. Here, we'll discover that we can describe complex numbers, $x + yi$, in an alternative geometric manner using just an angle and a length. Now, we can make this description precise by introducing a right triangle and resurrecting some trigonometry, but I promise you that we'll experience no trig tribulation, since a brief review of sines and cosines is all that we'll require.

Armed with this new way of writing complex numbers, we'll be ready to discover one of the most beautiful equations within all of mathematics, an identity that, in one equality, contains the five most important numbers: 0, 1, π, e, and i. This incredible result is known as "Euler's identity," is named after the great 18th-century mathematician Leonhard Euler. Although Euler is credited with first stating this identity, it was actually given to him and known to other people much earlier. This amazing identity, together with the Gelfond-Schneider Theorem that we discussed in Lecture Fifteen, will enable us to prove for ourselves that e^{π} is, in fact, a transcendental number. Thus, we leave the complex plane and our algebraic excursion into number, holding in our minds one of the most beautiful objects in nature, an equality that brings together the most important constants of our universe. We can all celebrate this formula while marveling at the beauty of both nature and number.

Here we focus on the five most important numbers in human history and see how they are, in fact, connected with one another. We first celebrate the two most important integers. We begin with unity, the Pythagorean number of reason, the number 1. Its mathematical importance comes from its unique role as the multiplicative identity. In other words, any number multiplied by 1 equals the any number. So, multiplying 1 by any number a preserves the identity of the number a. Next, we turn to 0, the number that took humanity hundreds of years to embrace as an actual number, but its true importance is the fact that it's actually the additive identity. If we take any number and add it to 0, we get the any number back again. Adding 0 to any number a preserves the identity of that number a.

We now honor the two most important real numbers, π and e. Beyond their transcendental qualities, we have seen that both π and e, through their nature, have allowed us to understand our universe in a deeper and richer way. Finally, we acknowledge the most famous number that is not real, $\sqrt{-1}$, also known as i. In our previous lecture, we discovered that the imaginary number i, in fact, has great importance, because we can use it to generate all the complex numbers, thus guaranteeing that we have solutions to all polynomial equations we could ever imagine.

Yes, we've studied and enjoyed the five most important numbers in mathematics, but now we wonder if mathematics and nature are so supremely beautiful as to bring together these five famous numbers into one profound equation. The amazing answer is yes. There is one equation that connects the numbers 0, 1, π, e, and i. In fact, that equation also captures the three fundamental arithmetical operations, addition, multiplication, and exponentiation. This fantastic formula is the following: $e^{\pi i} + 1 = 0$. Well, let's just take a deep breath here and take in this vision of number, all together in wonderful identity. You see the five numbers; you can see 0, 1, π, e, and i. You also notice that we have an addition, we have a multiplication (the π times i), and then the exponentiation ($e^{\pi i}$). So, we see it all in one equation.

So we can celebrate it, until we realize, exactly what does it mean to raise e to an imaginary power? So, of course, we have to make sense of this. What does it mean to raise e to a power that is imaginary? Well, we'll now dedicate ourselves to resolving this conundrum so that this amazing equation makes sense to us and then we'll see why this incredible formula really does hold. Along the way, we'll also discover many advances in our understanding of number, so numbers will certainly play a prominent role in proving this beautiful relationship.

To begin, let's recall how to visualize complex numbers in the complex plane. In the previous lecture, we saw how every complex number can be expressed in the form $x + yi$, where the x and y represent any real numbers at all. We can visualize these numbers not on the real number line, but as points in the complex plane. Remember, we saw that we had two sets of axes, a horizontal one, which is called the real axis, and then a vertical one that meets it perpendicularly, called the imaginary axis. They meet at right angles, and the entirety of the plane represents the plane of complex numbers. So for example, we could plot numbers by plotting their coordinates. So the number $4 + 3i$ would be arrived at by starting at the cross hairs, the origin, and then moving four units over in this horizontal direction, and then three units up in the vertical direction, and that point right there isolates the point $4 + 3i$. In general, we could look at $x + yi$, by moving over the appropriate number of units in the horizontal direction, x, and then moving up and down the appropriate number of units, y, and that point is $x + iy$.

We now look at this plane of complex numbers with a different eye. If we have the complex number, $x + yi$, marked on the plane with a point, then we can locate that point by measuring an angle and a length, and I want to describe how. Well, suppose that we have a dotted red line that's lying on the positive real axis, the positive, horizontal axis, and that line is actually hinged right at the intersection of the two axes, sometimes called the origin. So, it's like a spinner; it can go up and down like this, but actually it's going to only go counter clockwise, like in this direction, so it's only going to go this way. So, in other words, this dotted red line is more like a second-hand on a clock, but only going counterclockwise. Well, now let's imagine pivoting that red line up counterclockwise until it passes through the point that represents our complex number. Let's note that the red line will determine a specific angle. So, I just bring this line up from the horizon until I hit the point, and then once I hit that point, I can measure this angle.

Then we can actually measure the length from the cross hairs where the two axes meet out to the point along this red dotted line. So, I could measure any particular complex number, or point in this plane, by just giving the angle I go up and then how far I go out. And I can identify complex numbers in that fashion as an angle and a length. For example, let's consider the complex number $1 + i$. Well, first of all, let's plot it. So, if you imagine the plane right here, I go 1 unit over and 1 unit up, and so there's the point right there. Now, how would I measure that? Well, of course, I would start here at the horizontal axis, and I would tilt up until I get there and measure that angle. Well, since I went 1 over and 1 up, that point is exactly halfway in between the two axes which meet at 90 degrees, and so the angle must be half of 90 degrees, which is 45 degrees. So, I see the angle is, in fact, 45 degrees.

However, it's important here to remember that we are using the mathematical measure of angle, not the familiar degree measure, but as we saw in Lecture Thirteen, we're going to use the so-called radian measure of an angle. Radians contain the number π in it, and in fact, this is how we're going to get π into our famous formula. Once around, 360 degrees, equals 2π, halfway around, 180 degrees, equals π, one-quarter around, which is 90 degrees, equals $\pi/2$, and so therefore half of that, 45 degrees, would be $\pi/4$. So, that's how many radians, $\pi/4$ radians is that angle. Therefore, our angle is one $\pi/4$ radians.

Now the question is, what's the length from the center out to here, where I went one over and one up? Well, that length we actually computed in Lecture Eight, because notice that we actually have a square that's a 1-by-1 square, and so—a 1-by-1 square—this is, in fact, the diagonal, and we computed that to be $\sqrt{2}$. So, we see, in this case, we can describe this point here as "go up an angle of $\pi/4$ radians and then go out $\sqrt{2}$, and you hit the point."

So, any complex number $x + yi$ can be expressed, or at least its location can be expressed, in the complex plane as an angle and a length in this fashion. In fact, this angle and that red line segment can be used to form a right triangle in the complex plane by simply drawing a perpendicular line from our point down to the horizontal real axis. And all of the sudden we have a nice right triangle. Notice that, in this right triangle, the side of the horizontal equals the value x and the length of the vertical side of the triangle, in fact, is the value y. However, I want us to express these two lengths, the x and the y length, in terms of just the angle and the length of that red side, which is now known as the hypotenuse of this right triangle.

In order to express these lengths, we'll apply some tools from trigonometry. Now, we won't have to worry about all the technical details, so don't worry if you're not all up to speed on your trigonometry. But I do want us to briefly outline the basic ideas from trigonometry involving right triangles, and their angles, so that we can appreciate the formula. Well, back in our math classes, we were told that the ratios of various side lengths of a right triangle produce various trigonometric functions. For example, the sine of an angle of a right triangle is defined to equal the ratio of the length of the opposite side to the length of the hypotenuse. In other words, the shorthand way is, "sine is opposite over hypotenuse." The cosine of an angle, on the other hand, is defined to equal the ratio of the length of the adjacent side to the length of the hypotenuse, sometimes read "adjacent over hypotenuse."

In our case, to make things a little bit easier, let's now suppose that we have complex number $x + yi$ and that the right triangle it forms in the complex plane has hypotenuse of length 1. So, when we plot that point, the distance away from the cross hairs is 1. And let's call the angle that it makes with the horizontal axis A radians. So, this length is just 1, and this is A. Well, we can now see that the cosine of A equals the adjacent over the hypotenuse. The adjacent is x, and the hypotenuse is 1. So, we have x over 1, so it just equals x. The sine of A equals the opposite length, which is the y divided by the hypotenuse, which is 1. So, we see $y \div 1$, and we see it equals y. Well, this is really neat, because x equals the cosine of A and y equals the $\sin A$ in this example. So, we can now say this in a different way. We can say the complex number $x + yi$ is actually equal to $\cos(A) + i\sin(A)$, because of course, $\cos(A)$, we just saw, was x and $\sin(A)$ is y. We usually write $\cos(A) + i\sin(A)$. We write the $\sin(A)$ second only to not get it confused with the i and the A together, but it means i times $\sin(A)$. In other words, given any complex number $x + yi$ whose distance from the axes cross hairs is 1, there is an angle. If you measure that angle, and we call that angle A radians, then we see that $x + yi = \cos(A) + i\sin(A)$.

Now, here's the cool move. Let's now swing this observation around to create a degenerate triangle. Now, if we rotate our point around a circle counterclockwise, so just keep bringing it up, so here's the cross hairs and here's that point, length 1, and just keep rotating it around until it hits the horizontal axis, where's it going to hit? Well, it has length 1, so it's going to hit exactly at this side of the line, which is -1. So, it sweeps out and hits at -1. Well in this case, what's the angle that we swept out? Well, we went halfway around, and so that's going to be 180 degrees, or, in radians, we see it's π radians. Let me just note that our formula that we just derived still will hold even in degenerate triangles, even if the triangle kind of collapses on itself like it does here. So in this case, we see that we can say that the complex number, which is just the point -1, equals the cosine of the angle, which is π, plus i times the sine of the angle, which is π. So we see that $-1 = \cos(\pi) + i\sin(\pi)$.

Notice that hidden in this equality, we see the numbers 1, π, and i. Let's remember that we're trying to connect the numbers 0, 1, π, e, and i into one equation. So, we're just missing the numbers 0 and e, and now we have to try to get them into the picture. As a little aside, let me just say that this formula actually captures the spirit of Georg Cantor's work in writing functions as sines and cosines (sums of them) that we mentioned in Lecture Eleven.

To introduce the number e into our story, I want to bring back a formula that we saw for ourselves in Lecture Fourteen. It was a formula that expressed e to any power at all, e^x, and it said that e^x was an endless sum which read $1 + x^1/1!$ $+ x^2/2! + x^3/3! + x^4/4!$ and so forth. Again, don't worry about the details of how it looks, but notice the pattern that, in fact, the exponent and the denominator look the same, and that factorial means that I take that number and multiply by the previous natural number, and the previous natural number, until I get to 1. For example, 4! means $4 \times 3 \times 2 \times 1$, which equals 24.

Now here's the amazing moment, but this going to take just a teeny bit of algebra, and we're going get technical for literally a moment. Now, feel free to either follow every detail or just sit back and enjoy the big picture as you wish as the mathematical story unfolds. But in either case, the algebra details here will only last for a minute, I promise.

Using some deep ideas from calculus, we can find formulas for both the sine and the cosine that is similar to the one we have for e^x, and this is the key connection that links the trigonometric functions to our famous number e. First, for sine of A, we could write that as $A^1/1! - A^3/3! + A^5/5! - A^7/7!$ and so forth. Similarly, for the cosine of A, we could write that as $1 - A^2/2! + A^4/4! - A^6/6!$ and so forth. Notice there's a pattern here. First of all, in each case, the sines alternate from plus to minus, and in the sine of A, we see odd numbers appearing, and in the cosine of A, we see the even numbers appearing.

These formulas are extremely important, profound, and by no means obvious. Notice that, if we add those two formulas together, we almost get e^a, but not quite, since the negative signs get in the way and snare things up a little bit. So, we're very close, but this is where the number i comes to our rescue. Let's make sense out of the number $e^{\pi i}$. The way we'll do that is we'll plug πi in for x in our formula for e^x. We see something that's going to look a little bit unwieldy, but just watch it and enjoy it and watch how things, all of the sudden, get very simple after they get complicated.

We'd see $e^{\pi i} = 1 + (\pi i)^1/1! + (\pi i)^2/2! + (\pi i)^3/3! + (\pi i)^4/4!$, and the pattern continues forever. Now, we can actually simplify that a little, teeny bit, because when we have $(\pi i)^2$, for example, then we know that that actually equals $\pi^2 i^2$. We can actually square each of those and then multiply them together. So, we could rewrite $e^{\pi i} = 1 + \pi^1(i)^1/1! + \pi^2(i)^2/2! + \pi^3(i)^3/3! + \pi^4(i)^4/4!$ and so forth. All I did there was take the πi to a power and just split it up into π to the power, i to power, which is completely legal in the world of arithmetic.

Now, let's remember something fantastic: powers of i repeat. So we have i, and then $i^2 = -1$, $i^3 = -i$, and i^4 gets us to 1, and then we repeat. i^5 gets us i, and so forth. So, as we raise to higher and higher powers in i, we see i, -1, $-i$, 1, i, -1, $-i$, 1, and so forth. It repeats; it cycles. And if we use that fact and then replace all the is to powers by what they're equal to, we see the following. We see that our long expression becomes $e^{\pi i} = 1 + (\pi)^1 i/1! - (\pi)^2 / 2!$ (because i^2 is -1) $- (\pi)^3 i/3! + (\pi)^4/4!$ and so on. Again, it looks pretty complicated and threatening, but watch what happens now.

Notice that every other term in this sum has an i in it. So, let's reorganize the sum and write the first terms with no i's at the beginning, so first write the terms with no i's, and then write all the terms with one i at the end. And so, now, if we factor out the common factor of i from all the terms in the second sum, then we see something interesting. (By the way, we could factor out that i just using the good old distributive law in reverse. So, it could be done, it's not a problem.) But now look what we have. In the top sum we see only even numbers and the signs alternate plus/minus. In the bottom sum, we see only odd numbers, and again, the signs alternate plus/minus.

In other words, forget about all the derivation. Just look at these things visually, and you'll see that these are precisely the previous formulas that we saw for sine and cosine. So, the first actual sum is the sine and the second sum being multiplied by the i is the cosine. Well in this case, the A in our formula is π, since we have πs running around. Well, pulling all these ideas together, we see that $e^{\pi i} = \cos(\pi) + i\sin(\pi)$, again just using the formula that we derived before.

The algebra is over; you've survived, and now it's back to the fun. Let's now remember what we've learned from the right triangles and trigonometry. We saw for ourselves, by swinging that point way around, that $\cos(\pi) + i\sin(\pi) = -1$. We saw that in the complex plane. Well, this is wonderful, because if we pull all these discoveries together, we realize that $e^{\pi i} = -1$. And maybe you can see it coming. All we have to do is add 1 to both sides, and we come upon beauty: $e^{\pi i} + 1 = 0$.

We just derived for ourselves the most beautiful formula in all of mathematics and in all of nature. This is a moment to celebrate. So, feel free to celebrate, if you want. You've earned it. Thus, we proved that the five most important numbers come together in one identity involving the three most important arithmetic operations. This identity was first recorded in its modern form by Leonhard Euler in the 18th century, although English mathematician Roger Cotes had proved the result about 30 years earlier.

Euler was also the first to consider exponents with imaginary numbers as we did in this lecture, and this was in 1740. But as we remarked in an earlier lecture, as late as 1825, the great mathematician Gauss wrote that "the true metaphysics of $\sqrt{-1}$ remains elusive." So, we can see that, again, trying to wrap our minds around the imaginary number i takes an awful lot of time. Today, this formula is considered to be the most beautiful equation in all of mathematics.

The great 20th-century Nobel Prize–winning physicist Richard Feynman once called it "the most remarkable formula in mathematics." Nearly 100 years before that, the Harvard mathematician Benjamin Peirce, after proving this formula in a lecture, just the way we did in fact, said the following after he gave the proof to the group:

Gentlemen, that is surely true, it is absolutely paradoxical; we cannot understand it, and we have not the slightest idea what the equation means. But we have proved it, and therefore we know it must be the truth and we may be sure that it means something very important.

It's a beautiful, beautiful quote here where it's certain, it's mathematically correct, we proved it, but we have no idea what it means. Who knows?

However, given that all these wonderful numbers are coming together, it's got to mean something important, even though we don't know what it means yet. The truth is that not every formula must have a profound application in order for us to appreciate it. We can appreciate its beauty just as we'd appreciate a piece of music or a work of art. However, in mathematics, quite often what first appears as abstract and devoid of any practical value is later discovered to be the central piece of some important mathematical puzzle.

So, I thought we would actually close this lecture by seeing that Pierce's intuition was, in fact, correct. I'm going to now describe a tremendous number-theoretic consequence of this beautiful formula that was discovered nearly 100 years after Pierce made his famous remarks about "I'm sure it's important, but I don't know how." We'll actually use Euler's formula to prove that e^π is a transcendental number, which means that not only is e^π an irrational number, but we'll actually be able to know that it is not the solution to any polynomial equation involving integers. Now, there are two basic pieces to our argument. First is the famous identity that we just proved, $e^{\pi i} + 1 = 0$, although the version we'll look at is the one we did right before we actually proved it, we had $e^{\pi i} = -1$. And second, the important fact will be the Gelfond-Schneider Theorem.

So let's take a moment to recall the Gelfond-Schneider Theorem from Lecture Fifteen. It states that if we have an algebraic number that's not 0 or 1, and we raise it to an irrational algebraic number, then the result of this number raised to that number will be a transcendental number. Well, let's now show that e^π is transcendental indirectly. In other words, we'll assume that e^π is not transcendental, which implies that it's an algebraic number. Well, let's recall that from Lecture Sixteen we saw that i is, in fact, an algebraic, irrational number. Remember, i is one of the solutions to the polynomial $x^2 + 1 = 0$. So, it's algebraic, and since i is not a real, it must be irrational.

Given our assumption that e^π is algebraic, I can take e^π, an algebraic number, and raise it to the i, which is an algebraic, irrational number, and then by the Gelfond-Schneider Theorem, I know that this quantity, in fact, must be a transcendental number. But $(e^\pi)^i$ is actually the same thing as $e^{\pi i}$, which we just saw was -1, and -1 is an algebraic number. It's the solution, in fact, to $x + 1 = 0$, so it's certainly not transcendental. So, we just contradicted the

Gelfond-Schneider Theorem. So, what's the problem? Well, our assumption that e^π is an algebraic number must be false. That was the only thing we assumed. In other words, we have just proved that e^π is not algebraic, therefore it's transcendental.

Now, remember that we've already seen that both π and e are transcendental numbers. However, there are so many simple combinations involving π and e, as we noted in Lecture Fifteen, and for most of those combinations, we don't know if any of these numbers are irrational, let alone transcendental. But here we see a natural combination of π and e, e^π, for which we know the whole story, all because of our amazing identity.

So, we see that this amazing, beautiful formula, $e^{\pi i} + 1 = 0$, does have important applications after all. In particular, it allows us to discover that e^π is a transcendental number. Thus, it helps us see more detail into the jungle of numbers, but more importantly, we can look upon this formula with great delight and appreciate the beauty and the majesty that is the five most important numbers, the three most important operations, all together, singing in unison.

An Analytic Approach to Numbers
Lecture 18

Here we'll examine the real numbers from an entirely different vantage point. We'll construct the real numbers from the rational numbers in an *analytical* manner—that is, we develop ideas that are inspired by one of the most fundamental applications of numbers, measuring how close two objects are by computing the distance between them.

By the 1800s, there was a great desire among mathematicians to set the notions of mathematics—especially the ideas of calculus—on more rigorous ground. Intuitive and geometrical arguments became less attractive as a desire for precision grew. This desire led to a serious interest in finding the precise meaning of "real number." In 1817, the Czech mathematician Bernhard Bolzano saw this need for a precise definition of real numbers. Although he was not as influential as others to come, his work in this direction inspired some to call him the "Father of Arithmetization."

Mathematicians started with a world of rational numbers—which was completely understood—and wanted to use these familiar numbers to define all real numbers. In the 1820s, Augustin-Louis Cauchy developed a reasonable-sounding definition of real numbers as a limiting process using the rational numbers. Although his work was a major step forward, it lacked the rigor to satisfy some of his contemporaries. Karl Weierstrass, now known by many to be the "Father of Modern Analysis," declared all unending decimal expressions to be called *numbers*. This, for the first time, put the real numbers on solid footing.

Richard Dedekind and Georg Cantor (who was one of Weierstrass's Ph.D. students) were also working at making the real numbers precise. In the late 1800s, they articulated an idea that we have been taking for granted: Points on a line can be placed in a one-to-one correspondence with real numbers. This is called the Cantor-Dedekind Axiom. This idea can be made more precise and leads to a method of defining real numbers that is now known as "Dedekind cuts."

A modern view of real numbers further expands our notion of number. We find sequences of rational numbers for which the numbers get arbitrarily close to one another. Such sequences are called Cauchy sequences. Sometimes these numbers head toward a rational number and sometimes they do not (e.g., 1, 1.4, 1.41, 1.414, 1.4142, 1.41421, 1.414213…). In the previous example, we see a "hole" in which a number should reside; however, that value is not rational.

If we extend our notion of number from just the rational numbers to the rational numbers together with these undefined holes, then we produce the real numbers and the continuum that we can view as a line. This description represents our current precise view of real numbers. We say we *complete* the rational numbers and doing so produces the real numbers. The 19th-century German mathematician Leopold Kronecker strongly believed that all numbers should be derived from the natural numbers and zero. He thought that rational numbers, irrational numbers, and complex numbers were derived from false mathematical logic.

In all the 19th-century attempts to give a precise definition of real numbers, we encounter the notion of distance. Terms in a sequence of rational numbers get close together, implying a means of measuring closeness. We measure closeness by using the absolute value, denoted | |. The absolute value of a number is its distance from 0 (e.g., $|4| = 4$, $|-5| = 5$). To determine how close two rational numbers are, we take the absolute value of the difference between those two numbers. For example, the distance between 4 and 7 is 3, and this fact corresponds to: $3 = |4 - 7|$. The precise definition of the real numbers depends fundamentally on the notion of distance or, more specifically, on the absolute value.

An absolute value is any formula that observes certain rules. The absolute value of 0 must equal 0, and the absolute value of any nonzero number must be positive. The absolute value of a product of two numbers is equal to the product of their absolute values. For example, $|(4) \times (-3)| = 12$ and $|4| \times |-3| = 12$. The shortest distance between two points is a straight line; put

In all the 19th-century attempts to give a precise definition of real numbers, we encounter the notion of distance.

mathematically $|A + B| \le |A| + |B|$ (this is called the triangle inequality). For example, $3 = |5 + -2| \le |5| + |-2| = 5 + 2 = 7$.

There are other absolute values on the rational numbers that measure distance in an arithmetic sense. The p-adic absolute value is a new absolute value. To illustrate this new type of absolute value, let us consider the special case in which the prime p equals 3. We define the 3-adic absolute value (denoted as $|\ |_3$) first on the natural numbers as follows: Given a natural number N, the 3-adic absolute value of N equals the reciprocal of the highest multiple of 3 that divides evenly into N. For example, $|12|_3 = |2 \times 2 \times 3|_3 = \frac{1}{3}$, $|36|_3 = |2 \times 2 \times 3 \times 3|_3 = \frac{1}{9}$, and $|100|_3 = 1$. If the number N is not evenly divisible by 3, then its 3-adic absolute value is defined to be 3^0, which as mentioned earlier in this course, equals 1.

For any prime number p, we define the p-adic absolute value (denoted as $|\ |_p$) on the natural numbers as follows: Given a natural number N, the p-adic absolute value of N equals the reciprocal of the highest multiple of p that divides evenly into N. Note that $|0|_p = 0$; for negative numbers, we just ignore the negative sign. We can find the p-adic absolute value of fractions (that is, of rational numbers) by simply dividing the p-adic absolute values of the numerator and the denominator. For example, $\left|\frac{8}{45}\right|_3 = \frac{|8|_3}{|45|_3} = \frac{1}{\frac{1}{9}} = 9$. ■

Questions to Consider

1. Compute the following p-adic absolute values when $p = 2$: $|4|_2, |12|_2, \left|\frac{1}{6}\right|_2, |0|_2, \left|\frac{24}{25}\right|_2$.

2. Why were number theorists so obsessed with rigor and precision in defining the real numbers?

An Analytic Approach to Numbers
Lecture 18—Transcript

Here we'll examine the real numbers from an entirely different vantage point. We'll construct the real numbers from the rational numbers in an *analytical* manner—that is, we develop ideas that are inspired by one of the most fundamental applications of numbers, measuring how close two objects are by computing the distance between them. This important approach, first considered by Augustin-Louis Cauchy in the early 1800s and later made precise by Karl Weierstrass, Georg Cantor, and Richard Dedekind, not only led to a deeper understanding of real numbers, but also allowed scholars to make many ideas from other areas of mathematics, including calculus, more exact and arguments more rigorous. This seminal work was the inspiration and starting point for a modern branch of mathematics known today as analysis.

At its very core, finding the distance between two numbers requires us to find the so-called absolute value of the difference between those two numbers. Using this simple idea, if we assume that our entire "universe of number" is the collection of rational numbers, that is, the collection of fractions, then we discover that there are lists of numbers whose terms are getting closer and closer together, and yet the value they are heading towards is, in fact, not a number, that is, not a fraction. In other words, the collection of rational numbers has, in some sense, "holes" within it. We'll then construct the collection of real numbers, and thus expand our notion of number, by filling in these holes. This construction leads us to a hole-free, continuous real number line. We'll close this discussion with a very strange question. Are there other measures of distance, in other words, absolute values, on the rational numbers? The surprising answer will lead us to an extremely important, but entirely strange, new universe of number.

Let's open our discussion with a historical look at the construction of the real numbers. Now, as we discussed in Lecture Eleven, by the 1800s there was a great desire among mathematicians to set the notions of mathematics, especially the ideas of calculus, on more rigorous grounds. The intuitive and geometric arguments used in that day were becoming less and less attractive as a desire for precision grew. This desire led to a serious interest in finding the precise meaning of the term "real number." In 1817, the Czech mathematician Bernhard Bolzano saw this need for a precise definition of the real numbers. Although he was not as influential as others to come, his work in this direction inspired some to call him the "Father of Arithmetization."

The basic goal was to begin with the world of rational numbers, the world of fractions that we all understand, and then use these familiar numbers to rigorously define all the real numbers, including the irrational ones. In the 1820s, Augustin-Louis Cauchy, the prolific French mathematician who wrote over 780 journal contributions, developed a reasonable-sounding definition of real number as a limiting process using rational numbers. Let me try to explain this in my own words just so you get the flavor for the idea. You'll notice that it's actually slightly subtle, because it's not going to be quite right. He says, Let's take a look at a list of rational numbers, a list of fractions, which I'll just denote as terminating decimals that are getting closer and to each other. For example,

3.1
3.14
3.141
3.1415
3.14159

And so forth. You see this list coming out. In fact, what's actually happening here is I'm taking the successive digits of π, but let's look at this list of numbers. Notice that these numbers themselves are getting closer and closer to each other, for if you take the difference of any two consecutive numbers on this list, they'll differ in the far, far rightmost spot, which, of course, means it's getting smaller and smaller. These numbers are getting closer and closer together.

But unfortunately, if our world of number is just the world of rational numbers, then we run into a problem, because this limiting number that we're approaching is, in fact, π, which we know is irrational. So, what Cauchy said was, Well, all these numbers are getting closer and closer to each other so the number that these numbers are heading towards—let's just call that a real number. In other words, let's take the target that we're heading toward in this list of fractions and just say that target now is a real number. Well, this is a great idea, except for the fact that the moment he says, We'll take that target and call it a real number, which implies that we have the target in hand, which, at the moment, we don't. So in fact, this wasn't quite rigorous and precise enough.

Now, while his work was a major step forward, it did lack this rigor to satisfy some of his contemporaries. But soon afterwards, Karl Weierstrass, now known to many to be the "Father of Modern Analysis," put Cauchy's wonderful ideas on solid ground by, yet again, expanding the notion of *number*. And here's what he said. So, here's how he fixed the problem. He said, You know what? This collection of numbers are getting closer and closer together, but in fact, they're not heading toward anything. So, let's just declare, right now, the real numbers to be just anything that's an infinitely long decimal expansion. So formally, any time you put down a dot and then write down infinitely many decimal digits, and of course, we know that they're rational precisely when we have a periodic repeating part or a tail of just zeroes. But even if we don't have a tail of zeroes, I don't care, it's just an object. It's a formal object, and I'm going to declare that's what I mean by a real number. So, he fixes Cauchy's little, teeny issue by just declaring that that formal thing, that decimal thing, will be a number. He defines it to be so.

So, in fact, what Weierstrass does is he declares that all unending decimal expansions will be numbers, and for the first time, he puts the real numbers on solid footing. Now Weierstrass, by the way, was a very successful secondary school teacher before he started to publish important mathematical papers in the area that would later be known as analysis, and then he soon moved to university.

During this same period, two other German mathematicians, Richard Dedekind and Georg Cantor (Cantor, by the way, was one of Weierstrass's Ph.D. students), were also working on making the real numbers precise. In the late 1800s, they articulated what we've been taking for granted throughout this course—that the points on a line can be placed in a one-to-one correspondence with the real numbers. That is, when we look at the number line, the fact that we call it a number line is exactly what Cantor and Dedekind noticed, and in fact, sometimes this fact is known as the Cantor-Dedekind Axiom—that, in fact, the number line exists. As we noted in Lecture Nine, this connection was considered first by the 16th-century Flemish mathematician Simon Stevin, who considered a continuum of real numbers, which he so beautifully and poetically described as "flowing like a river of magnitudes." Here we see that this idea can be made very precise, and that's what these mathematicians of the 1800s were doing.

In fact, the Weierstrass version of real number can be made even more rigorous, and this leads to a method of defining real numbers that are known today as "Dedekind cuts." They were named in honor of Richard Dedekind, who derived the method, although we won't describe them here.

These 19th-century ideas have led to our modern view of the real numbers, and I want to outline how we consider the real numbers today within the world of mathematics. Let's just begin by starting with the view that number consists solely of the rational numbers, and let's study how those numbers cluster among themselves. We find sequences of rational numbers, that is, lists of fractions, for which those numbers are getting arbitrarily close to one another. That is, as we go down these lists of fractions, we see that the fractions are getting progressively closer and closer to each other. For example, consider the sequence 1/2, 1/3, 1/4, 1/5, 1/6, and so on, forever. Notice that, by just going down our list far enough, we can insure that all the subsequent fractions on this list are as close to each other as we wish, and so in fact, they're coming closer and closer to each other. Well, such sequences are called today Cauchy sequences, of course named in honor of Cauchy, who was the first to describe this idea.

Now, remember that our entire pretend world of number consists solely of the fractions, the rational numbers. Within this hypothesized world, Cauchy sequences either are heading toward a target, namely a number, a rational number, a fraction, or they're not heading toward a rational number. For example, the Cauchy sequence we just looked at, 1/2, 1/3, 1/4, 1/5, and so forth, is actually heading to the number 0. Notice that 0, being a rational number, is indeed a number in our current context. However, if we consider the Cauchy sequence of rational numbers expressed in decimal form in the following list, we're going to see something a little bit different. Let's consider:

1.4
1.41
1.414
1.4142
1.41421

And so on. Well, it turns out that these decimal fractions are getting closer and closer to each other, because they're agreeing further and further out, and yet they're not heading toward a fractional number. Their target is a non-repeating decimal, which means it's not a rational number, and that decimal is 1.41421356237, and it goes on (which, as an aside, is the decimal expansion of $\sqrt{2}$, a number that we've already proven is not rational).

Through this example, we actually see a hole in which a number really should reside, because these rational numbers are getting closer, and closer, and closer together, but where they're actually heading toward is missing, because it's not a fraction. Well, these holes, in fact, as you may remember, confounded the ancient Greeks. As we noted earlier in the course, the ancient Greeks viewed the rational numbers as a gapless list of numbers until they discovered the irrationality of $\sqrt{2}$. This conundrum remained unresolved until around 360 B.C.E., when Eudoxis offered his definition of irrational number. He defined irrational numbers in terms of an infinite list of rational numbers. Notice how beautifully this notion foreshadowed, 2,000 years earlier, the formal idea of Cauchy sequences of rational numbers. It's really remarkable when you think about it.

So we are faced with these holes, points around which rational numbers accumulate, but are themselves not numbers in this sense. If we extend our notion of number from just the rational numbers to the rational numbers together with these now-undefined holes, then we produce the real numbers and the continuum that we can now view as a line. In other words, with just the rational numbers, we have a Swiss cheese–like number line, riddled with holes. The holes are, in fact, the ever-present irrational numbers, and when we include them, that is, we fill in those holes, we produce the continuous, unbroken real number line that we know and love.

This perspective represents our current precise view of real numbers. In mathematical terms, we have some language for this. We say that we *complete* the rational numbers. That is, we fill in the holes. So, by filling in the holes and producing the real numbers, we say it's the "completion" of the rationals. That's just the language that we use in the math biz. In fact, Dedekind himself acknowledged this evolution of number that we just experienced. In fact, he wrote:

> Just as negative and rational numbers are formed by new creation, and as the laws of operating with these numbers must and can be reduced to the laws of operating with positive integers, so we must endeavor completely to define irrational numbers by means of the rational numbers alone.

So, we really get a sense that this is evolution, and here Dedekind himself acknowledged that we're pushing forward from the numbers we have, and the arithmetic, to this new world. We used the old numbers to build the new numbers in this fashion.

Now, this new idea was not unanimously embraced by all mathematicians of the math community of the time. The influential and respected 19th-century German mathematician Leopold Kronecker strongly believed that all numbers should be derived from the natural numbers and 0. He thought that rational numbers, irrational numbers, and even complex numbers were derived from false mathematical logic. In fact, his famous quote remains, "God made the integers, all else is the work of man." He meant, very much like the Pythagoreans, if you remember, where they thought of the natural numbers of God-given, we see that Kronecker, a millennia later, still has this point of view.

In fact, there's a great story. Remember that Lindemann was the mathematician who proved the amazing result that π is an irrational number. Well, in fact, when Lindemann offered his remarkable proof of the irrationality of π, Kronecker wrote him a letter. He wrote, "Of what use is your beautiful research on the number π? Why cogitate over such problems, when really there are no irrational numbers whatever?" It's remarkable. Here, he was receiving celebratory notes from all over the math community for this triumphant result about proving the irrationality of π, and here is good old Kronecker saying, "Well, they don't even exist. So why are you wasting your time with such things?"

Actually in the last part of this course, we'll see Kronecker's deep objections in an even more dramatic fashion against Georg Cantor in Cantor's revolutionary work on infinity. So, we'll see Kronecker later on in the course, and he's just as upset—in fact, more so. Today, all serious mathematicians can prove that irrational numbers exist and view them as real as the natural numbers themselves. So, the view of Kronecker is now old-fashioned, obsolete, and we've moved on.

I want to now focus on the main character in this story of the real numbers that, up to this point, has been totally overlooked. In the 19th-century attempts to give precise definitions for the real numbers, we encounter the notion of distance. Now when we say that the values in a sequence of rational numbers are getting close together, implicitly we must be measuring that closeness. Things are coming together; we should be able to measure that shrinking.

We measure closeness by using what's called the absolute value, usually denoted by two vertical lines. The absolute value of a number is merely the distance away from 0 that it is. So for example, |4| is just 4, because we measure the distance between 0 and 4. And | −5| is 5, because we just measure the distance from 0 to −5. For computational purposes, really, you don't even need to look at the number line. All you need to do is just look at the number, and if it's negative, strip off the negative sign, and that's the absolute value. To determine how close two rational numbers are, we take the absolute value of the difference between those two numbers. So, for example, the distance between 4 and 7 is 3, and this fact corresponds to |4 − 7|. Well, $4 − 7 = −3$, and |−3| is 3, so we arrive at the same answer.

The precise definition of the real numbers depends fundamentally on the notion of distance, or more specifically, on the notion of absolute value. Given its importance, I want us to consider the absolute value and make an absolutely surprising discovery. An absolute value, by definition, is any formula that observes certain rules. First, the absolute value of 0 must be

0. The distance from 0 to itself, of course, is 0. The absolute value of any nonzero number must be positive, because it represents distance.

Second, and this gets a little bit more arithmetical, the absolute value of a product of two numbers must always equal the product of their absolute values. So, let's look at an example. Let's find the absolute value of the quantity $4 \times (-3)$. Well, $|(4) \times (-3)| = |-12|$, and $|-12|$ is 12. Now, on the other hand, let's consider the $|4| \times |-3|$. So $|4| = 4$, and $|-3| = 3$. Then 4×3 is again 12, so we see that the absolute value of the product equals the product of the absolute values. Put informally, the absolute value is very sympathetic to multiplication. You can do the multiplication inside or outside, and you get the same answer.

Lastly, and most importantly, the absolute value conforms to the famous adage, "The shortest distance between two points is a straight line." We put this mathematically with an inequality that's called the triangle inequality, and it reads as follows. If we take the absolute value of a sum of two numbers, that quantity will always be less than, or possibly equal to, the individual sums of the absolute values. So put mathematically, we'd say $|A + B| \le |A| + |B|$. Let's look at an example to see this make sense. Let's let A be 5 and B be -2. Well, then $|5 + (-2)|$ would be $|3|$, which equals 3. On the other hand, let's find the absolute values individually. So $|5| = 5$, $|-2| = 2$, and $5 + 2 = 7$. And notice that 7 indeed is larger than 3, so $3 \le 7$, as the triangle inequality asserts. This inequality is called the triangle inequality, because it's related to the lengths of sides of triangles, but that's of no consequence here. The important thing is it allows us to think of the absolute value of a way of measuring distance.

In view of our mathematics studies in high school, we might guess that there is only one absolute value for the rational numbers. In other words, there's only one way to measure the distance between rational numbers. However, as we're about to see, this seemingly plausible assertion is, in fact, false. There are other absolute values for the rational numbers that measure distance, but in an arithmetic sense rather than a geometric sense involving a ruler.

These strange new absolute values are called *p*-adic absolute values. The letter *p* here stands for a prime number *p*, and they're called the *p*-adic absolute values because they measure the size of numbers based on how many times the prime number p divides evenly into that number. The basic idea is the more factors of p a number has, the smaller its p-adic absolute value. Well, to make this strange and very counterintuitive idea more precise, let's fix a prime number p in our minds. So, for example, let's consider the prime number 3. Although what we're about to do will work with any prime number at all, for now let's just describe the formula for the 3-adic value on the natural numbers.

We write this absolute value as $| \ |_3$. The subscript 3 tells us this is a 3-adic, not the usual absolute value. Here are the rules. If a natural number is not a multiple of 3, then its 3-adic absolute value is equal to 1. So, for example, $|2|_3 = 1$, because 2 is not a multiple of 3, and $|7|_3 = 1$, because 7 is not a multiple of 3. And even a number like 100—the $|100|_3 = 1$, because 100 is not a multiple of 3.

Moving on. If a natural number contains exactly one factor of 3 when it's written out as a product of prime numbers, then its 3-adic absolute value equals 1/3. For example, $|3|_3$ itself would be 1/3, one factor of 3. And what about 12? Well, 12 equals $2 \times 2 \times 3$—one factor of 3, so $|12|_3 = 1/3$. And what about a number like 300? Well, 300, if we were to factor it, is $2 \times 2 \times 3 \times 5 \times 5$. It's 3×100. Well, there's one factor of 3, so $|300|_3$ is just 1/3.

If a natural number contains exactly two factors of 3 when it's written as a product of prime numbers, then its 3-adic absolute value equals 1/9. For example, $|9|_3 = 1/9$, and $|36|_3 = 1/9$ because we have exactly two factors of 3 in 36 ($2 \times 2 \times 3 \times 3$). And 72, which is $2 \times 2 \times 2 \times 3 \times 3$—since there's two factors of 3, $|72|_3 = 1/9$.

More generally, given any natural number n, we define the 3-adic absolute value of n as the reciprocal of the highest multiple of 3 that divides evenly into n. So, it's what we've been seeing. So for example, let's compute the 3-adic absolute value of say 54. Well, first we write 54 as a product of prime numbers. So, $54 = 2 \times 3 \times 3 \times 3$. We see three factors of 3. So, we take the reciprocal of that, which would be 1/27, and so $|54|_3 = 1/27$. It takes a while to get used to this formula, by the way. This is very, very weird. Very weird—since normally we view 54 as a large number, and yet its 3-adic absolute value is very small, but that's because 54 contains several factors of 3. The more factors of 3 you have, the smaller the 3-adic absolute value will be.

More generally, for any prime number, let's call it p, we define the p-adic absolute value, which we write $|\ |_p$, on the natural numbers in a similar way. So, as follows: Given a natural number n, the p-adic absolute value of n equals the reciprocal of the highest multiple of the prime p that divides evenly into n, just as we've seen.

We'll also declare that the p-adic absolute value of the number 0 equals 0, just as before. And for negative numbers, we just ignore the negative sign, as we did before. So, for example, the 3-adic absolute value of –6 equals 1/3, because I just ignore the negative sign, take 6, factor it (2×3), I see one factor of 3, I take its reciprocal, and I see that $|-6|_3 = 1/3$.

Now, notice that, given this description, the p-adic absolute value of a number is never negative, because it's either going to equal 0 or it's a reciprocal of a product of p's, and that is always positive. Also, the p-adic absolute value is very sympathetic to multiplication. For example, let's find $|6 \times 18|_3$. Well, 6 $\times$ 18 = 108, and if we write that as a product of primes, we see that $108 = 2 \times 2 \times 3 \times 3 \times 3$. So, $|6 \times 18|_3 = 1/27$, because we see three factors of 3 in there. So, we write the reciprocal of $3 \times 3 \times 3$.

Let's compute individually $|6|_3$ and multiply it by $|18|_3$. So, $|6|_3$, we've already seen, is 1/3. And what about $|18|_3$? How many factors of 3 do we see? Well, two. So therefore, we see 1 / (3×3), or 1/9. When we multiply these two numbers together, we see $1/3 \times 1/9$ which equals 1/27. So, we see the same number again. So, the 3-adic absolute value of a product is equal to the product of the 3-adic absolute values. And similarly, in terms of addition, we can show that the triangle inequality also holds here, although we won't do it.

Anyway, what we're seeing here are some examples showing that the p-adic absolute value really does satisfy all the rules required in order to be an actual absolute value, and indeed, all these properties can be proven to hold true.

By the way, we can now find the *p*-adic absolute value of fractions, in other words, rational numbers, just by dividing. So for example, suppose we want to find $|8/45|_3$. All we do is take $|8|_3$ and divide it by $|45|_3$. So $|8|_3$ is just 1, because there's no factor of 3 in there at all. It's not a multiple of 3, so we just have an absolute value of 1. And what about 45? Well, $|45|_3$ is 1/9, because we have a 3 × 3 in its factorization. So when I divide these numbers, I see 1 ÷ 1/9. This is always a little bit tricky here, so we need to think about this. So, for (1/(1/9)), we have to actually take the reciprocal and multiply, if we remember our rules from a long time ago. And we see that the answer is 9. So, $|8/45|_3$ is actually 9, which is quite large. So, here's the moral: If we have powers of 3 in the denominator of a fraction, then the 3-adic absolute value will become large, and if we have a power of 3 in the numerator of the fraction, then its 3-adic absolute value will be small—very, very peculiar, but it's an arithmetical sense of distance.

As we'll see in the next lecture, this very strange *p*-adic absolute value leads to an even stranger notion of number—new numbers that exhibit properties that are stranger still. These numbers will actually defy our own intuition as to what "number" means, and so we'll actually be at a crossroads. We'll see that things about which we'll think, "they can't be numbers," and yet numbers they are.

A New Breed of Numbers
Lecture 19

Many of us today might find it difficult to empathize with the Pythagoreans' resistance to accepting the notion that the counterintuitive irrationals should be considered as numbers. Here we'll face numbers that will be as counterintuitive to us today as the irrational numbers were to the Pythagoreans several thousand years ago.

We begin with arithmetic absolute value. We recall the definition of the p-adic absolute value $|\ |_p$ on the integers, as follows: $|0|_p = 0$; if N is a natural number, then $|N|_p$ equals the reciprocal of the largest multiple of p that appears when N is written as a product of prime numbers (e.g., $|90|_3 = \frac{1}{9}$). We see that, because there are infinitely many prime numbers, there are infinitely many absolute values on the rational numbers. These new absolute values were discovered by the German number theorist Kurt Hensel in 1897. In the 1930s, Ukrainian mathematician Alexander Ostrowski proved that the only absolute values on the rational numbers are those that are equivalent to either the usual, familiar absolute value or one of these new p-adic absolute values.

Using this p-adic absolute value, we can measure the distance between two rational numbers. For example, if we fix $p = 3$, then the 3-adic distance between 5 and 2 is: $|5-2|_3 = |3|_3 = \frac{1}{3}$; thus, 3-adically, 5 and 2 are relatively close to each other. The distance between 0 and $\frac{1}{6}$, however, is: $\left|0 - \frac{1}{6}\right|_3 = \left|\frac{-1}{(2 \times 3)}\right|_3 = 3$; thus, 3-adically, $\frac{1}{6}$ and 0 are relatively far apart. The p-adic absolute value measures how many factors of p there are in a rational number $\frac{r}{s}$. The more factors of p in r, the smaller $\left|\frac{r}{s}\right|_p$ is; the more factors of p in s, the larger $\left|\frac{r}{s}\right|_p$ is. Closeness in this p-adic context means that the difference of the two rational numbers has a high power of p in its numerator.

Measuring the lengths of sides of triangles having rational-number vertices will show how p-adic value can offer greater structure than standard absolute value. We now consider the lengths of the sides of triangles formed by rational numbers (we will assume that we have not yet built the real numbers). If we consider a triangle having vertices 5, –1, and 7, then the lengths of the sides are the distances between these three numbers: $|5-(-1)| = 6$, $|5-7| = 2$, and $|-1-7| = 8$. Each side length has a different length; hence, this is an example of a scalene triangle.

We now attempt to find the 3-adic lengths of the sides of the previously described triangle. The lengths, measured 3-adically, are: $|5-(-1)|_3 = \frac{1}{3}$, $|5-7|_3 = 1$, and $|-1-7|_3 = 1$. In this case, we see that two sides have equal lengths; thus, this 3-adic triangle is isosceles. More generally, any triangle formed by three rational numbers will be an isosceles triangle when the lengths of the sides are measured p-adically. Consider the triangle having vertices 0, 3, and 12. If we measure the lengths of the sides 3-adically, we have: $|0-3|_3 = \frac{1}{3}$, $|0-12|_3 = \frac{1}{3}$, and $|3-12|_3 = \frac{1}{9}$. Again, we find that two sides have equal lengths. The more arithmetic p-adic absolute value offers greater structure than the usual absolute value.

Finding and filling in the p-adic holes with p-adic Cauchy sequences is challenging but follows a definite logic. Consider the infinite sum: $1 + 3 + 3^2 + 3^3 + 3^4 + 3^5 + \dots$. When measured 3-adically, this sum is heading toward a number—but which number? If we call the number N, then we see that

$$N = 1 + 3 + 3^2 + 3^3 + 3^4 + 3^5 + \dots$$

and

$$3N = 3 + 3^2 + 3^3 + 3^4 + 3^5 + \dots$$

Subtracting the two equations:

N	$=$	$1 + 3 + 3^2 + 3^3 + 3^4 + 3^5 + \dots$
$3N$	$=$	$3 + 3^2 + 3^3 + 3^4 + 3^5 + \dots$
$-2N$	$=$	1

We conclude that $N = \frac{-1}{2}$; when we measure closeness with the 3-adic absolute value, we see the very surprising fact that $-\frac{1}{2} = 1 + 3 + 3^2 + 3^3 + 3^4 + 3^5 + \dots$. By the same reasoning, we see that the terms in the infinite sum $3 + 3^2 + 3^4 + 3^8 + 3^{16} + 3^{32} + 3^{64} + \ldots$ are getting smaller and smaller. Moreover, the numbers $3, 3 + 3^2, 3 + 3^2 + 3^4, 3 + 3^2 + 3^4 + 3^8, 3 + 3^2 + 3^4 + 3^8 + 3^{16} \ldots$ are getting closer and closer to each other and form a 3-adic Cauchy sequence. We would like this sequence of numbers to head toward a number; unfortunately, because the exponents are doubling, this sum is not equal to a rational number or even a complex number.

We must extend our notion of number so that all 3-adic Cauchy sequences converge. We formally declare the sum $3 + 3^2 + 3^4 + 3^8 + 3^{16} + 3^{32} + 3^{64} + \ldots$ to be a new irrational number. This new irrational number is analogous to one of the irrational *real* numbers we found in Lecture 10: $0.101001000100001000001\ldots$.

If we consider the rational number but measure closeness with a p-adic absolute value, then we again see there are holes. When we filled in the holes using the usual absolute value, we extended our notion of number and discovered the irrational real numbers. In this p-adic context, if we fill in the holes using the p-adic absolute value, then we extend our notion of number and discover *new* irrational numbers, such as $3 + 3^2 + 3^4 + 3^8 + 3^{16} + 3^{32} + 3^{64} + \ldots$. This new collection of numbers is not the collection of real numbers and does not reside on a number line. It is known as the collection of p-adic numbers.

It may seem difficult, despite this mathematical explanation, to embrace the idea that p-adic numbers are actually numbers. Mathematicians have accepted this concept for more than a century. The p-adic numbers were first studied by Kurt Hensel in the late 1800s. They provide a purely abstract arithmetic notion of number that is extremely useful in studying integers, prime numbers, algebraic numbers, and even transcendental numbers.

Recently, physicists have used p-adic numbers to create ... new ideas about string theory and quantum mechanics.

It might appear as though these strange and foreign p-adic numbers have no utility in our everyday world or in any universe beyond the abstract universe of number. Recently, physicists have used p-adic numbers to create new models of space–time and new ideas about string theory and quantum mechanics. We see that these numbers might help us better understand the nature of our "real" universe. ∎

Questions to Consider

1. Consider the triangle having vertices 0, 5, and 25. Compute the lengths of the three sides of this triangle 5-adically and verify that the triangle is isosceles.

2. Give an example of a 5-adic number that is irrational. Describe how p-adic numbers in general challenge our intuition about the notion of number.

A New Breed of Numbers
Lecture 19—Transcript

Many of us today might find it difficult to empathize with the Pythagoreans' resistance to accepting the notion that the counterintuitive irrationals should be considered as numbers. Here we'll face numbers that will be as counterintuitive to us today as the irrational numbers were to the Pythagoreans several thousand years ago. Using an entirely new arithmetical measure of distance involving prime numbers, called the p-adic absolute value, we'll discover a very unusual world of numbers. This modern perspective, first discovered by Kurt Hensel in 1897, allowed for many important advances in the abstract theory of numbers. We'll see, for example, that when measured with this new notion of distance, all triangles are isosceles. We'll see a way of expressing rational numbers in a p-adic expansion that has reflections of the different bases discussed in Lecture Nine but, in fact, will be fundamentally different. While many of our previous observations about numbers will carry over into this new context, there will be several dramatic differences.

We'll see that, armed with this new measure of distance, the rational numbers still have holes, in other words, missing values. Here, as we plug up the holes and expand our notion of numbers, we do not build the real numbers, instead we produce a collection of very strange-looking numbers. These so-called p-adic numbers will appear as totally abstract consequences of arithmetic with no meaning in our physical world. However, we'll briefly describe how these seemingly unnatural numbers and their analysis, in fact, are at the core of describing some delicate ideas in quantum physics. Thus, on the one hand, we see again how nature fits, bringing together worlds that seem totally unrelated upon first blush. And on the other hand, we see the great danger in believing that, at any moment in our history, we've captured all the numbers.

Let's open what will be the very, very strange discussion of numbers by first revisiting the arithmetic absolute values from the previous lecture. There we saw a new absolute value called the p-adic absolute value. First, we select a prime number which we'll call p. Remember that the prime numbers are 2, 3, 5, 7, 11, 13, and so forth, the numbers that can't be broken down to the product of two smaller natural numbers. We then define the p-adic absolute value $|\ |_p$, and we define it on the integers by declaring that, first of all, if you plug in 0, $|0|_p = 0$, and if n is a natural number, then $|n|_p$ will equal the reciprocal of the largest multiple of p that appears when n is written as a product of prime numbers.

For example, if we fix the prime to be 3, then we can find $|90|_3$ by first noticing that if we factor 90, we see $2 \times 3 \times 3 \times 5$. So, we see two factors of 3, which equals 9. We take the reciprocal and discover that the $|90|_3$ will be declared 1/9. Remember that the more factors of 3 a number has, the smaller the 3-adic absolute value will be. Because we have such a p-adic absolute value for every single prime number p, and because we have shown that there are infinitely many prime numbers, we can now conclude that there are, in fact, infinitely many different absolute values on the rational numbers.

These new absolute values were discovered by the extremely talented German number theorist Kurt Hensel in 1897. Now, once we have infinitely many absolute values—we've gone from the usual one to infinitely many—we, of course, wonder, "are there other absolute values?" Well, the amazing answer is no. In the 1930s, Ukrainian mathematician Alexander Ostrowski proved that the only absolute values on the rational numbers are those that are equivalent to the usual one that we've always used in school, the familiar one, or one of these new p-adic ones. In other words, the only measures of distance, either arithmetic or using a standard ruler, on the rational numbers that satisfy the rules of being an absolute value, are precisely the familiar one and these new p-adic ones. That's it. There are no others—really amazing.

Using these p-adic absolute values, we can measure the distance between two rational numbers, in other words, the distance between two fractions. Let's look at an example. If we fix the prime number to be 3 again, then the 3-adic distance between 5 and 2 is just $|5 - 2|_3$. Well, $5 - 2 = 3$, and $|3|_3 = 1/3$. (We take the reciprocal.) So 3-adically, 5 and 2 are relatively close to each other when measured through this absolute value. However, the distance between 0 and 1/6 is what? Well, it would be $|0 - 1/6|_3$, which is just $|-1|_3 / |6|_3$. Well, what is $|-1|_3$? There are no factors of 3 in there, so it's just 1. And what is $|6|_3$? Well, 6 is 2×3 (one factor of 3). So we see 1/3, but we have to take the quotient, so we have 1/(1/3). If we take the reciprocal, we see the answer is 3. So 3-adically, 1/6 and 0 are relatively far apart. Their distance between each other is actually 3.

The p-adic absolute value measures how many factors of the fixed prime p there are in a natural number. The more factors of p there are in n, the smaller $|n|_p$ is. Closeness in this p-adic context means that the difference of two numbers is a multiple of a high power of the prime p. This measure of distance leads to some very strange geometric consequences which I want to share with you. To understand this new the geometry within the rational numbers, we'll assume for the moment—as we have in the past—that our entire notion of number consists solely of the rational numbers. That is, the only numbers in our mind's eye, for the moment, are the fractions.

Since we're measuring distance in this arithmetical sense, the fractions no longer line up on an orderly, increasing line. Instead, let's just visualize the rationals as an unordered jumble of numbers sitting in a bowl. In other words, we can consider all the fractions as tiny pieces of fruit inside an enormous bowl of translucent JELL-O. Believe it or not, if we haven't specified how we're measuring distance between two rational numbers, this is, in my opinion, probably the most accurate way to visualize rational numbers—as points suspended inside a chilled gelatin number dessert. Sounds delicious, doesn't it? Well, in fact, when I actually do research in number theory in this area, I actually visualize the rationals sitting in this fashion.

If we take any three fractions, we can imagine drawing three lines between those three fixed points in that JELL-O, and thus we'd see a shimmering triangle in our enormous JELL-O mold. Let's now consider the lengths of the sides of these triangles formed by the rational numbers. For example, if we consider a triangle having vertices at the points 5, –1, and 7—so remember, they're just sitting inside JELL-O, so 5 might be here, and –1 might be here, and 7 might be here, all mixed up—well, the lengths of the sides are the distances between these three numbers. Using our usual absolute value, we can figure out the distances. So, to find the distance between 5 and –1, we look at $|5 - (-1)|$, which equals 6. To find the distance between 5 and 7, we look at $|5 - 7|$, which would yield 2. Finally, to find the last length of the last side, we would look at the distance between –1 and 7, and we'd see that that distance would be 8. Now, notice that the image that we're looking at here is really not drawn to scale since we know that, with the usual absolute value, the rational numbers do line up perfectly on the real number line. So, we'd see −1, 5, 7, and in fact, you can even check and see that the far length is 8, which does equal 2 + 6. So in fact, the triangle, in some sense, collapses down.

But let's not worry about the image itself, let's just think abstractly. Now, we can just look at those three lengths and notice that each side has a different length. In other words, this is an example of a scalene triangle. That's what we call a triangle whose three sides have three different lengths. Well, let's now consider the exact same triangle, but now measure the lengths using the 3-adic absolute value rather than our usual familiar one. The lengths, measured 3-adically are, between 5 and −1, we would see $|5 - (-1)|_3$, which would be 6. And what's $|6|_3$? Well, 6 is 2×3—one factor of 3. We take the reciprocal of that, so we see that that distance is going to be 1/3. For the next length, we take the difference between 5 and 7 and take the 3-adic absolute value. That's going to be $|-2|_3$. Well, −2 has no factors of 3 in it, and so, in fact, there we see that $|-2|_3 = 1$. And what about the last length? Well, we're looking for the distance between −1 and 7, that would be $(-1) - 7$, which equals −8, and $|-8|_3$ would equal 1, since there's no factors of 3 in the number −8. Again, we notice that this image is not drawn to scale. But let's not worry about the image. Let's look at the actual numerical values which we have precisely.

In this case, however, we see that two sides actually have equal lengths. Thus, this 3-adic triangle is isosceles (the name we give triangles that have at least two side lengths that are equal). In fact, more generally, any triangle formed by three rational numbers will be an isosceles triangle when the lengths of the sides are measured with any 3-adic absolute value.

Let's illustrate this amazing fact with another example to see this again in action. This time, let's consider a triangle having vertices at 0, 3, and 12 (again, in our number Jell-O). So, there's three points. We've got a 0, a 3, and a 12 just swimming around in the Jell-O, locked in there. If we measure the lengths of the sides 3-adically, what do we see? Well, between 3 and 0, we see $3 - 0 = 3$. And 3-adically, that's 1/3. Between 12 and 3, we measure that distance, and we see 9, then $12 - 3 = 9$. So 3-adically, that's 1/9, since 9 is 3×3. (We take the reciprocal.) And the final length—the distance between 12 and 0 would be—well, $12 - 0$ is just 12 and $|12|_3$—factoring 12 we get 2 $\times 2 \times 3$. There's one factor of 3, and so I see that $|12 - 0|_3 = 1/3$. And look again. We see that two lengths have the same measure—in this case, 1/3. So, again we see that two sides have equal lengths.

This is very, very counter-intuitive and extremely hard to digest, that all number triangles are isosceles when measured using these p-adic absolute values—really, really hard to swallow, certainly much harder to digest than the JELL-O itself, but a beautiful fact to try to wrap our minds around. What we're witnessing here is that these new, more arithmetic p-adic absolute values place a simpler and more symmetric geometric structure on the rational numbers than the structure inherited by the usual absolute value. Again, we see this wonderful recurring theme that what first appears standard is really unusual, and what first appeared to be exotic is really the norm. Remember that out of all the infinitely many absolute values on the rational numbers, there's only one usual one that we've seen in school, and then the infinitude of what remains is, in fact, these p-adic ones. And so, what we see here is that the scalene triangles are now the peculiar ones in light of the fact that all the other absolute values give rise to isosceles triangles. So, it really forces us to re-think what is familiar and what is the norm, because within the numerical context, we see that's really the exception.

We'll now discover that, just as we saw with the usual absolute value in the previous lecture, when we built the real numbers, the rational numbers still have holes when we measure distance with these p-adic absolute values. Let's look at an example of a p-adic Cauchy sequence. In other words, a sequence of rational numbers that are getting closer and closer to each other when measured with the p-adic absolute value, but are not heading toward any rational number as a limiting value. Now, this is going to look very, very strange. I want to warn you right out of the get-go, so I want to warm up to it by considering a very weird looking sum of numbers. It's going to be an endless sum of numbers. It's going to look very peculiar, let's keep an open mind together. So, I want to consider the endless sum that begins $1 + 3 + 3^2 + 3^3 + 3^4 + 3^5$, and so on endlessly.

Now of course, we look at this and it appears that this sum equals infinity since the terms are getting bigger and bigger forever. But how are we measuring size? Suppose we are using the 3-adic absolute value. Then the terms are actually getting smaller and smaller. In fact, they're approaching 0. For example, look at the term $3^4 = 81$. Its 3-adic absolute value would be its reciprocal, $1/81$, which you can see is very, very small. Well, while this is not at all obvious, let me just say that when measured 3-adically, number theorists can prove that this sum, in fact, is heading toward a number. Well, which number is it?

We can now apply the same idea we used to show that the decimal number 0.99999 forever equals 1. If we adopt that same strategy, we can figure out what this sum is actually approaching. So, let's call this sum s. So, we give it a name. So, $s = 1 + 3 + 3^2 + 3^3 + 3^4$, and so on forever. Now our trick will be to multiply both sides of this by 3. Now on the left-hand side, we'll be left $3s$ (because we multiply s by 3, and we get $3s$). But what happens on the right-hand side? Well, we have to multiply every term in the sum by 3. When we multiply the 1 by 3, we get 3. So in some sense, we could line up that 3 underneath the 3 we have above. When we multiply the second term, which is 3 by 3, we now get 3^2. And 3^2 multiplied by 3 gives us 3^3, and 3^3 multiplied by 3 gives us 3^4. What we notice is that basically the role of each term sort of shifts down by one. So once we shift down, we see that they line up perfectly over their counterparts above. If we now subtract the two equations, what do we see? Well, we see that endless list of powers of 3 cancel out, because they line up perfectly forever, since there's no last term on either list. But we do have that 1 on the top, and $1 - 0$ is 1. And then I see s, and I'm subtracting $3s$. Well, $s - 3s = -2s$. So, I have $-2s = 1$. And so if we solve, we can just divide both sides by -2, and we see that $s = -1/2$.

That is, when we measure closeness with the 3-adic absolute value, we see the very disturbing fact that $-1/2 = 1 + 3 + 3^2 + 3^3 + 3^4 + 3^5$, and so on forever. Well, this equation is so disturbing to us and counterintuitive that it seems totally impossible. It just feels wrong, because our intuition with distance between numbers is so firmly planted in the real number line and the usual absolute value. By the way, if we use the usual absolute value, indeed that previous expression is complete nonsense. But the moment we think about the real numbers and we think about the number $1 + 3 + 9$, and so forth—we see that list going out this way—it's hard to imagine that that list is actually approaching the number −1/2, which is over here. Well, the moment we look on the number line, then we're assuming we're using the usual absolute value. We have to free ourselves, in some sense, from the shackles of the real number line and let our minds drift to different notions of absolute values.

When we use the 3-adic absolute values, nothing lines up. With the 3-adic absolute value, that equation is both provably correct and, believe it or not, sensible. Now, how is that possible? Well, in the 3-adic JELL-O, when we add $1 + 3 + 3^2 + 3^3$, and so forth, all those numbers are getting closer and closer to −1/2. Remember that, in the JELL-O, where the points are just sort of frozen in space, what we're seeing is that if we take 1, and then add it to 3, and then add it to 3^2, and it to 3^3, and so forth, what we're heading toward is the number in the JELL-O that happens to be −1/2. Well just to see this fact, let's just consider the sum $1 + 3 + 3^2$. In other words, let's just truncate that sum, not consider the endless list, but instead just consider up to 3^2. Well, $1 + 3 + 3^2$ is just $1 + 3 + 9 = 13$. Let's see how close 13 is to this target, −1/2, 3-adically. So, we look at $13 − (−1/2)$ and measure that difference 3-adically to see the distance between the two. Well $13 − (−1/2)$ is $13 + 1/2$, and $13 + 1/2$ is the same thing as $26/2$ $+1/2$, which is $27/2$. Well, what's $|27/2|_3$? The denominator is 2 and has no powers of 3 at all. And 27 is a power of 3, and so $|27/2|_3 = 1/27$. So, look how close $1 + 3 + 3^2$ is to the number −1/2. They're within 1/27 of each other when measured 3-adically. And so in fact, if you continue this process and let the sum drift off endlessly, we actually hit −1/2 in that limiting case. So, really counterintuitive—it takes an awful long time to make this sensible—but we see that it really is correct in this context.

Well, by the same reasoning, we could see individual terms in this endless sum. Let's consider $3 + 3^2 + 3^4 + 3^8 + 3^{16} + 3^{32} + 3^{64}$ and so on. Notice that the exponents here are doubling each time. Well notice that each of these terms, in fact, are again getting smaller and smaller and heading toward 0, 3-adically. Just as with the previous example, if we look at the successive finite truncations of the sum, then we'd see the numbers:

3

$3 + 3^2$

$3 + 3^2 + 3^4$

$3 + 3^2 + 3^4 + 3^8$

And so on. And in the eyes of the 3-adic absolute value, these numbers are actually getting closer and closer to each other. For example, what's the 3-adic distance between the fourth and the fifth numbers on this list? If we subtract them, we see all the terms cancel except that last term, 3^{16}. And so, what's the 3-adic absolute value of 3^{16}? Well, it's $1/(3^{16})$, which is a really tiny number. The difference in those two numbers is very small. Those two numbers are very close together 3-adically.

In fact, these numbers form what is called a 3-adic Cauchy sequence, a list of numbers that continue to get closer and closer to each other when measured 3-adically. As with the Cauchy sequences we saw in the previous lecture, we would like this Cauchy sequence of numbers to head toward a target, to head toward a number. But unfortunately, because the exponents of the 3s are doubling each time, this sum is not equal to a rational number or even a complex number. The algebra trick that we used to show that the previous sum equalled $-1/2$ is not going to work here. This fact means that we need to extend our notion of number so that all 3-adic Cauchy sequences converge. In other words, formally declare the sum $3 + 3^2 + 3^4 + 3^8 + 3^{16} + 3^{32} + 3^{64}$, and so forth—let's just call that a number just like Weierstrauss did with the infinitely long decimal expansions. We just call this a number. This, in fact, is a new irrational number.

This new irrational number is analogous, in some sense, to one of the irrational real numbers that we actually found in Lecture Ten, the decimal that begins 0.101001000100001, and so forth. Remember, we kept adding longer and longer runs of zeroes. Well, here our doubling the exponent each time is the corresponding idea to the number we saw from Lecture Ten. But our new irrational number is not part of the real number line, and in fact, it doesn't have a decimal expansion. Thus, we see that if we consider the rational numbers and now measure closeness with a p-adic absolute value, then we again see that we're faced with holes, points in which fractions are bunching up around, like going around a drain, but the target itself is not a fraction.

When we filled in these holes using the usual absolute value, as in the previous lecture, then we extended our notion of number and discovered the irrational real numbers. In this p-adic context, if we fill in these new holes using a p-adic absolute value, then we extend our notion of number and discover new irrational numbers such as $3 + 3^2 + 3^4 + 3^8 + 3^{16} + 3^{32}$, and so on forever. This new collection of numbers is neither the collection of real numbers nor the collection of complex numbers, and neither resides on a number line nor even on a plane. This new collection of numbers is known as the collection of p-adic numbers.

Now, if you remember from my gelatin vision of number, only the tiny bits of fruit, the fractions suspended in the JELL-O, were considered numbers. Now, in essence, we are extending our notion of number so that we can view the JELL-O itself, together with the points of fruit, as the entire collection of p-adic numbers. We filled in all the space. These new numbers are so strange upon first inspection, and so counterintuitive, that it is difficult to see these infinite-looking things as numbers. Look, we're adding $1 + 3 + 3^4$, and so on. It looks infinite. It's hard to imagine that this really is a sensible notion of number.

Once again we see that, as we travel further into the outer depths of mathematics, our notion of number is challenged. Now, imagine how the Pythagoreans must have felt when their notion of God-given natural numbers and their ratios were challenged with the discovery of $\sqrt{2}$. They couldn't fathom that quantity as number just as here it's very difficult for us to fathom a number like $3 + 3^2 + 3^4 + 3^8 + 3^{16}$, and so forth forever, as a number. And yet it is, and it's our job to see if we could somehow modify our view of number to embrace such a strange looking object.

These strange p-adic numbers provide a purely abstract arithmetical notion of number that is extremely useful in the study of integers, the prime numbers, algebraic numbers, and even transcendental numbers. In fact, the p-adic numbers were artfully applied by Princeton mathematician Andrew Wiles, who, in the mid-1990s, provided a proof of the so-called Fermat's

Last Theorem, a 350-year-old open question about the natural numbers left unanswered by the famous French mathematician Pierre de Fermat. The question was one of the most famous long-standing open questions in mathematics. The *p*-adic numbers played an important role in that final solution.

Thus, we see that these foreign and strange numbers allow us to actually better understand our familiar numbers. It might appear as though these strange and foreign *p*-adic numbers have no utility within our everyday world or in any universe beyond the abstract universe of number. Although, again, this new idea of number might enable us to better empathize with the struggle that the Pythagoreans had with irrational numbers. However, this lack of practical utility of the *p*-adic numbers in our real world appears to be an outdated point of view. Recently, physicists have used *p*-adic numbers to create new models of space-time, and new theories of string theory, and quantum mechanics. So, we see that these foreign numbers might help us better understand the nature of our real universe.

They certainly challenge our notion of what numbers should mean and what numbers should look like, because, at first blush, these objects seem totally infinite. In fact, there's a sensible, reasonable notion of number right behind all this theory. These numbers that look strange at first, in reality, are just as iron-clad as all the numbers we've seen up to this point.

The Notion of Transfinite Numbers
Lecture 20

Here, in the last part of our course, we'll journey beyond numbers and contemplate the enormous question: What comes after we have exhausted all *numbers?*

There are two uses of numbers: ordinal and cardinal. We can place objects in a certain order and refer to them by their placement in the ordering as first, second, third, and so forth. These numbers (1st, 2nd, 3rd, 4th, 5th, 6th ...) are known as ordinal numbers. A different use for numbers is as a means of enumerating; that is, counting how many elements (or members) there are in a collection. These numbers (0, 1, 2, 3 ...) are known as cardinal numbers. The number of elements in a collection is called the cardinality of the collection. We can extend each of these notions of number to infinity. Mathematicians have devoted a great deal of study to what are called transfinite ordinal and transfinite cardinal numbers.

As early as the 5th century B.C.E., scholars contemplated infinity.

As early as the 5th century B.C.E., scholars contemplated infinity. Around 450 B.C.E., Zeno considered paradoxes involving an infinite number of steps. In the 4th century B.C.E., Aristotle concluded that infinity did and did not exist. He recognized that the counting numbers had no end but also believed that an infinite object could not exist in the real world because it would be boundless. Sometime between the 4th and 1st centuries B.C.E., the Jains studied mathematics extensively and believed there were different types of infinity. In a work published in 1638, Galileo observed that the natural numbers (1, 2, 3, 4 ...) could be placed in a one-to-one correspondence with the perfect squares (1, 4, 9, 16 ...). His one-to-one correspondence was: $1 \Leftrightarrow 1$, $2 \Leftrightarrow 4$, $3 \Leftrightarrow 9$, $4 \Leftrightarrow 16$, and so forth (here, the double arrow, $\Leftrightarrow$, indicates the one-to-one pairing of numbers from each collection). Galileo observed that an infinite collection (the natural numbers) had been put in one-to-one correspondence with a proper subcollection of itself. He thought this seeming paradox was one of the challenges provoked by infinity.

The notion of one-to-one pairings resurfaced two centuries later. The 19th-century German mathematician Georg Cantor was the first to put the notion of infinity on a firm foundation. He needed this precision for his work in function theory and number theory. His revolutionary ideas were grounded in a very simple reality: We cannot use ordinary counting methods to understand infinity. Suppose now that we cannot count to 5. How would we know that the number of fingers on our left hand equals the number of fingers on our right hand? We can pair them up in a one-to-one fashion. This one-to-one pairing—comparing quantities rather than counting them—was humankind's first attempt at grasping the idea of quantities. Cantor showed that this ancient basic idea is the key to unlocking the mysteries of infinity.

Returning to Galileo's observation about infinity, we say that a one-to-one correspondence between two collections is a way of pairing the elements of two collections so that every element from the first collection is paired with exactly one element of the second collection, and every element of the second collection is paired with exactly one element of the first. We say that two collections have the same cardinality if there is a one-to-one correspondence between the two collections; that is, the collections are equally numerous. It is easy to see if finite collections have the same cardinality; we simply count and see if the counts are equal. For infinite collections, we need to describe a one-to-one correspondence to verify that the collections have the same cardinality.

We move from studying individual numbers to studying the size of collections of numbers. Let N denote the entire collection of natural numbers—that is, $N = \{1, 2, 3, 4 \ldots\}$—and let E denote the collection of all even natural numbers; that is, $E = \{2, 4, 6, 8 \ldots\}$. Do these two collections have the same cardinality? We recall Euclid's accepted "common notion" from 2,000 years earlier: "The whole is greater than the part." We compare natural numbers with integers. Let Z denote the entire collection of integers; that is, $Z = \{\ldots -4, -3, -2, -1, 0, 1, 2, 3, 4 \ldots\}$. Do the collections of natural numbers and the collection of integers have the same cardinality? The answer is perplexing on first inspection.

We compare natural numbers with rational numbers. Let Q denote the entire collection of rational numbers; that is, $Q = \{\text{every fraction}\}$. Do the collections of natural numbers and the collection of rational numbers have the same cardinality? The answer surprised many people. ∎

1. How do you define *infinity*? What does infinity mean to you and what images does the word evoke in your mind?

2. Imagine a collection of objects. Suppose we remove some members from this collection to produce a second collection; is it possible that these two collections have the same cardinality? Does your answer change if you further assume that the initial collection is finite?

The Notion of Transfinite Numbers
Lecture 20—Transcript

Here, in the last part of our course, we'll journey beyond numbers and contemplate the enormous question: What comes after we have exhausted *all* numbers? We'll begin with a distinction between two basic types of number, ordinal and cardinal, and then wonder how to extend these ideas past all numbers into the realm of the infinite. Galileo wrestled with the infinite, and in 1632, observed some perplexing paradoxes. In 1851, a tract authored by Bernhard Bolzano first attempted to study infinite collections. While others made contributions to this area, it was the pioneering and highly controversial work of Georg Cantor in the late 1800s that laid the foundation for our understanding of infinity. Cantor's realization was that "counting" infinite quantities leads us nowhere. Instead, he returned to our ancient ancestors and their first notion of number. He dared to compare.

Thus, we come full circle and revisit, from the opening of this course, the idea of a one-to-one correspondence between two collections. If the individual objects from two collections can be paired up in a one-to-one fashion, we declare these two collections to have the same cardinality. To illustrate this ancient idea within this abstract context, we'll consider several scenarios in which we'll compare different infinite collections of numbers. These examples will challenge our intuition and demonstrate that what we know about finite collections might not carry over to infinity. They'll also confirm one of our intuitively rock-solid notions: Infinity comes in just one size—one unending size. However, in the lecture that follows this one, we'll be forced to reconsider even this intuitively sensible-sounding opinion and, yet again, have the wonderful opportunity to retrain our intuition.

Well, we open with the two most basic applications for numbers in our everyday lives. We can place objects in a certain order and refer to them by their placement in this ordering as first, second, third, fourth, and so on. These numbers—first, second, third, fourth, fifth, sixth, and so on—are known as the ordinal numbers, because they're numbers that give order.

A different use for numbers is, as we have seen throughout history, as a means of enumerating. In other words, counting the size of a collection.

These numbers—zero, one, two, three, and so forth—are known as cardinal numbers. The number of elements in a collection is called the cardinality of that collection.

Now, both of these views of number can be extended to infinity. In fact, there's been much study in what are called infinite ordinal and infinite cardinal numbers, although "numbers" should really be in quotes. These are sometimes referred to as "transfinite" numbers. Here in this last part of our course, we'll explore infinite cardinal numbers.

Now, before moving on, this is a wonderful opportunity to challenge ourselves to ask, "What is our own definition of infinity? What does infinity mean?" Well, while we think about what it means to us personally, I thought I would share with you the thoughts of some others that came before us as they tried to wrap their minds around this mysterious idea.

The 20th-century British philosopher Bertrand Russell once wrote about how philosophers themselves would talk about infinity. He writes, "If any philosopher had been asked for a definition of infinity, he might have produced some unintelligible rigmarole, but he would certainly not have been able to give a definition that had any meaning at all." So, here's the philosopher Bertrand Russell saying that philosophers would have no idea.

The great 20th-century German physicist Albert Einstein once wrote, "Two things are infinite, the universe and human stupidity. And I'm not sure about the universe." Great quote.

The 15th-century English statesman Thomas More once wrote, "By confronting us with irreducible mysteries that stretch our daily vision to include infinity, nature opens an inviting and guiding path toward a spiritual life."

I once asked a math class of mine at Williams what was their notion of infinity, before we actually touched on this subject in depth, and there was a student in the class, whose name was Joe, who raised his hand said, "Infinity is a mother's love," to which the entire class broke up in gales of laughter. And it was just a very funny moment. And then about a month later, his parents actually came to visit him on campus, and they actually attended the class which he was in, this math class. And so when the parents were there, of course, I had to embarrass poor Joe and say to the mother, "By the way, we talked abut infinity a month or so ago. I asked what the definition was, and Joe said it was a mother's love." And as you can imagine, his mother was beaming. So it was a great story.

The 18th-century French writer Voltaire once wrote, "The only way to comprehend what mathematicians mean by infinity is to contemplate the extent of human stupidity." Well, perhaps Einstein's point of view was inspired by Voltaire. But anyway, I guess we will walk into the extent of human stupidity, because we're about to take on his vast notion of infinity and try to make some sense of it.

So, given these thoughts, how can infinity be rigorously defined or understood? Maybe by definition we should say that infinity is that which is beyond human comprehension. Well, we will see that these vague ideas of infinity can be replaced by a very rigorous notion which then can be analyzed as we analyze the numbers.

So, let's begin our exploration by taking a look at some early ideas involving infinity. As early as the 5th century B.C.E., scholars were contemplating infinity. Around 450 B.C.E., Zeno considered paradoxes involving an infinite number of steps. One such paradox was his argument to prove that motion is impossible and, thus, an illusion. And he argued as follows. He considered an arrow that was going to be shot to a target. In order for the arrow to get from the archer to the target, it must first reach the halfway point. Well, in order to reach the halfway point, it first has to reach half of that distance, or a quarter of the way. And in order to reach the quarter-way point, it has to reach half that distance, which is an eighth of the way, and so on.

So as this goes on, what do we see? Well, we see that the arrow must pass through infinitely many points in a finite amount of time, which Zeno argued was impossible. Thus, he concluded that motion was merely an illusion.

Aristotle actually once wrote, "That which is in locomotion must arrive at the halfway stage before it arrives at the goal." So, again we see that getting to the halfway point is really an issue. How do we deal with that?

Well, this paradox was finally put to rest using the concepts of calculus, the rigorous, mathematical study of motion. In fact, Aristotle himself, in the 4th century B.C.E., concluded that infinity both existed and didn't exist. He recognized that the natural numbers had no end, but also believed that an infinite object would not exist in the real world because it would be boundless. So you can see the dichotomy. In theory it should exist, but does it exist in practice? Aristotle struggled.

As we noted in Lecture Three, sometime between the 4th and 1st centuries B.C.E., the religious sect known as the Jains in India (the group that might have inspired Pythagoras to explore numbers) studied the mathematics of number extensively. They actually believed that there were many different types of infinity. They argued that, in fact, there would be (like on a number line, for example) a positive infinity, a negative infinity. There was the infinity of area of the plane, and they even considered time a type of infinity. So there were four types of infinity in the Jains' view of infinity.

In a work published in 1638, Galileo observed that the natural numbers—1, 2, 3, 4, and so forth—could be placed in a one-to-one correspondence with the perfect squares. Those are the numbers 1, 4, 9, 16, 25, and so forth. Now, let's just think—to see this correspondence that he come up with—let's view those perfect squares as red numbers, so we can just distinguish them from the numbers that are the natural numbers.

So, his correspondence was he paired the number 1 with the perfect square 1. He paired the number 2 with the perfect square 4. The number 3 was paired with the red 9. The number 4 was paired with the red 16, and so forth. (And here, by the way, a double arrow just indicates the one-to-one pairing, how we're paring up the collections.) So, thus, he observed that an infinite collection, in this case, the natural numbers, had been put in a one-to-one correspondence with a proper subcollection of itself—namely, just the perfect squares.

Notice that Galileo's observation ran counter to the sensible sounding ancient common notion of Euclid, which we actually discussed in Lecture Twelve, that stated that the whole is greater than the part. Well, he thought this seeming paradox was just one of the challenges provoked by infinity.

In fact, Galileo wrote that infinity, by its very nature, is incomprehensible to us. So, he actually viewed that maybe the view of infinity should be that which is incomprehensible. Well, in order to make infinity comprehensible to us, we must first acknowledge that infinity is not a number, and thus we cannot count infinite collections. The 19th-century German mathematician Georg Cantor was the first to adopt this point of view and place the notion of infinity on a firm foundation. Recall that we first met Cantor in Lecture Eleven, in which we examined his Cantor set. Remember that dusting of numbers on the number line of special numbers between 0 and 1 that formed a beautiful, self-similar fractal collection of numbers. There we also learned that he required a very precise view of infinity in order to further pursue his work in function theory and in number theory.

His revolutionary ideas were grounded in a very simple reality: We cannot use ordinary counting methods to understand infinity. In order to inspire Cantor's insight, let's return to our understanding of 5. How would we know that the number of fingers on our left hand equals the number of fingers on our right hand? Well, this seems like a ridiculous task. By counting 1, 2, 3, 4, 5. Then 1, 2, 3, 4, 5. Both are 5; we're done.

Suppose that we weren't able to count up to 5. We've seen in previous lectures that some cultures counted 1, 2, 3, many. So, suppose you couldn't count to 5. How would we know if, in fact, there are more digits on this hand than this hand, or if, in fact, each hand is equally digited? Well, how would we do that? We would just do a very natural thing. We would just do this. We would pair them up in a one-to-one fashion. And notice that I can see that every finger is associated with exactly 1 and only 1 finger from the other hand and vice-a-versa. And there's no finger left untouched. Therefore, I see that the number of fingers on one hand is the same as the number of fingers on the other hand, even though I have no idea what that number is.

I want us to recall from Lecture Two that this one-to-one pairing, this comparing quantities rather than counting, was humankind's first attempt to grasp the ideas of quantities. Cantor's tremendous insight was that this ancient, basic idea was the key to unlocking the mysteries of infinity. So, it's wonderful that, in our course, we come full circle back to the very dawn of counting, and that will take us out to the frontiers beyond, into infinity.

To formalize this idea, we want to say that a one-to-one correspondence between two collections is a way of pairing the elements of the two collections so that every element from the first collection is paired with exactly one element from the second collection. And every element of the second collection is paired with exactly one element of the first collection. Just like this.

We say that two collections have the same cardinality if there is a one-to-one correspondence between the two collections, that is, the collections are equally numerous. Informally, we're saying that they have the same size. We don't want to say that they have "the same number of elements," because that won't work when we talk about infinite collections. So, we say "the same the cardinality."

It's easy to see that if finite collections have the same cardinality, then we could just simply count, and we'd see that those counts are equal. So 1, 2, 3, 4, 5, and 1, 2, 3, 4, 5. They're equal. But for infinite collections, we need to describe a one-to-one correspondence to verify that the collections have the same cardinality. So, that's going to be our test. We see two collections. If we want to know if they are equally numerous or not, we have to find a one-to-one correspondence.

Let's now consider some examples to help solidify our intuition on infinity and this idea of a one-to-one correspondence. First, let's compare the natural numbers with just the even numbers. In other words, we're moving now from the study of individual numbers to the study of the size of a collection, an aggregate, of numbers. Now, we often use N to denote the entire collection of natural numbers. And so, formally, we'd say, "N equals"—and then we'd have a brace, which is this curly type bracket—and then we'd list 1, 2, 3, 4, and so forth and then close the curly brace. So we write $N = \{1, 2, 3, 4,...\}$.

And the way you'd read that—the curly brace means a collection, so we'd say, in this case, "N equals the collection of numbers 1, 2, 3, 4, and so on." That's the collection. So N represents the entire aggregate.

Let's write E for the collection of all the even natural numbers. That is, E will be the set, the collection, of 2, 4, 6, 8, and so forth. And we'd write it in a similar way. Well, do these collections have the same cardinality? Well, at first it appears not, since the even numbers are only half of the entire collection of natural numbers. But we can't be fooled by such logical-sounding distractions. Here, there's only one measure: Does there exist a one-to-one pairing or not?

Let's try to find a one-to-one pairing between the numbers in these two collections. So, we take the number 1 from the natural numbers, and let's just pair it up with the number 2 from the even numbers. We take the number 2 from the natural numbers, and why don't we connect it, or pair it up, with the number 4 from the even numbers. Similarly, we'll take the number 3, and we'll pair it up with the number 6 from the even numbers. 4 and 8 will be matched up, and in general, we can actually give a little formula for this—for the rule, anyway: If we take the natural number n, we would pair it up with its double, namely 2 times n.

Is there any number that's left unpaired from the natural numbers? No, every natural number is paired up. Is there any even number that's not paired to a natural number? For example, let's think of the even number 102. Is 102 paired up with some natural number? Sure. Namely 51, because when we take 51 and we double it, we get 102. Well, therefore, we see that these two collections have the same size. They have the same cardinality. This realization is surprising, since we're realizing that, in some sense, half of infinity is the same size as infinity. Yet again, we see a counterintuitive example to the Euclid common notion from 2,000 years earlier that the whole is greater than the part.

In fact, this observation may have inspired Richard Dedekind, a good friend and close colleague of Cantor's, to define an infinite collection to be, in effect, any collection that can be put in a one-to-one correspondence with just a part of itself. In a mathematical sense, this is our first precise definition of what we mean by an infinite collection. And in fact, now I can quickly prove that I do not have an infinite number of digits on my left hand. Here's the proof. I have to show you that it's impossible to put these digits in a one-to-one correspondence with just some of them. Well, certainly I can't take these digits and put them in a one-to-one correspondence with just my thumb because I'd be missing a whole bunch of other fingers that would not be paired up. Nor could I do it with just two fingers, because if I try with two fingers, I wouldn't have anything else left for these, or three, or four. The only one that works is five. That means that this collection must be a finite collection. Well, we now see that Galileo actually held the key to measuring infinite collections by means of a one-to-one pairing, but viewed it as a paradox. Namely, seeing the whole as the same size as the part perplexed him. And thus, again, unfortunately, he let this key idea slip through his fingers.

Let's now compare the collections of natural numbers with the collection of the integers. Now in mathematics, we traditionally use Z to denote the entire collection of the integers. Now remember, the integers are all the natural numbers, together with 0, and then all the negatives of the natural numbers. So, we'd write, "Z equals the set of the natural numbers, 0, and then all the negative natural numbers." Why do we use the letter Z? Well, Z might stand for *zahlen*, which is German for "number or quantity," and that's exactly where it comes from.

What are we doing here? We have the natural numbers that go 1, 2, 3, 4, and so forth, and now, in some sense, we're doubling it, because we're throwing in all the negatives as well. So, does the collection of natural numbers have the same cardinality as the collection of integers? Well, let's just try it. But I'm going to warn you right now that we're going to fail. But remember, failing is always good, because it leads to insights. See, on the one hand, we can say, "Well, no, this doesn't work," because I'll imagine the natural numbers on top, and I'll imagine all the integers here. I'll take 1 and pair it with 1, 2 with 2, 3 with 3, 4 with 4, and so forth, all the way out. And what do I see? I see that 0 and all the negative numbers have no pairs. And so, therefore, I see that this is not working. This is not a one-to-one correspondence. Well, yes. It's not a one-to-one correspondence, but that just means that that particular try didn't work. It doesn't imply that a one-to-one correspondence doesn't exist some other way.

Let's try it again and be a little bit more clever. This time—I'm calling the integers "red numbers" to distinguish them from the natural numbers—let's take the red 0 and pair it up with 1. Let's take red 1 and pair it up with 2; red –1 and pair it up with 3; red 2 and pair it up with 4; red –2 and pair it up with 5; red 3 and pair it up with 6; red –3 and pair it up with 7, and so forth. What we're doing here is we keep swinging back and forth from the positive to the negative integers like the hand on an old fashioned metronome. We just keep counting and counting and counting. So in this case, we see that, yes, there is a one-to-one correspondence because every natural number has been paired up with exactly one integer, and every integer, in turn, has been paired up with exactly one (and only one) natural number. So, we see that, in fact, we have a one-to-one pairing here. We actually are pairing the red positive numbers with the black even numbers, and the red negative numbers with the black odd numbers. In fact, that's the rule.

So, we're trying here to compare stuff, and we're seeing all sorts of counterintuitive things. If we take half of a collection that's infinite, it's the same size of infinity. If we try to double the infinity, we see it's the same size. Well, let's now compare the collection of natural numbers with the collection of rational numbers, all the fractions, all the ratios. We write Q for the entire collection of rational numbers. In other words, we say, "Q equals the collection of all fractions." The letter Q might stand for quotients, but actually I'm not sure about that, and I'm not sure if anyone really knows for sure. But the tradition in mathematics is to use a Q to denote the rational numbers.

Does the collection of natural numbers have the same cardinality as the collection of rational numbers? Now, the answer surprised many people. Because, if you think about it, you say the rationals must be much larger, because between any two counting numbers (like, for example, 1 to 2) there are infinitely many fractions, as we have seen. In fact, we proved that the fractions are dense. So, they are infinitely many right in between there, and so this would look like almost taking infinity and multiplying it by infinity. We took half of infinity, it didn't change the size. We doubled infinity, it didn't change the size. Now, in some sense, we're taking infinity and multiplying it by infinity, that must change the size, wouldn't you think?

The surprise is that these two collections, in fact, have the same cardinality. In other words, we can find a one-to-one correspondence between all the rational numbers and all the natural numbers. Now, this is a little bit tricky, and so, to inspire this particular one-to-one pairing, let's just first consider a cloud of dots.

So, we have this cloud of dots. And the question is, how would we count how many dots we have? Well, what we could do, of course, is what we've done when we were kids. We would just connect the dots, and as we connect the dots with little line segments, we could write a number at each dot so that connecting the dots makes sure that we hit every dot and hit every dot exactly once. So, we count—for example, I'll do it right now—1, 2, 3, 4, 5, 6, 7, 8, 9, 10. So, I see that we actually have 10 dots. And notice that those dots, and those lines connecting them, sort of form a snake, and that snake actually is snaking through these dots and shows us that we actually have a one-to-one pairing between the numbers from 1 to 10 and the dots. Right? We see that every dot was hit once, and we didn't count any dot twice. So, this snake helped us.

Let's now extend this idea to the cloud of ratios. To make room for the negative rational numbers in our list, we're first not going to worry about the negatives, but instead we'll just take a look at the positives. So, here's what we're going to do. Let's back up here and think about it. We're going to make a big array. The big array—on the top, I'm going to list all the counting numbers, but I'll start with 0. So, I'll say 0, and then 1, 2, 3, 4, and so forth. Then, down here, I'm going to list all the natural numbers—1, 2, 3, 4, and so forth, all the way down. Then the way I'm going to fill up this array is I'll take this number, and I'll look at the ratio of this number to this number, and I'll put it right here. So for example, 3/5. I go over to the number 3, and I go down to 5, and there's 3/5. So, I fill this grid with fractions. In fact, if you notice, every positive fraction is on this infinitely long chart. Well, now what do I do? Now of course, the fractions appear quite often. They repeat. For example, 2/2, 3/3, 4/4 are all just the same number, the number one. So, when we go through this cloud of rational numbers, we have to make sure that we're never going to count something that we've already counted.

But now we're just going to proceed like with the cloud of dots. I start in the upper top, and I associate that first rational number with the number 1. Then I slide over to the next number and associate it with 2. Then I slide down and associate that with the number 3, then slide over—4, 5. Now, if I get to a number that I've already seen, I just pass over that number and go to the next one. And I just keep zigzagging through this chart. And each time I zigzag, I'm counting—1, 2, 3, 4, 5, 6, and so forth. It's like taking the number line, and bending it up, and snaking it through. We see a one-to-one correspondence, for every natural number is used on this chart (because the chart is endless) and every fraction's been hit. For example, let's think about 35/22. Well, if we go over 35 and then down 22, there's that rational number. And this snake, you can see, in a finite number of steps, will eventually pass through that. And it will be associated with some natural number. So, we have this one-to-one correspondence.

Now, what about the negatives? Well, we now take our positive fractions, and we just shift them over so that they're paired up with the even numbers. In other words, let's slide the rational numbers down so that they're just paired up with every other natural number. And so, now the odd natural numbers have no partners, and in this pairing, what we can do is just plug in the negative of every fraction next to its neighbor. So, we just sort of fill in the other half with the negatives, and so we actually see a one-to-one correspondence in this way.

So again, if you try to find a formula for this, it's very, very tricky, but we can visualize this very easily by snaking through and determining that, in fact, we have a one-to-one pairing. Every natural number is paired with a particular rational number, and conversely, every ratio is paired with a natural number so that the size is the same. These two collections have the same size. In some informal sense, infinity multiplied by infinity still yields the same size of infinity. We haven't upped anything.

As we'll see in the next few lectures, infinity holds many more profound surprises. As the great early 20th-century mathematician David Hilbert once wrote:

The infinite. No other question has ever moved so profoundly the spirit of man. No other idea has so fruitfully stimulated his intellect. Yet no other concept stands in greater need of clarification than that of the infinite.

That clarification, and that wonderful journey to infinity, will be the concluding topic of this wonderful class.

Collections Too Infinite to Count
Lecture 21

Infinity, just as the numbers that came before, comes in different sizes.

Countable collections lead us to challenge our notion of infinity. We say that the cardinality of a collection is countable if it has the same cardinality as the collection of natural numbers or if the collection contains only finitely many elements. The collection of cards in a deck of playing cards is countable because the number of cards in the deck is a finite number: 52. Each of the following collections of numbers is a countable collection: the collection of even numbers, the collection of integers, and the collection of rational numbers. It seems obvious that any infinite collection should have the same cardinality as any other infinite collection, but is this sensible-sounding assertion mathematically correct?

Cantor's ingenious insight was to look at one-to-one correspondence in a new way. Suppose that the collection of real numbers is countable. We will just consider the interval of real numbers between 0 and 1. We will assume that the cardinality of natural numbers is the same as the cardinality of all real numbers between 0 and 1. We thus assume that there *does* exist a one-to-one correspondence between the collection of natural numbers and the collection of real numbers between 0 and 1. What might such a one-to-one correspondence look like?

Cantor used the digits of real numbers appearing on our assumed correspondence to generate a real number. He focused on the digits that lie along the "diagonal" of the column of real numbers. He then created a real number that does *not* appear on our list. Each digit of this new real number was selected so that it differed from the corresponding digit along the "diagonal" digits. Cantor then argued that this new number is a real number between 0 and 1 that never appears in our assumed one-to-one correspondence. Our assumption that a one-to-one correspondence exists between these two

The collection of real numbers between 0 and 1 is a larger collection than the collection of natural numbers.

collections is, thus, false. We are forced to conclude that the collection of real numbers between 0 and 1 is a larger collection than the collection of natural numbers; there is no way to pair them all up. For this reason, there can be no one-to-one correspondence between natural numbers and the entire collection of real numbers. Cantor's method of proof is now known as Cantor diagonalization.

Cantor discovered that infinity comes in more than one size. Informally, the infinity that represents the cardinality of real numbers is larger than the infinity that represents the cardinality of natural numbers. The real numbers are *not* countable. We call collections that are not countable "uncountable."

There were numerous reactions to this startling reality from the mathematics community. In 1831, Carl Friedrich Gauss protested vehemently against Cantor's position. In 1906, the French mathematician Henri Poincare wrote that there was no actual infinity; he saw the Cantorians as being trapped by contradictions. Leopold Kronecker was a powerful opponent of Cantor's work, which established the existence of infinitely many irrational numbers (in fact, uncountably many). Kronecker did not view irrational numbers as natural. Given that Kronecker was perhaps the most powerful mathematician in Germany during his lifetime, his opposition to Cantor had an enormous impact. Carl Weierstrass, Bertrand Russell, and David Hilbert were impressed with Cantor's work and defended Cantor to his detractors.

Despite the support of these mathematicians, Cantor struggled throughout his career. Cantor spent his entire career at the University of Halle, a less prestigious institution than he felt he deserved. He desired a position at the University of Berlin, the premiere German institution, but the department chair was Kronecker. He imagined Kronecker's distress were he to obtain a position in Berlin. Cantor was extremely confident about the truth of his work on infinity. The controversy inspired by his work took a toll.

Bearing Cantor's work in mind, we must revisit irrational and transcendental numbers. We start with the cardinalities of the collections of rational numbers and irrational numbers. Recall that the collection of rational numbers is a countable collection. Because real numbers are uncountable, we conclude that the collection of irrational numbers is uncountable. This discovery reflects back to our previous discussions of irrationals; again, we see our recurring theme of familiar and exotic numbers. By a modified diagonalization argument, we can establish the fact that the collection of all algebraic numbers is countable; thus, we discover that the collection of transcendental numbers is uncountable.

Recall that the Cantor set is the collection of real numbers between 0 and 1 whose base-3 expansions contain only 0s and 2s. In view of our previous discoveries from Lecture 11, we deduce that this collection of numbers is uncountable. In this way, we discover that there must be transcendental numbers in the Cantor set. ∎

Questions to Consider

1. How could mathematicians of the day be so opposed to Cantor's ideas leading to different sizes of infinity?

2. The list below attempts to show a pairing between the natural numbers and the real numbers. Use Cantor's idea of diagonalization to produce the first seven decimal digits of a number that will not be on the list.

   $1 \Leftrightarrow 0.27364810\ldots$

   $2 \Leftrightarrow 0.01926573\ldots$

   $3 \Leftrightarrow 0.22937510\ldots$

   $4 \Leftrightarrow 0.61100029\ldots$

   $5 \Leftrightarrow 0.71099058\ldots$

   $6 \Leftrightarrow 0.29384655\ldots$

   $7 \Leftrightarrow 0.56478392\ldots$

   $\ldots$

Collections Too Infinite to Count
Lecture 21—Transcript

In this lecture, we offer one of the most counterintuitive and surprising facts about our world. Infinity, just as the numbers that came before, comes in different sizes. This incredible discovery was first made by Georg Cantor in 1874. Here, we'll tell his story and his struggle to have his outrageous, but totally correct, mathematical ideas accepted by the mathematical community. His wondrous argument, at once simple and subtle, proves that the collection of real numbers is a greater infinity than the collection of natural numbers. While the natural numbers can be listed in an orderly fashion and, in some sense, can be counted, we cannot count the real numbers. Collections that are larger than the natural numbers are known as *uncountable* collections.

Well, we'll then connect our discussion on the sizes of infinity with our earlier discussions on the likelihood that a real number, selected at random, will be irrational and transcendental. We'll describe how both the collections of irrational numbers and transcendental numbers are so large that they are, in fact, uncountable as well. We'll also connect our observations with the previous discussions of the Cantor set we had, and we'll show that that while, on one hand the set is a very dusty collection of numbers and is sparse, on the other hand, it's actually quite robust. In fact, the Cantor set, we'll see, is uncountable as well. Once again, we open our minds to the reality that there are at least two sizes of infinity, and once we do that, we see the nature of our universe in an entirely new light.

Well, let's open what will be the most mathematically dramatic and shocking lecture of this entire course with a totally innocuous riddle. Why is it so easy to count the digits on my right hand? Well, let's see. First of all, I can do it. Let me prove that to you—1, 2, 3, 4, 5. Now, why was that so easy? Well, one answer is that they all lined up in an orderly fashion, or perhaps I should say ordinally fashion, because were talking about order. There's a first, second, third, fourth, and fifth. Well, notice that the collection of natural numbers, although they're infinite in size, can be easily and endlessly lined up in an ordinally fashion: 1 is the first, 2 is the second, 3 is the third, and so on. Well, I want us to now formalize this trivial observation and see where it leads.

We say that a collection is countable if that collection contains only finitely many elements, or has the same cardinality as the collection of natural numbers. In other words, a collection is a countable collection, or we say, "It's countable," if we can line up its elements from that collection so that there's a clear first element, then a second, a third, a fourth, a fifth, and so on. That is, I can list the elements in this collection even if our list is endless. For example, the collection of cards in a deck of playing cards is countable because, in fact, the number of cards in the deck is finite. It's a finite number, it's 52, and we can count them just off the deck. The first, second, third—so we have 1, 2, 3, 4, … 49, 50, 51, 52, and there we can see we can actually count them. We can list them all.

Applying this new notion to our previous observations from the previous lecture, we can now say that the collection of even numbers is a countable collection, because we found in that lecture a one-to-one correspondence between it and the collection of natural numbers. So, it fits the bill. The collection of integers is also a countable collection, because remember there what we did was we took the positive integers and the negative integers, and we sort of shuffled them together and then we were just able to list them one after the other and pair them up with the natural numbers. Even the collection of rational numbers is a countable collection, because we found a one-to-one correspondence between it and the collection of natural numbers. In other words, we were able to produce an orderly list that contains every rational number, and that was how we snaked through that large array of fractions.

This idea brings us to a very natural question. Is the collection of real numbers countable? Well, we know that the collection of real numbers is infinite, so we're asking, Is the collection of real numbers a countable collection? Which means, Is there a one-to-one correspondence between the collection of real numbers and the collection of natural numbers? Well intuitively, it seems obvious that infinity is infinity, after all. In other words, any infinite collection should have the same cardinality as any other infinite collection. Certainly, every infinite collection that we've considered thus far has had the same cardinality as the natural numbers. The big question now is, Is the previous sensible-sounding assertion mathematically correct?

This question captured Cantor's imagination and led to his ingenious insight. In a shocking turn of events, Cantor was able to prove that there cannot exist a one-to-one correspondence between the natural numbers and the collection of real numbers. In other words, Cantor proves that the real numbers are too numerous to be paired up with the natural numbers. The collection of real numbers is not a countable collection. Its cardinality is actually a larger infinity than the countably infinite collection of natural numbers.

Let's now follow in Cantor's logical footsteps and deduce this astounding fact for ourselves. Cantor offers an indirect proof. In other words, he assumes that there is a one-to-one correspondence between the natural numbers and the real numbers, and then shows that this assumption leads to a contradiction, a logical impossibility. Therefore, his assumption must have been false, and so the collection of real numbers is larger than the collection of natural numbers.

Let's see how Cantor's argument unfolds. To simplify the notation, we'll just consider the real numbers between 0 and 1 on the number line and show that this collection of numbers is actually larger than the collection of natural numbers. In other words, instead of working with the entire real number line and all the numbers on it, we'll just restrict ourselves to the points between 0 and 1 and show that this collection is not countable. I mean, of course, if the collection of real numbers in this small interval is not countable, then certainly the collection of all the real numbers must not be countable as well.

So, Cantor's assumption, which we plan to prove is wrong, is that the cardinality of the natural numbers is the same as the cardinality of all real numbers between 0 and 1. This assumption implicitly implies that there *does* exist a one-to-one correspondence between the collection of natural numbers and the collection of real numbers between 0 and 1. Now, let's just recall that the real numbers between 0 and 1 can be expressed as an unending decimal expansion starting with "0-point." For example, 0.56982, and so forth—that's one number, one real number, between 0 and 1—and they all have that same flavor; they start with "0-point" and then an endless run of digits.

What might such a one-to-one correspondence between the natural numbers and the real numbers between 0 and 1 look like? Well, it would look like a list of real numbers, one matched up or paired with each natural number. Cantor is assuming that some such list exists, although we have no idea of the particulars of the numbers appearing in that list. Now just to give an illustration of what this presumed list might look like, let's consider a particular attempt.

Now, we'll use this example just to demonstrate the idea of Cantor's general argument. So, you see here a table where, in the first column, we see the counting numbers, the natural numbers (1, 2, 3, 4, 5, and so on), and then paired up with each, in the second column, we see this list of real numbers represented in decimal form.

Cantor now uses the digits of these real numbers appearing in his assumed one-to-one correspondence to generate a real number that is definitely absent from his list. Now, in order to generate this missing number, Cantor focuses on the digits that lie along the "diagonal" in the column of real numbers. Then he's going to describe his missing number by giving its decimal expansion. So, he describes this number's digits systematically, one digit at a time. Each digit of this missing real number is selected so that it differs from the corresponding digit along the "diagonal" digits from his real numbers in the chart, in this assumed one-to-one correspondence.

Let's take a look at the example. For this particular example, let's see what Cantor might do. We look at the number paired up with the natural number 1, and we see it's 0.8472651, and so forth. Cantor only cares about the very first digit, which is an 8. He'll now, in the number that he's constructing, just switch it to make sure that the digit that he writes down is different from 8. In particular, let's say he writes down a 5. So, he starts his number 0.5. Now we need the second digit of our allegedly missing number. Where do we go? We look at the number paired up with the natural number 2; it's 0.5000, and so forth. I look at its second digit, and I see that it's a 0, and I want to write a digit that's not a 0, so I'll just put down a 5 again. So, now I see my second digit of my number is 5, and I continue this process. To find the third digit of the number we're constructing, we look at the number associated, or paired up, with the number 3. It happens to be 0.3333, and so forth. I look at its third digit, I see that it's a 3, and so now I'll want to switch it to make sure that our digit doesn't equal 3, and so I'll make it a 5 again. So far, our number is 0.555. What should our next digit be? Well, I look at the fourth number on our list, 4, which is paired up with 0.1075963, and so forth. I look at the fourth digit of that real number which, notice, is a 5. So, I'll write down a digit that differs from 5, and I'll just pick 8. So, now my fourth digit is 8. For the fifth digit, how do I select that? I look at the number paired up with the number 5, which happens to be, in this example, 0.01000. The fifth digit is a 0, and so what do I do? I'll make sure that our digit is not a 0. I'll select 5 again, and then I repeat this. For the sixth digit, I look at the number paired up with 6. I see 0.216008. I see that the sixth digit there is a 5. I'll switch that, and I'll make that an 8. And I do this and continue forever. So, I've built a real number between 0 and 1.

Now, Cantor argues that this new real number, that is between 0 and 1, never appears in our assumed one-to-one correspondence. Now, why not? Let's think about this. If it appears somewhere on my list, then that means that all the digits must be perfectly lining up somewhere on that chart. Could it be in the first spot? Could the number we just built be paired up with the number 1? No, because we know that it differs at least in the very first decimal spot. The number paired up with 1 has 8, and we switched ours, it's a 5. So, it can't be that number. We don't even care about the rest of the digits; we know it can't be, because of that. Could it be paired up with the number 2? No, because we know that the number paired up with the number 2 will differ, at least in the second spot, with our number, because we picked our second digit to be different from the second digit of the second number. Could this number actually be on the chart somewhere further down, like in the millionth spot? Well, no, because of the way we systematically developed this number. If we go down to the millionth decimal spot, we see that whatever digit is there, our digit in that same spot will differ. Therefore, this number is not that number paired up with 1,000,000, and this is true for all the numbers. This number is genuinely not on our chart.

Notice that this argument works no matter what the real numbers are and which ones we've written in the right column. And so, we can always look at the diagonal digits and create a real number by simply switching the corresponding digits so that the newly constructed number is a real number that's definitely not on our list. Thus, his assumption that there was a one-to-one correspondence between these two collections is false. In other words, Cantor proved that any such list of real numbers will be incomplete. Therefore, we're forced to conclude that the collection of real numbers between 0 and 1 is a larger collection than the collection of natural numbers. It's impossible to pair them up in a one-to-one fashion.

Thus, Cantor proves that it's impossible to have a one-to-one correspondence between the natural numbers and the real numbers. Well, Cantor's method of proof is now known as "Cantor diagonalization," because we use those diagonal digits to switch them in order to build a missing number from that list. This straightforward argument is so subtle that it takes a long time to understand and digest it completely. However, once we see that it's correct, we're forced into a totally counterintuitive realization. Cantor discovered that infinity comes in more than one size. Informally, the infinity that represents the cardinality of the real numbers is a larger infinity than the infinity that represents the cardinality of the natural numbers. In other words, the real numbers are *not* countable.

Cantor proved this startling result in 1874, and he called the collection of real numbers the "continuum." Today, we call collections that are not countable "uncountable." So, we can now say that the collection of real numbers is uncountable. Well, as we struggle to comprehend this counterintuitive idea that surrounds infinity, it is perhaps comforting to realize that we're not alone. The mathematics community itself struggled with and resisted these same ideas when it was first grappling with them.

In 1831, the great Carl Friedrich Gauss wrote:

> I protest against the use of infinite magnitude as something completed, which is never permissible in mathematics. Infinity is merely a way of speaking, the true meaning being a limit which certain ratios approach indefinitely close, while others are permitted to increase without restriction.

So, here he was thinking about heading up to infinity by going off the horizon. This was his view.

In 1906, one of France's greatest mathematicians, Henri Poincaré, wrote, "There is no actual infinity, that is what the Cantorians have forgotten and have been trapped by contradictions." So, Poincaré was, in fact, dramatically opposed to this idea. In fact, he is said to have actually said of Cantor's ideas that the ideas were a "grave disease." Here, we see Cantor, in some sense in Poincaré's mind, has infected mathematics—a very dramatic thought, if you think about it, over a mathematical principle, but it was so outlandish that it was it hard to believe.

Now, Poincaré was not alone in this point of view. As we'll see in just a moment, Cantor's work established, among other things, the existence of infinitely many irrational numbers, in fact, uncountably many irrational numbers. Now as we noted in Lecture Eighteen, Leopold Kronecker did not view the irrational numbers as natural at all. Remember that he wrote the famous line "God created the integers, the rest is the work of man," reminiscent of the Pythagoreans, in fact. Even though Cantor was a former student, Kronecker called him a "corrupter of youth" and a "scientific charlatan." Really brutal words and emotionally charged, because he just did not want to accept the arguments that Cantor was offering.

Now, Kronecker was perhaps the most important and powerful mathematician in Germany during his lifetime, so his opposition to Cantor had an enormous impact. In fact, when Cantor published his paper showing that there were more real numbers than natural numbers, he gave the paper an innocuous title, in part, some suggest, so as not to draw Kronecker's attention to it during the review process. So, the title that Cantor gave the paper was "On a Property of the Set of All Real Algebraic Numbers," which really says almost nothing at all. But the idea was that it would fall on to Kronecker's desk, and he would just look it over very quickly and say, Okay, that's fine; we can just publish that—and not read the details. And in fact, that's exactly what happened, and it was published, although if Kronecker would have looked a little bit closer, he would have certainly never allowed the publication of this very controversial work.

Well, of course, Cantor did have some supporters. Karl Weierstrass, another great German mathematician of the day, and the Ph.D. advisor for Cantor, was impressed with Cantor's original and interesting and dramatic work, and he encouraged Cantor to publish it and he defended Cantor to his detractors. Richard Dedekind, another important German mathematician at the time, was also a longtime friend and supporter, as well. Once the mathematics community had the time to absorb Cantor's groundbreaking ideas, it was finally able to embrace them.

In 1917, Bertrand Russell wrote that Cantor had "conquered for the intellect a new and vast province which had been given over to Chaos and Night." So, here Cantor is combating the chaos and the darkness and is shining the light on infinity. So, again, dramatic, but now on the other side—dramatic support. Also during this period, the great mathematician David Hilbert had said, "No one shall expel us from the Paradise that Cantor has created." So, here we go from corrupting youth and from a grave disease infecting mathematics to Paradise. You can see this dramatic shift once the math community has decided that it will, in fact, embrace these correct ideas of Cantor despite their own personal biases before. Hilbert also said Cantor's work was "the most astonishing product of mathematical thought, and one of the supreme achievements of purely intellectual human activity." So, Hilbert is putting this work on the very summit of all intellectual thought, which is, again, dramatic in a very positive way and justified, certainly in my opinion and I believe in the opinion of many.

Cantor himself felt the wrath of all the controversy surrounding his original work. Cantor spent his entire career at the University of Halle, a less prestigious institution than he felt he deserved. In fact, he desired a position at the University of Berlin, the top German institution, but the chair of the mathematics department was none other than Kronecker. He actually imagined Kronecker's distress, were he to obtain a position in Berlin. In fact, he wrote about this, and it's almost comical. So, he's able to take a light-hearted approach to this grave feud between the two. He writes of Kronecker's sense of what would happen if Cantor were to show up:

> I knew precisely the immediate effect this would have: that in fact Kronecker would flare up as if stung by a scorpion, and with his reserve troops would strike up such a howl that Berlin would think it had been transported to the sandy deserts of Africa, with its lions, tigers, and hyenas.

So, you really get a sense of the ruckus that Kronecker would kick up. While Cantor did know how Kronecker's reaction would unfold, apparently he was unaware that there were no tigers in Africa. But we'll let that go; it's still a wonderful quote.

Despite the enormity of all the controversy, Cantor was extremely confident and steadfast about the truth of his work on infinity. In fact, in 1888 he wrote:

> My theory stands firm as a rock; every arrow directed against it will return quickly to its archer. How do I know this? Because I have studied it from all sides for many years; because I have examined all objections which have ever been made against the infinite numbers.

So, here we see Cantor as a victim. He feels as a victim; everyone shooting at him, and he's going to repel all of these arrows that are being flung toward him, and he's going to stand strong. He's very confident in his work, which is difficult to do, you could imagine. If an entire community is against you, it's difficult to stand up and say, "No. I'm right, and I will fight for that belief." However, I should point out that the controversy inspired by his work continued to take a toll on Cantor, as we'll discuss in a later lecture. So, even though he was very firm in his beliefs, it did take a personal toll on life.

Let's now return to the notion of number and consider Cantor's work on some of the types of numbers that we've already explored in this course. Now, in the previous lecture, we established that the collection of rational numbers is a countable collection. In other words, we found a one-to-one correspondence between the rational numbers (the fractions) and the natural numbers. Now, let's remember that the real numbers come in two basic flavors. They're either rational (ratios) or they're irrational. Now since the entirety of real numbers is uncountable, we can conclude that the collection of irrational numbers is uncountable as well.

Let's think about why that's true. Let's do this indirectly like Cantor would do. Let's pretend that the collection of irrational numbers was, in fact, countable. Well, then we have the rational numbers, that's a countable collection, and the irrational numbers, we're assuming that's a countable collection. So we have two countable collections. And in fact, if you take a countable collection and a countable collection and put them together, we can interleave them like we did with the positive and the negative numbers of the integers to see that we can list them all. So, in fact, then their totality would also be countable. But we've already said the totality, in fact, are the real numbers; and they're uncountable, so that's a contradiction. Therefore, since the infinity of the rationals is so small and the infinity of the reals is so large, the infinity of what's left over, which are the irrational numbers, must be large as well. In particular, they must be uncountable.

This discovery amplifies our previous discussions of irrational numbers being the norm. Remember, again we see this recurring theme that the at-first exotic irrational numbers are actually abundant, while the more familiar rational numbers are, in fact, really rare. In fact, the irrationals totally dominate, because they're uncountable.

Going a step further by a modified diagonalization argument, it's possible to actually establish the fact that the collection of all algebraic numbers—that means like $\sqrt{2}$ and the number i, $\sqrt{-1}$, and so forth—they're countable as well. Thus, since the real numbers are either algebraic or transcendental—the opposite of being algebraic—we discover that the collection of transcendental numbers is uncountable. And it's by the same principal. We have the real numbers, an uncountable collection, broken up into two pieces, the algebraics and the transcendentals. The algebraics are countable. If the transcendentals were to be countable, then these two collections together would be countable as well. But the reals are, in fact, uncountable. Therefore, the transcendentals must also be uncountable. So, again we see that the transcendentals are very robust, whereas the algebraics ($\sqrt{2}$), which we know very well, are actually the exotic ones.

Let's conclude by returning to Cantor's original collection of numbers that we explored in Lecture Eleven, his Cantor set, or the so-called Cantor dust. Let's recall that the Cantor set is the collection of real numbers between 0 and 1 whose base-3 expansions contains only 0s and 2s. The digit 1 never appears. That was the defining trait of the Cantor set. In Lecture Eleven, we described the one-to-one correspondence that Cantor himself gave between the Cantor set and all the real numbers between 0 and 1. Therefore, these two collections have the same cardinality. In other words, the Cantor set is an uncountable collection of numbers.

What can we conclude from this particular discovery? Well, since we know that there are only countably many algebraic numbers, we now discover that there must exist transcendental numbers in the Cantor set. Let's think about this now. The Cantor set, Cantor showed in Lecture Eleven, is an uncountable collection. That means that it's larger than the collection of algebraic numbers, which is merely countably infinite. Well, is it possible that this large collection, which is larger than the collection of all the algebraic numbers, could this large collection actually contain only algebraic numbers? Well, of course not, because the algebraic numbers are small and this set is even larger. So, there must exist at least one number in the Cantor set that's not an algebraic number, because the set is too big. It can't fit in with all the algebraics. And so, therefore we actually have established indirectly that there must exist transcendental numbers in the Cantor set. In particular, numbers that when expressed in base 3 just using 0s and 2s, will never be the solution to any polynomial equation with integers. This is highly not obvious, and, even using the modern technology of today's mathematics, we don't know how to prove this result without going back to Cantor's amazing idea showing that the transcendental numbers, in fact, are uncountable.

Well, thus we see that not only does Cantor's incredible work allow us to slowly see a more accurate vision of infinity, in particular, that some infinities are larger than other infinities, but by applying his ideas, we are drawn into an even richer understanding and appreciation for the subtly of number.

In and Out—The Road to a Third Infinity
Lecture 22

In this lecture, we extend our reach even further into the stratosphere of infinity. ... That is, we'll find a third size of infinity.

The goal of this lecture is to construct a collection whose cardinality is greater than the uncountable cardinality of the collection of real numbers. To this end, we will follow Cantor's own creative path yet again and consider the abstract but attractive idea of power sets. We begin with an examination of sets. A set is an abstract collection of elements. For example, the collection that contains the Marx family—Chico, Groucho, Harpo—can be viewed as a set with three elements: {Chico, Groucho, Harpo}. Another example is the collection of the two most popular condiments in this country, ketchup and mustard: {ketchup, mustard}. The term *set* first appeared in 1851 in a work by the Italian mathematician Bernhard Bolzano entitled *Paradoxes of the Infinite*.

A subset of a set is any collection that comes from a given set. For example, {Chico, Harpo} is a subset of {Chico, Groucho, Harpo}. An empty set is a subset that contains no elements. The entire set is the other extreme example of a subset (called the improper subset).

Given a set, we define the power set of the set to equal the set whose elements are precisely all the subsets of the original set.

Given a set, we define the power set of the set to equal the set whose elements are precisely all the subsets of the original set. Our set of condiments, C = {ketchup, mustard}, has the following four subsets: the empty set, {ketchup}, {mustard}, and {ketchup, mustard}. These four sets are the elements of the power set of {ketchup, mustard}. We call the power set $P(C)$; that is, $P(C)$ = {empty set, {ketchup}, {mustard}, {ketchup, mustard}}. Let S = {Chico, Groucho, Harpo}. The power set of this collection, denoted as $P(S)$, is the collection of all subsets of S; that is, $P(S)$ = {empty set, {Chico}, {Groucho}, {Harpo}, {Chico, Groucho}, {Chico, Harpo}, {Groucho,

Harpo}, {Chico, Groucho, Harpo}}. Notice that the order in which we list the elements does not matter.

We compare the cardinality of a set with its power set. Returning to the previous examples, we notice that the cardinality of C is 2, the cardinality of $P(C)$ is 4; the cardinality of S is 3, and the cardinality of $P(S)$ is 8. The original set is smaller than its power set. Does this observation always hold?

Cantor's Theorem helps us answer this question. Cantor was able to apply his powerful insights into sets to prove that even if a set is infinite, its cardinality will always be smaller than the cardinality of its associated power set. This result is known as Cantor's Theorem. Cantor used his diagonalization idea to prove his result about power sets. We illustrate his argument with our example involving the three Marx brothers. We assume that, contrary to what we wish to establish, the set S and its power set $P(S)$ have the same cardinality. Given this assumption, a one-to-one correspondence between the elements of the set S and the elements of the set $P(S)$ must exist. For example:

Chico $\Leftrightarrow$ {Chico, Groucho, Harpo}

Groucho $\Leftrightarrow$ { Harpo}

Harpo $\Leftrightarrow$ {Chico, Groucho } Cantor observed that any element in S is either in a subset or not; there are exactly two possibilities.

Cantor modified his diagonalization idea to create a new subset of S. If the element in the left column was in the collection to which it is paired (the corresponding collection in the right column), then he did not include this element in his new subset. If the element in the left column was not in the associated subset in the right column, then he did include that element in his new set. In our example, his diagonalization method would produce the subset {Groucho, Harpo}. We know, by the way we produced it, that this subset does not appear in the right column. We can generalize this idea to see that no one-to-one pairing between a set S and its power set $P(S)$ can exist—we can always find a missing element of $P(S)$ whenever we try to pair up the elements from the two collections.

An infinity exists beyond the cardinality of the real numbers. Recall Cantor's discovery that the cardinality of the set of natural numbers is smaller than the cardinality of the set of real numbers; that is, the infinity of the set of natural numbers is a smaller infinity than the infinity of the set of real numbers. How can we find a collection whose cardinality is larger than the cardinality of the collection of real numbers? We can consider the power set of real numbers; that is, we can consider the collection of all subsets of real numbers. The power set of real numbers has a greater cardinality than the set of real numbers, which means there is a third size of infinity! ∎

Questions to Consider

1. Consider the collection of suits in a deck of cards: {♣, ª, ©, "}. How many elements are there in the power set of this collection? Can you list them all?

2. Do you believe there is a collection that has a greater cardinality than the cardinality of the power set of the real numbers, or do you think that the power set of the real numbers has the largest possible cardinality?

In and Out—The Road to a Third Infinity
Lecture 22—Transcript

In this lecture, we extend our reach even further into the stratosphere of infinity. Our goal is to construct a collection whose cardinality is greater than the uncountable cardinality of the collection of real numbers. That is, we'll find a third size of infinity. To this end, we'll again follow Cantor's own creative path and consider the abstract, but attractive, idea of so-called power sets. A power set associated with a collection is the totality of all possible sub-collections of the original collection.

After shoring up our understanding of this new idea through several simple examples, we'll return to Cantor's diagonalization argument, which established that the cardinality of the real numbers is greater than the cardinality of the natural numbers. We'll apply it to prove that given any collection, the cardinality of its associated power set is greater than the cardinality of the original collection. We will then apply this principle, now known as Cantor's Theorem, to show that there exists an infinite collection that has greater cardinality than the collection of the real numbers. We'll prove that there are at least three different sizes of infinity.

So far in our journey through infinity, we've studied collections of numbers. Now we'll abstract that exploration and study collections, not of numbers, but of collections. In other words, we'll now study collections whose elements are themselves collections. Let's begin by informally describing what mathematicians mean by "set." By a "set" we mean "an abstract collection of elements." For example, the collection that contains the funny brothers of the Marx family, Chico, Groucho, and Harpo. We can view that as a set of three elements, and we denote that with that curly bracket, and we'd say, {Chico, Groucho, Harpo}. And the curly bracket would be read as "the set consisting of Chico, Groucho, Harpo." Notice that the order doesn't matter at all here. It's just the totality of all those things. Another example is the collection of the two most popular condiments in this country, ketchup and mustard. Again we can write that as {ketchup, mustard}. The order doesn't matter.

In fact, throughout our course, we've been working with sets of numbers, although we've referred to them as "collections." Now we'll adopt a more standard mathematical terminology and call them "sets." And so for example, the set of natural numbers is {1, 2, 3, 4, ...}; the set of integers is the collection of all the natural numbers, together with 0, and all the negatives of the natural numbers; the set of rational numbers equals all the fractions; the set of real numbers..., and so on. So, we'll use the word "set" to mean "collection."

The word "set" appeared for the first time in mathematics in 1851 in a work authored by Italian mathematician Bernhard Bolzano, considered by many to be the father of set theory. The work was entitled *Paradoxes of the Infinite*. Now, in his text he offers the very first definition of a set. In fact, he writes, "I call a set a collection where the order of its parts is irrelevant and where nothing essential is changed if only the order is changed." So, again you see it's a collection where order doesn't matter. Now, it turns out that the concept of a set, or a collection, is much more subtle than it would first appear. This observation was made famous by Bertrand Russell in 1901. He illustrated the subtly with an amusing paradox involving a barber. Today, this paradox is actually known as the Barber's Paradox.

Here we go. Suppose there's a small village in which every man shaves, but there's only one barber in this town, whose name is Pierre. Let's further assume that there's a strange law declaring that this barber must shave all men, and only those men, who do not shave themselves. Let's just think about this for a second before we go on in the paradox. So, we have a small village. There's one barber. Each man must shave, and either he can shave himself or he must be shaved by the barber. And the barber only shaves those men who decide not to shave themselves. That's the law that was decreed for some strange reason.

Let's now consider the set of all men from this village who do not shave themselves. Well here's the question for us to contemplate: Is Pierre the barber in this set? Well, if he's in this set, that means that he doesn't shave himself, which then by law means that he, as the barber, must shave himself. In other words, if Pierre is in the set, then he is not in the set, which is impossible. On the other hand, if Pierre is not in this set, that means that he shaves himself, but by law, he's not allowed to shave those men who shave themselves, thus he cannot shave himself. In other words, if Pierre is not in this set, then he must be in this set, which, again, is impossible. In other words, this law is such that no matter what Pierre does—he either shaves himself or doesn't shave himself—whatever he does, he's breaking the law. Well, that's a paradox.

So, what's giving rise to this funny paradox? Well, the answer is that a set must be "well defined." That is, for the set to truly be a mathematical set, it must be clear what things are in the set and what things are not in the set. The barber scenario is a paradox because we have a set of men for which Pierre can be neither in the set, nor not in the set, and that's impossible. Mathematicians today would say this collection is not well defined and would not consider it a set. This is why the area of set theory, in fact, has so many subtle points. Well, Cantor's work was fundamental in creating the subject that's known as set theory, and actually laid the groundwork for further progress in topology, calculus, and fractals.

Given the notion of a set as a collection of objects, we'll now consider sub-collections of collections, which we'll call "subsets." So, a subset of a set we define to be "any collection that comes from the original set." So, for example, {Chico, Harpo} is a subset of {Chico, Harpo, Groucho}, because those two elements come from this larger set. So, a subset means just a smaller group of things from there. The empty set is the subset that contains no elements at all, and usually we denote that by just two curly brackets with nothing in there— 0, which of course we saw was a big problem in trying to understand 0 as a number. So, here we see {}, the set that consists of nothing. The entire set is another extreme example of a subset. For example, the set {ketchup, mustard} is a subset of itself, {ketchup, mustard}, because all the elements from {ketchup, mustard} are from {ketchup, mustard}. So, the whole collection is another example, an extreme example, of a subset.

Armed with the idea of subsets as subcollections of a set, we'll now gather together all possible subsets of a set and view them as one collection. More precisely, we'll now raise the level of abstraction by considering collections whose elements are themselves a collection. That is, we consider sets consisting of sets—a little weird sounding, but let's think about it. Given a set, we define the "power set" of that set to equal "the set whose elements are precisely all the subsets of the original set."

Let's consider a tasty example—our set of condiments. Let me call it c, so $c = \{$ketchup, mustard$\}$. Well, it has the following four subcollections: $\{\}$, that's always a subcollection; just ketchup alone, so $\{$ketchup$\}$; $\{$mustard$\}$; and then, of course, $\{$ketchup, mustard$\}$, the other extreme, naming all of them. These four subsets are the elements of the power set of $\{$ketchup, mustard$\}$, and we denote the power set by using the letter P, for "power." So, $P(c)$, or P of the condiments, would be $\{\{\ \}, \{$ketchup$\}, \{$mustard$\}, \{$ketchup, mustard$\}\}$. So, you see four elements in this big power set. By the way, again, remember the order in which we list these things doesn't matter at all. Well, notice that the power set represents the four possible toppings that we could put on a burger when faced with a rather skimpy condiment bar containing just two options, ketchup and mustard. And so, in some sense, the power set depicts all possible ways of selecting from the set. So, with respect to a hamburger, you could either put nothing on top of it, you could put just ketchup, just mustard, or both. And those are the elements of the power set.

Let's now consider a more comical collection to illustrate the idea. Let's let s denote the set of brothers Chico, Groucho, Harpo. Well, the power set of this collection, which is denoted as $P(s)$, or the power set of s, is the collection of all subsets of s. In other words, the power set is going to be a big set. It starts, of course, with $\{\}$, so no brothers; then $\{$Chico$\}$; then $\{$Groucho$\}$; then $\{$Harpo$\}$; then $\{$Chico, Groucho$\}$ together; then $\{$Chico, Harpo$\}$ together; and then finally $\{$Groucho, Harpo$\}$ together; and then one last set, which is all of them, the entirety, $\{$Chico, Groucho, Harpo$\}$.

Now notice that in this power set what we're doing here is we're looking at all the ways of taking subcollections of them, taking them off into a room. We either take none of them, we take them one at a time, we take them in pairs, or we take all three. So, again notice that the order in which we list these elements, or these brothers, really doesn't matter at all.

Now that we've seen a couple of examples of power sets, let's compare the size of a set with the size of its associated power set. Reflecting on these examples, we see that the set of condiments has cardinality 2—there was {mustard, ketchup}—and its power set has cardinality 4. We also notice that the set of Marx Brothers has cardinality 3, while its power set has cardinality 8. Well, what do we notice? We see that the original set in both of these examples is smaller than its power set. But does this observation always hold? In particular, the truly interesting question is, Does this observation hold for infinite sets? Well, we'll discover that the answer is yes.

This amazing fact is now known as Cantor's Theorem, and it states that the cardinality of a set is always smaller than the cardinality of its associated power set. Cantor was able to apply his powerful insights into sets to prove that even if a set is infinite, its cardinality will always be smaller than the cardinality of its associated power set—really a remarkable, stunning theorem. Now, in fact, Cantor used his diagonalization idea from the previous lecture to prove his result about the cardinality of power sets, and this is standard in mathematics. When you have a great idea that allows you to prove a theorem, what you try to do is, after you have the theorem, you move on from that and ask, Can I take the same idea, that germ of an idea, and actually prove much more? This is beautifully depicted by Cantor's work, which proves what's called Cantor's Theorem, which is a dramatic extension of Cantor's original result showing that the reals are greater than the natural numbers, and now in this context, with just the same idea of proof. That's the power of thinking about how we prove theorems and whether we can extend them.

Now, I want to illustrate his argument with our example involving the three Marx brothers. Of course, in this case, it's easy to see that the set of Marx brothers is smaller than its power set, since we listed all the elements and saw that the set of brothers has size 3 while the power set has size 8. So, of course, that's easy. However, let's apply Cantor's argument with this simple set so we can see how his idea can be applied to any set, especially infinite sets. Let's assume that, contrary to what we wish to establish, the set s and its power set have the same cardinality. Now, remember that we're letting s represent the set of the three Marx brothers, and $P(s)$, or the power set, is the collection of all the subsets of s.

Given this assumption, there must be a one-to-one correspondence between the elements of the set s and the elements of the power set. In particular, a one-to-one correspondence between the individual brothers and the collections of brothers. Now, just as an example, just to sort of set this on solid ground, let's just consider one. So, let's take Chico and pair him up with the collection {Chico, Groucho, Harpo}. Let's take Groucho and pair him up with the subset {Harpo}. And finally, let's take Harpo and pair him up with the subset {Chico, Groucho}. So, you can see that in the first column we have the elements, the brothers Chico, Groucho, Harpo, and then each one is paired up with a particular, different subset.

Cantor observed that for any particular subset of the set s, any given element from the set s, there are exactly two possibilities: Either that given element is in the given subset or not in the given subset. In other words, every element is either in, or not in, any particular subset. So, given an element and given a subset, we can look at this element and ask, is it in the subset? And the answer is either yes or no. It's kind of like the binary arithmetic we looked at with just 0 and 1, yes/no. So, we see this binary arithmetic almost happening here, at least reflections of it. There's just a quantum decision. You're in or you're out, yes or no.

Armed with this basic idea and the assumed one-to-one pairing between elements of s and the subsets of s, Cantor then modified his diagonalization idea from the previous lecture to describe a subset of s that is definitely missing from the list of pairings. Well, by finding a missing subset, we'll show this pairing is not a one-to-one correspondence, because some subset was left off the list. Well, his method for describing this missing subset requires us to consider each element from the set s and to decide if we're going to include it in this subset, or not include it in this subset. So, we make a yes or no decision with each element of our set s.

What's Cantor's method for systematically making these decisions to insure that the subset we're creating is not on our list? Well, the elements of the set s are listed in the left column of our correspondence. So, let's look at the first element. In our particular example, we see Chico. Now, Chico is paired up with a subset. In our example, it's {Chico, Groucho, Harpo}, so the entirety of the three brothers. Cantor modifies his switch the first digit idea from his diagonalization argument. In particular, he would say, Well, the brother Chico is paired up with the subset {Chico, Groucho, Harpo}. Is Chico in this subset? If the answer is yes, then I will not include Chico in the subset I'm creating. And if the answer is no, Chico is not in the subset associated with Chico, then I will include Chico in the subset that I'm creating. In other words, Cantor will do the opposite. If Chico is in the first subset on the list, then he will not include Chico in his subset, and conversely, if Chico is not in the first subset on the list, then he will include Chico in his subset.

In our particular example, we actually see that Chico is, in fact, inside the first subset (because that consists of all three brothers), so we'll not include Chico in our subset. Let's now move to the next element in the left column. In this case, we see the brother Groucho. We now have to decide if we want to include Groucho in our subset or not include him in our subset. How do we decide? Well, we see and we look at the subset associated with Groucho. If Groucho is an element of that subset, then we will not put him in our subset, and if Groucho is not an element of that subset, then we will include him in our subset. In other words, we again do the opposite.

In our specific example that we see, we see that Groucho is paired up with the subset that just consists of Harpo. In other words, Groucho is not in that subset, therefore we include Groucho in the subset that we're building. We're always doing the opposite. So far in our special set that we're building, we don't include Chico, but we do include Groucho. Thus, our subset will not match the first subset and will not match the second subset, because they at least differ in one element consisting of an element or not. Now, what about Harpo? Well, we look at the subset paired with Harpo on our list. Do we see Harpo in that subset? In our specific example, we see no, that subset contains only Chico and Groucho. Therefore, we do the opposite, and we include Harpo in the subset we're constructing. Well, we've gone through every element from our set *s* of Marx brothers, and so we see that we've built the subset {Groucho, Harpo}.

Now, Cantor claims that this subset must always be missing from our pairing list. Now, how do we know that for sure, in general? Well, can our subset be the subset paired with Chico? Well, no, since we saw that Chico is in that subset, and by our construction, we know that Chico is not in our subset. In other words, these two cannot be the same subset. Well, can our subset be the subset paired with Groucho? Well, no, since we actually see that Groucho is not in that subset, and by our systematic construction, we know that Groucho is in our subset. In other words, those two cannot be the same subset. Can our subset be the subset paired with Harpo? Well, no, since we see again that Harpo is not in that subset, and we know that, by our construction, we put Harpo in our subset. In other words, those two cannot be the same subset.

Therefore we find that the subset we've just built is definitely not on our list, because it differs in at least one element with each of the elements there. Now, of course, if we tried to build another list like this, we could use the same exact idea of looking along the diagonal and making the opposite decision about whether to include each element or not. This would again create a subset not on our list. So, no matter how we paired elements of s with subsets of s, we see we can always find a subset missing from that list. Thus, we conclude that there is no one-to-one pairing between the elements s and the subsets of s. We always have more subsets left over.

As before, we can generalize this idea to show that no one-to-one pairing between any set s and its power set can exist. Remember, the power set is just the totality of all the subsets of s. We can always find, no matter how we try to pair them up—the subsets with the elements of s—missing elements from the power set whenever we try to pair them. In other words, the power set of a set—remember, that's the collection of all subsets—is always larger than the original set s. Hence, we've just established the validity of Cantor's Theorem.

We can now apply this beautiful and counterintuitive result of Cantor's to show that there exists a collection that's even larger than the collection of all real numbers. Let's warm-up to this by first recalling Cantor's discovery that the cardinality of the set of natural numbers is smaller than the cardinality of the set of real numbers. Put more informally, just for us, the infinity of the set of natural numbers is a smaller infinity than the infinity of the set of real numbers. Well, how can we now find a collection whose cardinality is larger than the cardinality of the collection of real numbers?

Cantor's Theorem gives us an answer. We can consider the power set of the real numbers. In other words, we can consider the collection of all subsets of real numbers. Now, this is a really, strange, strange collection to visualize, because remember that the real numbers can be represented by the points on a number line, all the infinite points on a number line, which we saw, in the last lecture, as uncountable sets. So, it's teeming with lots and lots of points, and now we're going to consider a new set, not the set of the points individually, but the collection of all subsets of these points. Well, this contains all sorts of things. For example, one element in the power set would just be the natural numbers all taken together. That's a subset. Or all the numbers between the interval 0 and 1—that's also a subset. That becomes one element of the power set, because remember, the power set consists of subsets. All the integers put together—0, 1, 2, 3, –1, –2, –3, and so forth—that entirety is just one element in this power set. And there are lots of other subsets that we could imagine, taking a bunch of points and maybe a couple of intervals all together and mixing that up and scrambling. I mean, it's hard to really wrap your mind around the totality of all these sub-collections and for good reason, because in fact, we're about to see that this is a very, very, very large set. The point is that we are looking at a set whose elements are precisely all the sub-collections of points from the real line. So the primes, the collection of primes, form one element in the power set of the reals. It's really hard to wrap your mind around a collection that's so vast that all the primes together make up just one point in this power set.

The power set of the real numbers has a greater cardinality than the set of real numbers. That's what Cantor proved. He said that if you take any set at all and look at its power set, the power set will always be larger. So, in this case, what we're seeing is we just found a third infinity. Right? This incredible consequence of Cantor's mathematical logic is totally vexing and requires much contemplation. Right? What are the implications of different sizes of infinity? I mean, here we're seeing that we took the counting numbers, the natural numbers—they're a certain size of infinity, and countable. Then we saw that the real numbers, all the points in our number line, are an even larger infinity, and then, by Cantor's Theorem, if we look at the collection of all subsets of points on a real line, that's even larger still. So, now we see three different levels of infinity, three different sizes of infinity.

How do we interpret that? How do we wrap our minds around that? I mean, how does it even affect our view of the world, our view of the universe, and even our personal and spiritual beliefs? We need to think about how this new realization, once we grapple with it, fits in. Cantor himself felt that his work on infinity would have an impact far beyond the confines of mathematics. He believed that his theory would have implications within the realms of philosophy and religion. In fact, he believed that his set theory was shown to him by God. In 1896, in fact, he wrote, "From me, Christian philosophy will be offered, for the first time, the true theory of the infinite." And the way I interpret that is that Cantor is saying, Well, you know, Christian philosophy maybe talks about God as being infinite, and Cantor is saying, Well, okay, you can believe that, but now this is really the true theory of the infinite, and so it really is looking into this big collection of religious beliefs and saying, Well, here's how to do it mathematically correctly.

Independent of our individual views and beliefs, the first time we encounter Cantor's revolutionary ideas, we're at once surprised and mystified. Only over time can we embrace this new mindset, one that includes the reality of different sizes of infinity.

Infinity—What We Know and What We Don't
Lecture 23

Are there infinitely many different sizes of infinity? Is there a largest infinity, a "mother of all infinities," if you will? Is there a cardinality of numbers that is actually between the cardinality of the real numbers and the natural numbers, or is the cardinality of the real numbers the next infinity after the infinity of natural numbers?

Once our simplistic view of infinity collapses under the weight of Cantor's mathematical arguments, other previously unimaginable questions arise. Are there infinitely many infinities? In the previous lecture, we considered Cantor's argument that showed that the power set of any set has a greater cardinality than the original set. By applying Cantor's result, we see the answer to this question is yes. If we consider the power set of the power set of the collection of real numbers, then we have produced a collection having greater cardinality than the cardinality of the power set of the real numbers; thus, we have just produced a fourth size of infinity. If we repeat this process, we discover that there are infinitely many different sizes of infinity.

Is there a largest infinity? Another intriguing question is to wonder if there is one all-encompassing infinity. The answer is no; if we had such an infinite collection, then we need only consider its power set to produce a collection that, by another application of Cantor's Theorem, must be larger.

Is there an infinity between the sizes of the natural numbers and the real numbers? We recall that the cardinality of the natural numbers is smaller than the cardinality of the real numbers. This question leads to the Continuum Hypothesis, which states that the cardinality of the collection of real numbers—the cardinality of the continuum—is the next-to-largest infinity after the countable collection of the natural numbers. Cantor was the first to pose the Continuum Hypothesis and worked very hard to prove it, but he was unable to resolve the issue. David Hilbert listed the Continuum Hypothesis as the first challenge in his list of 23 challenges at the turn of the 20th century.

For Cantor, the Continuum Hypothesis was almost an obsession. This obsession, along with his combative relationship with Kronecker and others, had a serious and negative impact on his life. Between 1874 and 1884, Cantor published his seminal papers on infinity, including six papers that formed the foundation of modern set theory. In 1884, he suffered his first major depression and was plagued by mental health problems for the remainder of his life.

During the last 30 years of his life, Cantor was in and out of mental clinics. His problems grew with the loss of his mother in 1896 and his youngest son in 1899. Though he continued to do valuable research, his focus on mathematics declined. At one point, for example, Cantor turned his energy to proving that Francis Bacon was the author of Shakespeare's plays. After finally seeing the mathematical community acknowledge and celebrate his incredible contributions to mathematics, Cantor died of a heart attack in a sanatorium in 1918 at the age of 73.

The Continuum Hypothesis, a seemingly straightforward assertion, has a very strange resolution.

The Continuum Hypothesis, a seemingly straightforward assertion, has a very strange resolution. The statement, in fact, resides outside the domain of mathematics; that is, it can be shown to be neither true nor false within the narrow confines of mathematics. The issue is extremely deep and involves advanced work in logic and set theory. In 1940, Kurt Gödel showed that the Continuum Hypothesis cannot be disproved using tools and theorems from mathematics. In 1963, Paul Cohen showed that it cannot be proved using mathematical machinery. We thus say that the statement is *independent*. Two independent launches into the galaxy of mathematics reveal that the world of mathematics would look the same if we assume that the Continuum Hypothesis is true or if we assume it is false. The study of number and the notion of comparing sizes of collections have brought us to one of the cliffs of mathematics itself. ■

1. Suppose a friend said to you that she had an infinite collection of which she was very proud because it was so vast. How would you gently put her in her place by describing a collection that has greater cardinality than hers?

2. How have these lectures on infinity changed your view of this abstract mathematical notion? How have these discussions on different sizes of infinity challenged your beliefs and your view of the world?

Infinity—What We Know and What We Don't
Lecture 23—Transcript

Here in this lecture, we'll visit the frontiers of our understanding and discover what humankind knows and doesn't know about infinity. After an overview of what we've seen thus far in our travels through infinity, we'll turn to several important questions that we've yet to consider: Are there infinitely many different sizes of infinity? Is there a largest infinity, a "mother of all infinities," if you will? Is there a cardinality of numbers that is actually between the cardinality of the real numbers and the natural numbers, or is the cardinality of the real numbers the next infinity after the infinity of natural numbers?

In view of our previous discoveries, we'll be able to answer two of these questions. However, the remaining question has a shocking answer. This question, which preoccupied Cantor at end of his life, remained unsolved for decades. Here, we'll delve into the fascinating 20th-century story of that question, known as the Continuum Hypothesis. As we'll see, this question leads us to one of the highest cliffs of mathematics. Its unusual answer, found by Kurt Gödel and Paul Cohen in the mid-1900s, literally brings us to the very edge of that cliff. We'll take a moment to enjoy the incredible sight and then carefully take one step back to avoid falling off into the abyss.

Before we travel further into the endless wonder of infinity, I wanted us to reflect on where we've been and what we've seen. As demonstrated by the events of history, we've seen that it's a long and challenging undertaking to transform the counterintuitive theorems of infinity into reasonable sounding, believable, and—dare I say—intuitive notions. I know how difficult it is to wrap one's mind around these incredibly foreign ideas. Thus, before we move forward, I want us to take a momentary pause and take in the panoramic view of the mathematics of the infinite.

The first step in infinity is to admit that it's beyond all real numbers and, thus, we must immediately forego any desire to enumerate a collection containing infinitely many objects, but instead compare these collections with other such collections. If two collections are equally numerous, we say they have the same cardinality. That implies that even if the collections are infinite in size, there's a way of pairing up the elements from one collection with the elements in another collection so that no element is without a partner, and no element is partnered up with more than one other element. This idea is called a one-to-one correspondence and allows us to determine if one collection is larger, smaller, or equal in size to another.

Any infinite collection whose elements can be listed so that there's a first, second, third, fourth, fifth, and so on, without any element left unaccounted for, is said to be countable, or a countably infinite set. In other words, it has the same cardinality as the set of natural numbers, since we have an immediate, one-to-one correspondence. We pair the natural number 1 up with the first element, the natural number 2 up with the second element, the natural number 3 up with the third element, and so forth.

We've shown that the even natural numbers are countable, since we can list them and pair them with each natural number, which is equal to the double of the number. For example, 1 gets paired with 2×1, or 2, 2 gets paired with the even number 4, 3 gets paired with 6, 4 gets paired with 8, and so forth. Even the integers (that is, the collection of natural numbers, their negatives, and 0) are countable, since we can shuffle together the negative and positive numbers and then pair them with the natural numbers in this alternating fashion where we have the one-to-one correspondence between the integers and the natural numbers. The rational numbers (that is, the collection of all ratios or fractions) first appears to be much larger than the collection of natural numbers. But again, a systematic means of expressing those ratios (in this case, in an endless array of numbers) leads to a means of snaking a one-to-one correspondence with the natural numbers. Even the rational numbers are countable.

The situation changes dramatically when we turn to the collection of all real numbers, that is, all the points on the number line, or equivalently, all endless decimal expansions. It is provably impossible to create a one-to-one correspondence between the real numbers and the natural numbers. There are just too many real numbers. Cantor demonstrated this impossibility by assuming that there was such a one-to-one pairing between the natural numbers and the endless decimal numbers, and then using his idea of diagonalization, he successively switched digits along a diagonal of digits to construct a real number in decimal form that is, by its very construction, definitely not on the list. Hence, Cantor arrived at a contradiction, a logical impossibility, if you will, and thus his assumption was wrong. Therefore, we see that the real numbers form a larger infinite collection than the natural numbers. The real numbers aren't countable. We call collections that are not countable "uncountable," and so the real numbers are, in fact, uncountable.

Holding two different sizes of infinity, we wonder if there is a third. Extending his original idea, Cantor was able to generalize his result and produce what is now known as Cantor's Theorem. If we have any set, any collection, then the set of all its subsets is a larger infinity than the original given set. This remarkable theorem followed from a modification of Cantor's diagonalization argument, and we illustrated this generalized technique with an example involving the set of three Marx brothers. If we now consider the collection of all subsets of real numbers, then, by Cantor's Theorem, this vast collection must be larger than the collection of real numbers. Hence, we uncovered a third size of infinity, one even larger than the uncountable collection of all real numbers.

With time, these ideas can be made our own, but it does take time. Once our narrow view of reality, which stated that there's just one infinity, collapses under the weight of Cantor's mathematical arguments, other previously unimaginable questions soon arise. For the remainder of this lecture, I want us to explore some of these interesting new conundrums and their solutions.

Let's face our first question right off the bat. Are there infinitely many infinities? Great question. In the previous lecture, we considered Cantor's argument that showed that the power set of any set has a larger cardinality than the original set. We used this result to find a third size of infinity. Well, are there infinitely many different sizes of infinity? By applying Cantor's Theorem again, we see the answer is yes.

Let's consider the power set, of the power set, of the set of real numbers. Now, this collection is the set of all subsets of the power set of the real numbers. In other words—but this is not going to be clear—this collection is the set of all subsets, of the set of all subsets, of the real numbers. It's really a set-theoretic tongue twister, if you will. It's a collection whose elements are sets of sets. Well, by Cantor's Theorem, this new collection has a greater cardinality than the cardinality of the power set of the real numbers, because remember, Cantor proved that whenever you have a set, no matter how convoluted it may be, if you take the power set, it's bigger. So, if we take the power set of the power set of the reals, that's going to be larger still. Thus, we just produced a fourth size of infinity. So, the smallest infinity we considered was the set of natural numbers. Then the next largest infinity we found was the infinity of the real numbers, then larger still was the infinity of the power set of the reals. Now we are seeing an even larger infinity, the power set of the power set of the real numbers.

Well, if we just repeat this process, we discover that there are infinitely many different sizes of infinity. We can just take the power set of the power set of the reals, and take its power set, and build from there. This insight leads us to our second question: Is there a largest infinity? In other words, is there one all-encompassing "mother of all infinities" in which all other infinities reside? Well, given our previous observations, we see that the answer is no. If we had such an infinite collection, then we'd need only consider its power set to produce a collection that, by another application of Cantor's Theorem, must be larger still. Thus, just like the natural numbers themselves, there's no largest, or last, infinity. They keep growing and growing without bound, just like the natural numbers. This is an amazing, amazing fact which takes a while to become intuitive. So, we've gone from one infinity to infinitely many infinities.

We now come to the final question of our quest toward infinity: Is there an infinity between the sizes of the natural numbers and the real numbers? Now, let's first remember that the cardinality of the natural numbers is smaller than the cardinality of the real numbers. So, we now wonder, Is the cardinality of the real numbers the next infinity after the cardinality of the natural numbers, or is there a collection whose cardinality is strictly between those two infinities? Well, this question leads to what is known as the Continuum Hypothesis. The Continuum Hypothesis asserts an answer to this question. Specifically, it states that the cardinality of the collection of real numbers is the next largest infinity after the countable collection of the natural numbers. (And remember, Cantor has used the word "continuum" to mean the real numbers. That's why this is called the Continuum Hypothesis, hypothesizing that the continuum is the next infinity.)

Cantor himself was the first to pose the Continuum Hypothesis and worked very hard to prove it, but he was unable to resolve the issue. David Hilbert listed the Continuum Hypothesis as the very first challenge in his list of 23 that he posed in his keynote address at the International Mathematics Congress at the turn of the 20th century, as we mentioned earlier. For Cantor himself, the Continuum Hypothesis was almost an obsession. This obsession, along with his combative relationships with Kronecker and others, had serious and negative impacts on his life. Now, to place these issues in historical context and expand our understanding of Cantor's struggle, let's take a moment to consider some revealing moments from his personal life.

Between 1874 and 1884, Cantor published his seminal papers on infinity, including six papers that formed the foundation of modern set theory. But in 1884, he suffered his first major bout with depression and was plagued by mental health problems for the remainder of his life. During the last 30 years of his life, Cantor was in and out of mental clinics. While his professional struggles contributed to his difficulties, today he might have been diagnosed as bipolar manic depressive. His problems only intensified with the loss of his mother in 1896, and his youngest son in 1899.

Though he continued to do valuable research, his focus on mathematics did decline. In fact, for a brief period, Cantor turned his energy to proving that Francis Bacon was the author of Shakespeare's plays. Some have argued that exposing Bacon to the world corresponded to Cantor's desire to expose the ugly sides of Kronecker to the world. As it turned out, later in his life, it appears as though Kronecker and Cantor had some type of reconciliation. After finally living to see the mathematics community acknowledge and celebrate his incredible contributions to mathematics, Cantor died of a heart attack in a sanatarium in 1918. He was 73. Today, in the center of the University of Halle, there's a plaque that shows his face and an image suggesting his diagonalization method. Engraved we find the words of Cantor himself: "The essence of mathematics lies precisely in its freedom." It's difficult to know if Cantor was thinking about mathematics being the freedom to get away from his personal problems and personal demons or, in a different context, because Cantor himself preferred the term "free mathematics" to mean the more traditional term which we call today "pure mathematics." That is, the mathematics that are not modeling the world, but instead are understanding mathematics for mathematics's sake. In either case, the statement and the quote seems very fitting for such a great man who dealt with so many demons.

Let's now return to the story of the Continuum Hypothesis. The Continuum Hypothesis proposes that the cardinality of the continuum, in other words, the cardinality of the real numbers, is the next largest infinity after the countable collection of natural numbers. Well, this seemingly straightforward assertion has a very strange resolution. The statement, in fact, resides outside of the domain of mathematics. In other words, it can be shown to be neither true nor false within the narrow confines of mathematics and its standard axioms. The issue is extremely deep and involves advanced work in logic and set theory.

In 1940, the great Austrian logician Kurt Gödel showed that the Continuum Hypothesis cannot be disproved using the tools and theorems from mathematics. In 1963, Paul Cohen, who passed away in 2007, showed that it cannot be proved using mathematical machinery. So, the Continuum Hypothesis is provably not provable and not disprovable (and we can prove that). Thus, we say that the statement itself is *independent*. It actually resides outside of the world of mathematics and her axioms. Now, this is very difficult to appreciate and even to wrap your mind around, but an analogy can help make this idea a little bit more intuitive. Let me offer you my intuitive sense of what it means to have a statement that's outside of mathematics.

So, let's visualize this. Imagine that we have two spaceships that are going to be independently launched out into the galaxy of mathematics, and they will reveal the world of mathematics and how it looks. Well, let's think about how this will go. We have two spaceships, and we're going to have two mathematicians that are actually going to enter into these spaceships. In these spaceships, we will give them every single axiom of mathematics. Those are the self-evident truths upon which we start all of mathematics. So, those are the foundation of mathematics. We give them all of them, these facts. They're going to go out into outer space, and they're going to now spend years, and years, and years, and years, thousands of years, producing all the theorems of mathematics up to this point.

Now, what does that mean? Let's back up here and think about what that means. What it means is that they're going to go off and generate mathematics. However, before the first spaceship goes off, I approach the capsule and go in and say to the mathematician, Okay, you've got all your axioms here, but I'm going to give you one extra axiom, one extra fact that the other spaceship doesn't have, which you can use. That fact is that the Continuum Hypothesis is true. There is no infinity between the natural numbers and the real numbers. You can use that. And so, the mathematician astronaut is like, Oh great! I've got an extra one. That's fantastic. And so she takes off, goes out into the outer reaches of mathematics and starts to prove all the theorems that are possible using the axioms and this one extra axiom.

Then I walk over to the other spacecraft, and I walk into the capsule and say, Okay, you've got all the axioms, the same as the other person. You're ready to go, but I'm going to give you one more axiom that you can use. That axiom is that the Continuum Hypothesis is false. There does exist an infinity in between the natural numbers and the real numbers. You can use that fact. And so this astronaut mathematician is like, Yes! One more! That's great, and she takes off into the outer reaches of mathematics to describe all of mathematics that are possible.

So a few thousand years later, these two mathematicians returned from their voyage through the galaxy of mathematics, and they have these large tomes, these big books, of all the theorems of mathematics—one, then the other. And what do we see? We see that when we look down their list of theorems, they are identical. They agree exactly. All the theorems up to the point of the Continuum Hypothesis match up perfectly. Mathematics was unaffected by the truth or the falsehood of the Continuum Hypothesis. When they go off into space, each having an opposite point of view, it doesn't change anything at all. That's a visual way that helps me think about what it means to be independent. It means, in fact, that it doesn't change mathematics at all whether we assume it's true or not. All the mathematics up to that point would be proven correct whether we assume its truth or assume its falsehood. The Continuum Hypothesis can be either true or false.

Now, Gödel actually believed that the Continuum Hypothesis was false, and Cohen also leaned toward rejecting the Continuum Hypothesis, but we know it's independent of mathematics. I actually wanted to take a moment just to share with you the story of Paul Cohen, or at least the story as I heard it when I was growing up in the math community as a graduate student, because it's a great story, and it's probably true. As the story goes, Paul Cohen was a young mathematician. He'd just earned his Ph.D., and he was attending a seminar in set theory and logic, which is the area in which the Continuum Hypothesis resides. Now, he wasn't an expert in this area, his area was analysis, but he wanted to attend this seminar just for fun, if you can imagine such a thing. And so, he was this young Ph.D. sitting there in the audience, and apparently, as the story goes, he wasn't particularly impressed with what he heard. The theorems that were presented and the mathematics didn't sound particularly profound. Remember that Cohen wasn't an expert in this area, so in fact, he just didn't seem to appreciate this.

After the seminar, as the story goes, he approached some people in the audience and said, Well this didn't, you know, seem all that dramatic. What's the big question in set theory these days? And so this person, who was a little taken aback by the moxie of this young, brand new, newly minted Ph.D., said, Well the biggest question, of course, is the independence of the Continuum Hypothesis. (You see, because in 1940, Gödel proved one direction. He showed that you can't disprove it, but still it remained an open question whether you can prove that you can't prove it.) And Paul Cohen had the audacity to say, You know what? I'm going to go out, and try this problem, and work on it. And he did, and in fact, he solved it, which just goes to show you that sometimes it takes a lot of guts, and maybe almost arrogance, although I'm not implying that Paul Cohen was arrogant. But it takes a lot of guts to face these big, challenging questions. If we don't face those challenging questions, they'll never get resolved. We'll never answer difficult issues if we don't have the guts to go and face them.

Anyway, this young, brand new Ph.D. produced a solution. It was extremely complicated, took up pages, and pages, and pages of complicated advanced mathematics, logic, and set theory. And of course, it was so complicated that people didn't want to read it. Who wanted to go through and read this beginner's solution to a very well known, difficult problem that great minds haven't been able to solve, only to go through page after page to inevitably find the mistake that we're sure is there? People really did not want to read it, and they wouldn't look at it.

But Kurt Gödel himself got a copy of this manuscript. Now, Kurt Gödel was at the Institute for Advanced Study at Princeton, right next to Princeton University, and Paul Cohen was a visitor there at the same time. One afternoon they had tea. In mathematics, the community gets together, usually in the afternoon, and has tea. It's very civilized; they get together and have tea. There are also blackboards there, so you can do math, too, but generally you drink tea. This man, Kurt Gödel, walked up to this very young Ph.D. and quietly said to him over tea, "Congratulations. You've solved this problem." And that was the first time that the math community acknowledged that, in fact, he had produced a complete and correct solution. So, it took Gödel himself to read through the work to actually be able to articulate and say, "Yes, this is right. This young guy got it." Well of course, once he got Gödel's seal of approval, then everything fell into place, and people read it, and studied it, and did agree that it was indeed correct.

Just a little time later, Paul Cohen was awarded the Fields Medal, which is the equivalent of the Nobel Prize in mathematics. And he won this, the highest honor in mathematics given to a young mathematician, for his work on that. And so then he actually went and spent a good portion of his remaining years thinking about trying to conquer the Riemann Hypothesis, which we mentioned about primes in a previous lecture, and that wasn't nearly as successful. Now, I don't know how much of the story is true, but it's a wonderful story, and it does show us that, in fact, it's important to tackle hard questions and to be brave enough to take on things that seemed like greater minds weren't able to solve.

The study of number and the notion of comparing sizes of collections have brought us to one of the cliffs of mathematics itself. This is one of the edges of math, where we see a statement that is neither provable nor disprovable within the realm of mathematics. Now, I can't resist closing this chapter on infinity with the very moving words of the great 18th-century English poet William Blake. I wanted to share with you one of his poems that I believe captures an image of grasping the incredible ideas that surround the study of the infinite. So, while we sit back and we try to make sense of all these totally counterintuitive ideas and try to make them intuitively reasonable, while we struggle with that, I want us to think about the words of William Blake:

> To see a World in a Grain of Sand
> And a Heaven in a Wild Flower
> Hold Infinity in the palm of your hand
> And Eternity in an hour.

The Endless Frontier of Number
Lecture 24

Throughout human history, we've seen the notion of number evolve from a practical necessity to a creative art of nature.

A look back at our course takes us from the dawn of quantification through the struggle to communicate quantitatively, from number as an attribute to number as an object. Once we name abstract objects, they exist in our minds. Numbers have captured the imaginations of all people from all cultures throughout human history. It is an intellectual curiosity that brings humankind together.

We explored various perspectives, particularly an algebraic approach and an analytic one. An algebraic approach of solving polynomial equations allows us to discover i, the complex numbers, and the distinction between algebraic and transcendental numbers. An analytic approach of measuring distance allows us to give a precise definition of the numbers on a real number line and explore parallel universes of numbers, such as the p-adic numbers.

What is number? We have come to see an ever-evolving notion of number as we have journeyed along the intellectual paths toward an understanding of the theory of numbers. Every time we develop a new insight into numbers, it forces us to rethink our old notion of what number means. Numbers hold many surprises—often what first appears strange in our minds is, in fact, ordinary. Numbers have a power and import both within the universe of mathematical ideas and far beyond. Are numbers discovered or created?

> **Numbers have a power and import both within the universe of mathematical ideas and far beyond.**

Number theory today is something our early ancestors probably couldn't have imagined. There are many different areas of number theory. The two main branches of modern number theory are analytic and algebraic number theory. Analytic number theory employs the ideas of calculus to answer questions about numbers. Algebraic number theory focuses on the study of numbers that arise from solutions to polynomial equations with integer coefficients—the algebraic numbers.

Mathematics moves from observations to conjectures to rigorous proof. Mathematicians at once balance rock-solid truth and wild, unbridled creativity. Mathematicians must first stumble upon some structure or pattern, believe that this form is a general principle, then verify that this principle is valid through a clear, correct, and complete logical argument that establishes its validity.

A culture of number theory research exists, and within it, we find scholars who extend our boundaries of understanding, despite—and sometimes because of—their failures.

The study of number is not a science devoid of human emotion, and throughout our course we have seen great passion as humankind struggled with the idea of number. We have seen some very deep and abstract ideas within mathematics. We also have experienced the joy and pleasures generated by discovery, creativity, and imagination. We have witnessed for ourselves that within numbers we find beauty, elegance, grace, and mystery. Many questions remain, and we can all appreciate and contribute to the quest to conquer new frontiers ahead. ∎

Questions to Consider

1. In Lecture 1, you were invited to write a definition of *number*. How do you view your original answer now? *Extra Challenge:* Are numbers created or discovered?

2. As we have seen in this course, the notion of number evolves with humankind's intellectual development. What other great ideas have followed a similar evolutionary trajectory?

The Endless Frontier of Number
Lecture 24—Transcript

Welcome to what I'm sad to say is our last lecture together in our exploration of number. Here, we'll reflect back on our journey and discuss how numbers are, at once, useful tools and also abstract objects of the mind that allow us to understand our world and our lives with greater clarity. For the number theorist and enthusiast, far beyond numbers' utility, numbers are objects of independent beauty, intrigue, and curiosity.

In this lecture, we'll open with a sense of what number is. How, throughout human history, we've seen the notion of number evolve from a practical necessity to a creative art of nature. We'll consider the subtle structure of numbers, and explore the many means by which we can view them through their individual numerical personalities in order to attain a better understanding of their nuance. We'll then face the question, "What *is* number?" And we will explore how that notion evolved and, in fact, remains an ever moving target.

We'll then step outside of our historical journey and take a moment to consider number theory today. Moving the frontiers forward requires the logic and rigor that have underscored our entire exploration together. The ever present rigorous proof is both a science and an art. Mathematics, in general—and the theory of numbers, in particular—is truly a culture with customs and traditions. We'll offer a window into this modern world of mathematical research and how, in practice, those frontiers move outward.

Finally, we acknowledge and celebrate the human passion involved in moving the notion of number forward. The study of number is not a cold, austere discipline, but instead one that is teeming with imagination, emotion, and excitement. Humankind has been drawn to number like a moth to flame; we're unable to resist the attraction. The quest to understand and tame the notion of number has transcended human history and cultural divides. What makes this intellectual quest so universally appealing? Perhaps it is the realization that numbers have a life and a will of their own. Even though the abstract notion of number came from our imagination, once born, numbers conform to the laws of nature and mathematics. What surprising and intriguing plot twists lie ahead in the story of number? That enticing, unending journey will fuel the creativity and imagination of our descendants for the next thousand years.

Before moving on, I wanted to say a word about the structure of our journey. Perhaps there's still a question about this course. Was it a history course, or was it a math course? Well, of course, this was the story of number, and so plainly we frolicked among the delights of mathematics. So, then, why the history? Well, the history is the only means by which we can intuit the critical human element that gave rise to the ever-evolving notion of number. If we were just handed a laundry list of numbers—here are the rational numbers, here are the natural numbers, here are the irrationals, the reals, the complexes, the algebraics, the transcendentals, the p-adics, and so forth—then our view of numbers would be a jumbled collection of abstract definitions that would have no real meaning and would not tell any story.

The story of number is really the story of humankind's ever-evolving imagination, creativity, understanding, and curiosity. It's the history that breathes life into our story of number. So, was this a history or a math course? Honestly, I don't know, and I'm not really sure that we need to have our efforts here dropped into any particular pigeonhole. My hope is that we feel the synergy between the two; the history inspired the mathematics, and the mathematics amplified the history. It really was an incredible story. Over 30,000 years ago, long before there was a notion of number, Sumerian shepherds didn't want to lose track of their flocks, so they used pebbles to create a one-to-one pairing between the stones and the sheep. Later, the pebbles were pressed into clay, and the abstract writing of numbers began. After thousands of years of trying to express numbers, humankind finally came upon the now familiar ten symbols for our digits, and a place based system that allowed science and mathematics to race forward at breathtaking speeds.

With the Jains, Pythagoreans, and many others, numbers move from an attribute to an object. Once those objects are named, they exist in our minds and take on a life of their own. Numbers have captured the imagination of all people from all cultures throughout human history. It's one of the few intellectual curiosities that truly unite humankind. Of course, we've explored various perspectives on numbers. If we just count—thinking of numbers as Ahmes did in the Rhind Papyrus, as heaps—then we have the natural numbers, 1, 2, 3, 4, and so on. When we consider ratios of these numbers, we see the rational numbers. When we extend our imagination and our view of number, we can actually embrace negative numbers and 0. When we measure lengths, we discover irrational numbers such as $\sqrt{2}$ and π. When we measure growth, we find the number e. When we consider numbers as solutions to certain polynomial equations, we naturally uncover the number i (the square root of -1), the complex numbers, and the distinction between algebraic and transcendental numbers. When we use number to measure distance, our analysis allows us to produce a precise definition of the real numbers, and to explore parallel universes of numbers known as the p-adic numbers. When we invoke the Sumerian shepherd's idea of a one-to-one correspondence, we unlock the secrets of infinity and build an infinite tower of different infinities.

Now we arrive back at the very first challenge I offered all of us in the very first lecture. What is number? As we see now, there's no one all-encompassing definition that has any precise meaning. Every time we developed a new insight into numbers, it forced us to rethink our old notion of what number means. In fact, so often, the numbers that at first seemed common and abundant turned out to be the exception and extremely rare, while those numbers that first appeared as exotic and peculiar turned out to be the norm and, in fact, dominated the ever-expanding number world.

We saw this with the rational numbers. We started off with just the natural numbers, and then we looked at ratios, and we built the rational numbers, which seemed so familiar and so common to us that we see them everywhere, even in our everyday lives. And the Pythagoreans stumbled upon $\sqrt{2}$ and showed it was irrational. What did this imply? Well, this implied that there was this different notion of number. But these numbers seemed strange, foreign, and peculiar. Only later in our exploration did we realize that, in fact, those numbers were the more common. On the real line, if we were to pick a number at random by randomly generating digits by rolling a fair, ten-sided die, with probability 100% we would produce a screed of digits that never repeats, that, in fact, is irrational. The irrationals are, in fact, the more dominant, and the familiar rationals turn out to be the exotic.

We then saw this again with the algebraic numbers, where we studied these numbers, $\sqrt{2}$ or the number i, $\sqrt{-1}$. They became familiar to us. Then we discovered numbers like π and the number e, and we saw that these are actually transcendental numbers that don't conform to any polynomial equations, for they're not the solution to any polynomial equation with integer coefficients. Well, these seemed so foreign and strange to us that it was difficult for us to really visualize these as numbers. Yet we saw later that if we were to pick a number out at random, with probability 100% it would be transcendental, and we would never, in fact, accidently stumble upon an algebraic. That followed from Cantor's great work that the transcendental numbers are uncountable, while the algebraic numbers are an infinite set that are actually countable.

Then again we come across this when we look at the real numbers. We have the real number line that's been with us for centuries, and we're comfortable with it, and we can understand points on it. Then we're introduced to a visualization of the complex numbers called the complex plane, where we have imaginary numbers running around this entire plane. If we were to take a pin, close our eyes, and put it down on a plane, what do you think the likelihood is that we will hit that one, particular horizontal real axis, the number line? With probability 100% we'll be off that line. The real, familiar numbers to us are, in fact, the exception. The complex numbers dominate the complex plane.

These moments in which our initial intuition runs counter to reality are moments in which we can grow, for it's these very moments that feed the life of the mind. We can retrain and fine-tune our intuition so that what first appeared strange will later appear commonsensical. This refinement of understanding is the responsibility of every open-minded, learned individual. Numbers evolve alongside humankind's intellect and imagination. Numbers have power and import within our physical universe and our everyday world. Given this intriguing balance between their utility in nature, and their abstract countenance within our imagination, I now offer a second conundrum here to bookend the very first conundrum I offered at the very opening of this course. That conundrum is, are numbers created or discovered? I'll leave this question with you to contemplate and discuss with others, but that's the question that really is one that deserves some contemplation.

We've seen a panoramic of number that began over 30,000 years ago and took us right into the 20th century. However, before we close this course, I would like to say a few words about the modern number theory of today which, I'm delighted to report, is alive and well and as vibrant as ever. Today, within the universe of number, we see many specialized areas. These include elementary number theory, combinatorial number theory, the geometry of numbers, Diophantine analysis, arithmetic geometry, cryptography, and probabilistic number theory. These areas, for the most part, fall into two very broad branches of study. These two main branches are known as analytic and algebraic number theory. I wanted to highlight each and place these areas of interest today within the historical context of number that we've already seen together.

Analytic number theory is an area that exploits the ideas of calculus. We noted, in an earlier lecture, that calculus is the study of motion and change. Here we use the mechanics of this study to deduce facts about numbers. For example, one of the jewels of analytic number theory is the prime number theorem. This result, as we discussed in Lecture Seven, gives an estimate for how many prime numbers there are up to any given point. Remember that the primes are the fundamental building blocks, or atoms, of the natural numbers, because every natural number greater than 1 is either a prime number, or is a product of primes. The study of how well the estimate in the prime number theorem approximates the number of primes up to any given point leads to one of the most important unanswered questions in all of mathematics today. This question has been around since 1859 and is known as the Riemann Hypothesis.

The Riemann Hypothesis is named after Bernhard Riemann, an extremely influential German mathematician who first stated the conjecture. Basically, the heart of the issue is to describe all the solutions to a certain equation. What makes this question so difficult is that the equation itself has an endless sum of terms, each term of which has the unknown x appearing. Moreover, there are infinitely many numbers that actually satisfy this equation. In other words, there are infinitely many solutions to this equation, so we just can't just list them all and be done with it. The Riemann Hypothesis asserts that every one of the numbers that satisfies this special equation exhibits a certain property. Now, if we could show that the Riemann Hypothesis holds, then we would actually be able to say more about the placement of the primes within the natural numbers. The surprising fact is that the numbers that are solutions to Riemann's endless equation are, in fact, complex numbers (numbers that contain the imaginary number i, which equals $\sqrt{-1}$). In other words, if we could better understand the nuances of the seemingly imaginary complex numbers, then we'd be able to say more about the grounded, concrete prime numbers. Again we see an example in which different types of numbers and notions all hang together.

In fact, today, there's some very exciting progress in understanding these subtle issues. Mathematicians in this area are now exploiting a surprising connection between the study of primes and what are called random matrices. Random matrices are mathematical objects that were originally applied to better understand the quantum behavior of large atoms in physics. Thus, it seems wonderfully fitting that these mathematical objects might also unlock the mysteries of the basic atoms of numbers.

Moving now to the other main branch of modern number theory, algebraic number theory is an area that focuses on the numbers that are the solutions to certain polynomial equations. As we've seen in Lecture Sixteen, these numbers today are called algebraic numbers. For example, $\sqrt{2}$ is an algebraic number, because it's a solution to $x^2 - 2 = 0$. Another famous algebraic number we studied was i, since it's a solution to $x^2 + 1 = 0$, a different polynomial. It turns out that, using algebraic numbers, we can generalize the collection of integers. Remember that the integers are precisely the natural numbers, together with 0 and all the negatives of the natural numbers. Within the integers, we find the prime numbers, and we saw that all integers greater than 1 can be expressed uniquely as a product of primes. Again, the primes are the atoms of the integers, and thus if we understand them better, we understand the integers better.

One way to generalize the integers is to define the collection of all numbers of the form integer + integer times i. This new collection is known today as the Gaussian integers, named after Carl Friedrich Gauss. So for example, $5 + 7i$, $-4 + 3i$, and $9 - 2i$ are all examples of Gaussian integers. Well, within this collection of numbers, we can also identify the fundamental multiplicative atoms. That is, we can study the Gaussian primes, those Gaussian integers that cannot be written as a product of two smaller Gaussian integers. For example, $2 + i$ and $2 - i$ can be shown to be Gaussian primes. Well, what about 5? Is 5 (which we can think of as $5 + 0i$) a Gaussian prime? Well, we know it's an ordinary prime, but let's carefully multiply our two new primes ($2 + i$ and $2 - i$) together. Now, to multiply those two numbers together, we actually have to use the distributive property that we mentioned in Lecture Twelve to perform the multiplication correctly, but don't worry about it. I'll just do it for you and you can just relax and enjoy it. It's our last lecture after all.

We take $2 + i$ and multiply it by the quantity $2 - i$. I'll first take the second 2 and multiply it through by the 2 and the i. And so, I see $2 + i$ all times 2 minus—and now I'll take that i and multiply it through by the $2 + i$ and see $2 + i \times i$. Now if I distribute that little 2 through, I see 2×2, which is 4, and then I see plus $2 \times i$ which is $2i$. And then I see minus $2 \times i$ (which is $2i$), so I have $-2i - (i \times i)$, which is i^2. So what do I see? Well, I see 2×2, which is $4 + 2i - 2i - i^2$. Well, notice that $+2i - 2i$ cancel, add to give 0, so I'm only left with $4 - i^2$. Well, i^2 is actually -1, so we have $4 - (-1)$, which is $4 + 1$, or 5. So, 5 is no longer a prime in this setting. It's the product of two other Gaussian primes. Just as with the regular prime numbers, number theorists have proven that Gaussian integers also can be factored uniquely into Gaussian primes. However, there are other generalizations of integers in which this unique factorization property no longer holds. Thus, we discover, yet again, an entire new universe of number. This one, at once, has reflections of the regular integers and yet is dramatically different.

In algebraic number theory today, some of the unanswered questions of interest involve studying subtle issues surrounding this lack of unique factorization into primes. Studying such strange, new integers and strange, new sense of prime numbers actually allows us to better understand our usual primes in a richer way. In fact, we can actually use the ideas from algebraic number theory to devise algorithms that will allow us to factor very large numbers and test to see if large numbers are, in fact, prime. These results have several practical applications within the modern study of cryptography, the study of codes, and thus even have national security implications. Number theory research today remains as important as ever and is still going strong.

Throughout the course, I've shared my vision of how number theorists, and how mathematicians in general, move the frontiers of our understanding forward. I suggested that our understanding of mathematics is like a large copper orb, and we mathematicians are inside with ball peen hammers, tapping away on its surface, producing dimple upon dimple as, very slowly, the orb expands and grows. Here, in the last few minutes, I wanted to briefly describe this modern community of mathematical scholars and how, in practice, we extend our boundary of understanding. Mathematicians stumble around in the dark, trying to uncover some structure or pattern by examining simple things very, very deeply. We study the mathematical works of others and ask how can we extend or generalize them. We create questions. We also fail.

Mathematicians are expert failures. Failing is the most important thing we can do. In our research, we're failing approximately 99% of the time. It's true. Every failed attempt, though, brings to light some otherwise unseen subtlety in an issue and provides the opportunity to try something else. After countless false starts, we finally discover some general principle or see a pattern that leads us to a conjecture. Next, we struggle to find a logical and rigorous proof of this conjecture, and then write up our results in the form of a research article. These are usually very cryptic-looking papers that are teeming with notation and symbols and can even intimidate other mathematicians. It's true.

Now we're ready to present our work to the mathematical community. We do this informally through lectures at seminars, colloquia, and addresses at conferences, but we do this formally by submitting our paper to a scholarly research journal. The editor will select several experts in the subject and ask those mathematicians to review the work and assess the results. These anonymous referees study the work and declare if the arguments are correct and complete, and if the results are significant enough to warrant publication. When our paper is accepted for publication and it appears in the literature, it has the endorsement of the mathematical community, and the results contained within it are now viewed as theorems of mathematics for others to study and explore. Captured in that published work is our creativity, our imagination, and our original contribution that moves the boundary of humankind's understanding, ever slightly, a bit forward.

This is the study of number. It's a very human struggle filled with both passion and imagination. Within numbers, we find beauty, elegance, grace, and mystery. Perhaps more importantly, the exploration into number is a liberating entertainment. We can discard the kid gloves of reality's restrictions and let our minds play where they will. The theory of numbers is about what can be thought, what can be imagined, and what can be dreamed. And it all leads to new truths. Exploring the deep consequences of simple ideas takes us on a journey of startling sights and unexpected insights.

We compare flocks of sheep and numbers are born. We keep track of inventories by pressing pebbles into clay, and a symbolic language of numbers graces the page. We look upon the spirals of nature and we see Fibonacci numbers. We search for the atoms of the numbers, and we uncover an infinitude of primes. We measure a diagonal, and we discover the irrational numbers. We look beyond $\sqrt{2}$, and we find the transcendental values π and e. We transcend the real line and the complex plane, and we find ourselves suspended in the vexing world of the p-adic numbers. We look beyond all numbers, and we travel to a universe of infinitely many infinities.

The history of numbers paints a story about casting off the constricting coil of bound thought. The ever-evolving world of number is full of wonder, but our guide is the clear principle of following creative ideas to their logical conclusions. We can all embrace the strategy of mathematical thinking to guide us in our everyday lives as well. We've seen here just a glimpse of the vast richness that our minds can create. Numbers and our imaginations have no bounds, no ends, no finish line. Every horizon reached opens new horizons more glorious still. Thank you very much.

Timeline

30,000 B.C.E. .... Paleolithic peoples in Central Europe and France use notched bones as counting tools.

c. 4000 B.C.E. .... Sumerians use clay tokens (*calculi*) to represent quantities of different items.

c. 3500–3000 B.C.E. .... Sumerians record numeral symbols on clay tablets (an abstraction from the token system used previously); Babylonians develop base-60 numeral system; Egyptians use pictographs for numbers.

c. 3000 B.C.E. .... Sumerians develop cuneiform, a system of writing that includes distinctive numbers.

c. 2000 B.C.E. .... Babylonians and Egyptians use fractions.

c. 2000–1650 B.C.E. .... Babylonians apply the Pythagorean Theorem to approximate $\sqrt{2}$.

c. 1650 B.C.E. .... Rhind Papyrus is written (copied by the scribe Ahmes), showing extensive Egyptian calculation techniques, including an approximation to π of 3.16.

c. 1400 B.C.E. .... Chinese use base-10 numeral system.

1000 B.C.E. .... Chinese book gives the first record of a magic square.

c. 540 B.C.E. .... Pythagoras founds his school and proves the Pythagorean Theorem; the Pythagoreans and the Jains in India begin the earliest explorations of abstract properties of numbers; later, the Pythagoreans are confounded by the irrationality of $\sqrt{2}$.

c. 400 B.C.E. .... The Hindu numerals 1–9 begin to develop in India.

300 B.C.E. .... Euclid presents his axiomatic method for geometry in *Elements* and proves the infinitude of primes and the irrationality of $\sqrt{2}$, while his common notions form the basis for modern arithmetic; Babylonians have a symbol for zero as a placeholder.

c. 100 B.C.E. .... Chinese solve equations with negative numbers.

c. 250 C.E. .... Mayans use a base-20 numeral system, including symbols for zero as a placeholder.

c. 650 C.E. .... Brahmagupta understands zero as a number, not just a placeholder, and uses negative numbers systematically; Bhaskara I uses the symbol 0 for zero.

c. 800 C.E. .... Arab mathematicians adapt and promote the use of Hindu numerals.

1202... Fibonacci brings knowledge of Islamic mathematics to Europe, encourages the adoption of the base-10 numeral system, and writes on the Fibonacci sequence.

1350... Oresme offers the first clear treatment of fractional exponents.

1489... The first appearance of + and − signs occurs in a German arithmetic book by Widman.

1545... Cardano introduces the idea of the square root of a negative number, leading to the discovery of complex numbers.

1585... Simon Stevin offers the first written reference to the number line and the first full account of the decimal expansion of numbers.

1614... Napier gives the first reference to the number e.

1632... Galileo discovers an apparent paradox concerning infinite sets.

1637... Descartes develops superscript (exponential) notation for powers of numbers and begins using x, y, and z to denote unknown quantities.

1713... Bernoulli approximates e using a formula for compound interest.

1727... Euler is the first to name e.

1740...Euler is the first to consider imaginary numbers as exponents.

1761...Lambert shows that π is irrational.

1799...Gauss shows that every polynomial equation has its solutions within the complex numbers.

1815...Fourier shows that e is irrational.

1820...Cauchy defines the real numbers using infinite sequences of rational numbers and a limiting process.

1844...Liouville constructs the first example of a transcendental number.

1873...Cantor shows that the infinity of real numbers is larger than the infinity of natural numbers. Hermite shows that e is transcendental.

1877...Cantor proposes the Continuum Hypothesis.

1882...Lindemann shows that π is transcendental.

1883...Cantor defines the set of real numbers between 0 and 1, known as the Cantor set.

1891...Cantor proves that the cardinality of a set is always smaller than the cardinality of its power set (now known as Cantor's Theorem).

1896.. Hadamard and de la Vallée Poussin independently prove the prime number theorem.

1897.. Hensel defines the p-adic absolute value.

1900.. Hilbert poses 23 questions at the Second International Congress of Mathematics in Paris as a challenge for the 20th century.

1902 .... Hensel defines the p-adic numbers.

1909.. Borel introduces the concept of normal numbers and shows that the chance a random real number is normal in base 10 is 100%.

1940 .... Gödel establishes that the Continuum Hypothesis cannot be disproved within the standard axioms of mathematics.

1963.. Cohen establishes that the Continuum Hypothesis cannot be proved within the standard axioms of mathematics.

2006.. The 44th Mersenne prime is found. It equals $2^{32582657} - 1$ and has 9,808,358 digits.

Glossary

abacus: A calculation device in which beads representing numbers are strung on wires, allowing for speedy arithmetic.

absolute value: The distance of a real number from zero on the number line.

additive identity: Zero is the additive identity because $a + 0 = a$ for any number a.

additive inverse: The additive inverse of number a is $-a$ because $a + -a = 0$, which is the additive identity. For example, the additive inverse of 5 is -5, and the additive inverse of -17 is $-(-17) = 17$.

additive system: A numeral system in which symbols represent specific values. A number represented by a collection of symbols equals the sum of the values of the individual symbols.

algebra: The branch of mathematics that studies equations, their solutions, and their underlying structures.

algebraic number theory: The branch of number theory that studies numbers that are solutions to certain polynomial equations.

algebraic numbers: The collection of all numbers that are solutions to nontrivial polynomials with integer coefficients.

algebraically closed: Complex numbers are algebraically closed because every polynomial equation with coefficients from the complex numbers has all its solutions within the complex numbers.

"almost all": A portion of a collection is said to be "almost all" of that collection if, when an item is selected at random from the entire collection, the chance of choosing something inside that portion is mathematically 100%.

amicable numbers: Two numbers are amicable if each is equal to the sum of the proper divisors of the other.

analysis: The branch of mathematics that generalizes the ideas from calculus, especially notions of distance and continuous change.

analytic number theory: The branch of number theory that studies integers (especially primes) using ideas from calculus and analysis.

axiom: A fundamental mathematical statement that is accepted as true without rigorous proof.

Babylonian: A dominant culture in Mesopotamia during much of the 2nd millennium B.C.E. The Babylonians developed a true place-based numeral system in base 60 and approximated $\sqrt{2}$ to seven decimal places.

Barber's paradox: Suppose there is a town in which all the men shave and a lone barber shaves exactly those men who do not shave themselves. The question is: Who shaves the barber? If he does not shave himself, then he *must* shave himself, but if he does shave himself, then he *must not* shave himself. This paradox is attributed to Bertrand Russell.

barred gate: A symbol for 5 consisting of four vertical slashes crossed with a single diagonal slash.

base-2 numeral system: A positional system using only the numerals 0 and 1, with the value of the digit equal to itself times a power of 2. For natural numbers, the rightmost position is the face value of the digit. Each position to the left has a higher power of 2. Also called the binary system.

base-3 numeral system: A positional system using only the numerals 0, 1, and 2, with the value of the digit equal to itself times a power of 3. For integers, the rightmost position is the face value of the digit. Each position to the left has a higher power of 3. Also called the ternary system.

base-10 numeral system: A positional system using the numerals 0, 1, 2 … 9, with the value of the digit equal to itself times a power of 10. For integers, the rightmost position is the face value of the digit. Each position to the left has a higher power of 10. Also called the decimal system.

base-60 numeral system: A positional system using the numerals from 0 to 59, with the value of the digit equal to itself times a power of 60. For integers, the rightmost position is the face value of the digit. Each position to the left has a higher power of 60.

binary: See **base-2 numeral system**.

Botocudos: An indigenous tribe from what is now eastern Brazil. The Botocudos have a primitive knowledge of numbers; their language does not include names for numbers beyond 2.

calculi: The Greek word for "pebbles," *calculi* were pebbles or clay tokens used to represent numbers or for basic counting.

calculus: The branch of mathematics that studies continuous processes and instantaneous rates of change based on precise measures of distance.

Cantor-Dedekind Axiom: Points on a line can be placed in a one-to-one correspondence with real numbers.

Cantor set: The set of real numbers between 0 and 1 whose base-3 expansions contain only the digits 0 and 2.

Cantor's Theorem: The cardinality of the power set of a set is always larger than the cardinality of the set itself.

cardinal number: A number that represents the size of a collection. Also called cardinality.

cardinality: The cardinality of a set is a quantity that represents the size of the collection. Also called the cardinal number.

Cauchy sequence: An infinite list of numbers that get arbitrarily close together.

coefficient: In a polynomial, a coefficient is the number multiplied by an unknown power (e.g., in the polynomial $27x^8 + 7x^3 - 8x$, 27 is the coefficient of x^8).

complex numbers: The collection of all numbers of the form $x + yi$, where x and y can equal any real number and i is the square root of -1.

complex plane: A representation of the complex numbers consisting of a plane with horizontal (real) and vertical (imaginary) axes meeting at a right angle at a point called the origin.

Congress of Mathematics: One of the largest and most important conferences for mathematicians in the world, it has been held approximately every four years since 1897. The Congress of 1900 was marked by David Hilbert's announcement of 23 open questions, which included the Continuum Hypothesis.

conjecture: A mathematical statement thought to be true but for which a rigorous proof has not yet been found.

continuum: The collection of real numbers.

Continuum Hypothesis: Cantor's conjecture that there is no size of infinity between the cardinality of the natural numbers and the cardinality of the real numbers.

cosine: The cosine of an angle of a right triangle is the quotient of the length of the side adjacent to the angle divided by the length of the hypotenuse.

countable: A set is countable if it is finite or there is a one-to-one correspondence between its elements and the natural numbers.

counting numbers: The collection of numbers 1, 2, 3, 4, 5, and so on. Also called the natural numbers.

cuneiform: One of the earliest forms of writing, invented around 3000 B.C.E. by the Sumerians.

decimal expansion: The representation of a number in base 10. A decimal point separates the places representing (to the left) 1s, 10s, 100s, and so on, and (to the right) the $1/10^{th}$s, $1/100^{th}$s, and so on.

dense: Rational numbers are dense within real numbers because between any two distinct real numbers, there is at least one rational number.

diagonalization: The method invented by Georg Cantor to prove that the cardinality of real numbers is greater than the cardinality of natural numbers.

distributive law: The arithmetic law that states $a \times (b + c) = (a \times b) + (a \times c)$ for numbers a, b, and c.

e: The fundamental parameter in the measure of growth. The value of e is 2.71828... and is equal to the limiting value of the expression $(1 + \frac{1}{n})^n$ as n grows without bound.

Egyptian fraction: A fraction with numerator equal to 1.

element: A member of a collection.

empty set: The collection containing no elements.

equation: An expression that sets two quantities equal. For example, $2 + 2 = 4$ and $x^2 - 2 = 0$ are equations.

Euler's formula: $e^{\pi i} + 1 = 0$.

exponent: A superscript following a number or variable (e.g., the number 3 in the expression $2^3 = 2 \times 2 \times 2$ is an exponent).

factor: A natural number m is a factor of a integer n if m divides evenly into n.

factorial: For a natural number n, n factorial is the product of the numbers from 1 up to and including n. Denoted by $n!$.

Fibonacci numbers: The sequence of numbers 1, 1, 2, 3, 5, 8... in which each number after the first two is equal to the sum of its two predecessors.

finite: A set is finite if the number of elements in the set is equal to a natural number.

fractal: An object that exhibits both infinite detail and self-similarity; as portions are repeatedly magnified, more and more detail is revealed and patterns are repeated at different scales.

fundamental theorem of arithmetic: Every natural number greater than 1 can be written uniquely—up to reordering—as a product of prime numbers.

Gaussian integers: The collection of numbers of the form $a + bi$, where a and b can equal any integer and i is $\sqrt{-1}$.

Gaussian prime: A Gaussian integer that cannot be written as the product of two smaller Gaussian integers.

Gelfond-Schneider Theorem: If an algebraic number not equal to 0 or 1 is raised to an algebraic irrational power, then the result is a transcendental number.

glyph: See **pictograph**.

Goldbach conjecture: Goldbach's conjecture states that every even number greater than 4 equals the sum of two primes.

golden ratio: The number $\dfrac{1+\sqrt{5}}{2}$.

Hilbert's problems: The list of open questions David Hilbert posed at the Congress of Mathematics in 1900. He considered them to be the most important open questions in mathematics for the 20th century.

Hindu-Arabic numerals: The numerals 0, 1, 2, 3, 4, 5, 6, 7, 8, and 9, developed from symbols used by Hindu mathematicians and brought to the West through Arab use.

i: $\sqrt{-1}$.

I Ching: An ancient Chinese system of philosophy and prediction based on a collection of symbols, called trigrams and hexagrams, that mimic a binary numeral system (though the symbols did not represent numbers).

imaginary numbers: The collection of numbers that are multiplies of *i*.

infinite: A set is infinite if it is not finite.

infinity: An abstract mathematical concept based on unbounded or unending quantities.

integers: The collection of numbers consisting of natural numbers (1, 2, 3 ...), together with all their negatives and 0.

irrational numbers: The collection of all numbers that are not rational.

isosceles triangle: A triangle with at least two equal sides.

Jains (or Jana): A religious community in India dating back to 600 B.C.E. The Jains studied numbers extensively and even posited the existence of several sizes of infinity. Along with the Pythagoreans in Greece, they were one of the first groups to study numbers as abstract objects.

Lengua: An indigenous tribe from what is now Paraguay. The Lengua's number vocabulary included many words reflecting body parts.

limit of four: A conjecture that the human brain can deduce that a collection of items has four or fewer objects in it without actually counting the objects. For collections of five or more objects, most people have to truly count the items, however quickly, to determine how many there are.

logarithm: The exponent to which a base must be raised to produce a given number (e.g., the base-10 logarithm of 1000 is 3, because $10^3 = 1000$). When the base is *e*, the logarithm is called the natural logarithm.

magic square: A square array of the numbers 1, 2, 3...n^2 in n rows and n columns so that the sums of each row, each column, and the two diagonals are all equal.

Maya: A civilization that thrived in Mesoamerica (present-day Central America and Mexico) from c. 1800 B.C.E. until 900 C.E., with continued presence until around 1600. The Mayans had a place-based numeral system in base 20 and were one of the earliest groups known to use a symbol for zero.

Mersenne prime: A prime number of the form $2^n - 1$.

Mesopotamia: The ancient region between the Tigris and Euphrates Rivers in what is now southern Iraq. Civilizations that flourished there are sometimes collectively called Mesopotamian.

multiplicative identity: The number 1 is the multiplicative identity because $1 \times a = a$ for any number a.

multiplicative inverse: The multiplicative inverse of a nonzero number a is its reciprocal, $\frac{1}{a}$, because $a \times \frac{1}{a} = 1$, the multiplicative identity (e.g., the multiplicative inverse of 5 is $\frac{1}{5}$, and the multiplicative inverse of $\frac{1}{2}$ is 2).

natural numbers: The collection of numbers 1, 2, 3, 4, 5 ...; also called the counting numbers.

nonrepeating expansion: A number expansion in any base is nonrepeating if it is not periodic.

normal number: A number is normal if it is normal in base 10 and that property holds analogously in expansions in all bases.

normal number in base 10: A number is normal in base 10 if its decimal expansion contains equal proportions of the digits 0, 1, 2 ... 9, as well as equal proportions of the two-digit expressions 00, 01, 02, 03 ... 99, as well as equal proportions of all three-digit, four-digit, five-digit expressions, and so on, for all finite-length expressions.

number: An ever-evolving mathematical concept involving quantity, measurement, and their abstractions and generalizations.

number line: A representation of the real numbers; a line extending endlessly in both directions, with a point marked as 0 and at least one more point, usually 1, marking the unit of length. Each point on the line corresponds to a real number according to its distance from 0, with points to the right of 0 denoting positive numbers and points to the left of 0 denoting negative numbers.

number theory: The area of mathematics that focuses on the properties and structure of numbers.

numeral system: A consistent system of symbols and notation for representing numerical values.

one-to-one correspondence: Two collections are said to be in a one-to-one correspondence if each item from one collection is paired with exactly one item from the other collection and vice versa.

ordinal number: A number that represents the position of an item in an ordered list (1st, 2nd, 3rd...).

***p*-adic absolute value**: The p-adic absolute value of a natural number m is $\frac{1}{p^n}$, where p is a specified prime number and n is the largest power of p that divides the number m.

***p*-adic numbers**: The collection of all numbers in which all Cauchy sequences of all rational numbers approach a number in this collection under the p-adic absolute value.

pebble jar: A clay container made to hold number tokens or pebbles as a record of quantity.

perfect number: A number that is equal to the sum of its proper divisors; that is, the sum of all natural numbers less than that number that divide it evenly.

periodic expansion: A number expansion in any base is periodic if, eventually, the digits to the right of the decimal point fall into a pattern that repeats forever. Also known as repeating expansion.

pi: The ratio of the circumference of a circle to its diameter. Pi is denoted by the Greek letter π and equals 3.14159....

pictograph: A character, drawing, or symbol used especially in early civilizations in Mesopotamia and Egypt. Also called a glyph.

place-based system: See **positional system**.

polynomial: An expression involving a single unknown (usually denoted by x) in which various powers of the unknown are multiplied by numbers, then added (e.g., $3x^2 - 17x + 5$ and $27x^8 + 7x^3 - 8x$ are polynomials).

positional system: A numeral system in which the position of each symbol determines its value. Also known as a placed-based system.

power set: The power set of a collection is the set containing exactly all subsets of the particular collection.

prime factorization: Calculation of all the prime factors in a number.

prime number: A natural number greater than 1 that cannot be written as the product of two smaller natural numbers.

prime number theorem: The number of primes less than or equal to a particular natural number n is approximately $\frac{ln(n)}{n}$, where $ln(n)$ denotes the natural logarithm of n. As n increases without bound, the number of primes less than n gets arbitrarily close to $\frac{ln(n)}{n}$.

proof: A sequence of logical assertions, each following from the previous ones, that establishes the truth of a mathematical statement.

Pythagorean Theorem: $a^2 + b^2 = c^2$, given a right triangle with side lengths a, b, and c (with c the longest length—the hypotenuse).

Pythagoreans (Brotherhood): The community founded by Pythagoras in the 6th century B.C.E. in what is now southern Italy. Members studied mathematics and philosophy, believing that numbers were fundamental to all reality. Along with the Jains in India, they were one of the first groups known to study numbers as abstract objects.

quadrivium: The four primary subjects studied by the Pythagoreans: arithmetic, geometry, music, and astronomy. These are considered by many to be the basis for the modern liberal arts.

radian measure: The measure of an angle equal to the length of the arc that the angle subtends on a circle of radius 1.

radix: The symbol, usually a period, that separates whole number digits from fractional digits in the decimal (or other base) expansion of a number.

ratio: A quantity that compares two measurements by dividing one into the other.

rational numbers: The collection of numbers consisting of all fractions (ratios) of integers with nonzero denominators.

real numbers: The collection of all decimal numbers, which together make up the real number line.

reed-stem stylus: A hollow reed with one end cut at an angle; used to mark wet clay with symbols for numerals and, later, cuneiform markings.

repeating expansion: See **periodic expansion**.

Rhind Papyrus: A papyrus scroll purchased by Scottish Egyptologist A. Henry Rhind in 1858. Dated to c. 1650 B.C.E., the scroll was copied by the scribe Ahmes from an original at least 200 years older. The Rhind Papyrus has been a critical document in understanding early Egyptian mathematics.

Riemann Hypothesis: A conjecture involving the complex number solutions to a particular equation. If true, the Riemann Hypothesis has important implications about the distribution of prime numbers. A prize of $1 million has been offered for a complete proof.

Roman numerals: A largely additive numeral system used in the Roman Empire employing capital letters as numeral symbols, including I, V, X, L, D, C, and M.

sand table: A calculation device in which columns of pebbles or impressions were made in sand to do arithmetic; thought to be a precursor of the abacus.

set: A well-defined collection of objects whose members are called elements.

sine: The sine of an angle of a right triangle is the quotient of the length of the side opposite the angle divided by the length of the hypotenuse.

solution: Given an equation involving an unknown, a number is a solution to the equation if substituting that value for the unknown yields a valid equation.

square root: The square root of a number is a number that, when multiplied by itself, yields the first number.

square root of 2: $\sqrt{2}$, which equals 1.414....

subset: A collection is a subset of a set if every element in it is also an element in the set.

Sumerian: The earliest known civilization to inhabit the region of Mesopotamia in what is now southern Iraq. Dating primarily from about 5000 to 2000 B.C.E., the Sumerians invented a base-60 numeral system and cuneiform writing.

tally stick: A notched stick used for keeping records, especially in commerce, without using a specific numeral system.

ternary: See **base-3 numeral system**.

tetractys: The arrangement of 10 dots in a triangle, with rows of 1, 2, 3, and 4 dots. This figure and the number 10 had great significance to the Pythagoreans.

theorem: A mathematical statement that has been proven true using rigorous logical reasoning.

tokens: Molded clay or shaped stones used to represent quantities of particular items in days of early counting, beginning around 4000 B.C.E.

transcendental numbers: The collection of all numbers that are not algebraic.

Twin Prime Conjecture: The conjecture that states that there are infinitely many twin primes. Two prime numbers are twin primes if their difference is 2.

uncountable: A set is uncountable if it is not countable.

unique factorization: Every natural number greater than 1 can be written as a product of prime numbers in only one way, up to a reordering of the factors. This product of primes is the unique factorization of the number.

Zeno's paradox: A scenario that suggests the impossibility of motion. When an arrow is shot toward a target, it must first reach the halfway point; before that, it must reach the point halfway to the halfway point; before that, it must reach the halfway point and so on. Because the arrow must travel infinitely many points before it even gets halfway to the target, it never gets there; thus, motion is impossible.

zero: The size of a collection having no members.

Biographical Notes

Ahmes (c. 1680 B.C.E.–c. 1620 B.C.E.). The Egyptian scribe who wrote the Rhind Papyrus—one of the oldest recovered mathematical documents. Though the text was authored 200 years earlier, Ahmes's copy has been critical in revealing work in early Egyptian mathematics and the important role of the scribe as a teacher and preserver of knowledge.

Archimedes (c. 287 B.C.E.–c. 212 B.C.E.). A Greek physicist and engineer, as well as a mathematician, Archimedes made many contributions to number theory and geometry. He calculated excellent approximations to π given the arithmetic tools of the time, and he proved that the area of a circle is π times the radius squared.

Bernoulli, Jacob (1654–1705). This Swiss mathematician was the first to approximate the value of e, having recognized this fundamental constant as a limiting value in a process of computing compound interest on an investment. One of eight mathematicians in his family, he made many contributions to the theory of probability.

Bhaskara I (600 C.E.–680 C.E.). This Indian mathematician is credited with the first use of what are now called the Hindu-Arabic numerals, including the symbol for zero.

Bhaskara II (1114–1185). Representing perhaps the height of mathematical knowledge of the 12th century, this Indian mathematician made many contributions to the study of equations and pondered infinite quantities. His work *The Gem of Mathematics* includes many story problems allegedly written to challenge and entertain his daughter.

Borel, Emil (1871–1956). This French mathematician was a pioneer in probability and a related area of analysis called measure theory. He introduced the concept of normal numbers as a measure of the randomness of the decimal expansion of real numbers.

Brahmagupta (598 C.E.–665 C.E.). The first known written record that acknowledges zero as a number was the work of this Indian mathematician and astronomer. He also understood many fundamental rules of arithmetic and the solving of equations.

Cantor, Georg (1845–1918). A German mathematician of Russian heritage and a student of Weierstrass, Cantor established much of the early fundamentals of set theory. Between 1874 and 1884, he created precise ways to compare infinite sets, establishing the existence of infinitely many sizes of infinity, as well as infinitely many irrational and transcendental numbers. The controversy stirred by his work, along with bouts of depression and mental illness, caused him great difficulties later in his life, and he died in a sanatorium.

Cardano, Girolamo (1501–1576). An Italian physician and mathematician, Cardano was the first to consider square roots of negative numbers, calling them "fictitious" numbers. He published solutions to the general cubic and quartic equations, acknowledging that the results themselves were credited to others.

Cauchy, Augustin-Louis (1789–1857). This French mathematician was particularly concerned with rigor and precision. He began the effort to prove the results of calculus rigorously and developed a definition of the real numbers using a limiting process involving rational numbers. The resulting infinite lists of numbers are called Cauchy sequences.

Cohen, Paul (1934–2007). An American mathematician at Stanford University who proved in 1963 that the Continuum Hypothesis cannot be proven true within the standard axioms of set theory. For this work, he won the Fields Medal in 1966, the closest award mathematics has to a Nobel Prize.

Dedekind, Richard (1831–1916). This German mathematician made significant contributions in abstract algebra and number theory and did fundamental work on the real numbers and infinite sets. He was an important friend and supporter of Georg Cantor during the time when Cantor struggled to have his work on infinite sets accepted.

Diophantus (c. 210 C.E.–c. 290 C.E.). This Greek mathematician lived in Alexandria, Egypt, where he wrote one of the earliest treatises on solving equations, *Arithmetica*. Though he considered negative numbers to be absurd and did not have a notation for zero, he was one of the first to consider fractions as numbers. In modern number theory, Diophantine analysis is the study of equations with integer coefficients for which integer solutions are sought.

Euclid (c. 325 B.C.E.–265 B.C.E.). A mathematician of Alexandria, Egypt, Euclid's major achievement was *Elements*, a set of 13 books on basic geometry and number theory. His work and style is still fundamental today, and his proofs of the infinitude of primes and the irrationality of $\sqrt{2}$ are considered two of the most elegant arguments in all of mathematics.

Euler, Leonhard (1707–1783). A Swiss mathematician and scientist, Euler was one of the most prolific mathematicians of all time. He introduced standardized notation and contributed unique ideas to all areas of analysis, especially infinite sum formulas for sine, cosine, and e^x. The equation known as Euler's formula, $e^{\delta i} + 1 = 0$, is considered by many to be the most beautiful in all mathematics.

Fibonacci, Leonardo de Pisa (c. 1175–1250). An Italian mathematician, Fibonacci traveled extensively as a merchant in his early life. Perhaps the best mathematician of the 13th century, he introduced the Hindu-Arabic numeral system to Europe and discovered the special sequence of numbers that bears his name.

Gauss, Carl Friedrich (1777–1855). A German commonly considered the world's best mathematician, Gauss is known as the "Prince of Mathematics." He established mathematical rigor as the standard of proof and provided the first complete proof that complex numbers are algebraically closed, meaning that every polynomial equation with complex coefficients has its solutions among complex numbers.

Gelfond, Aleksandr (1906–1968). In 1934, this Russian mathematician answered the seventh of Hilbert's famous questions posed in 1900. The question was also answered independently in 1935 by Theodore Schneider;

thus, the result is called the Gelfond-Schneider Theorem. Gelfond began teaching at Moscow State University in 1931, continuing there until the day he died. The Gelfond-Schneider Theorem can be used to show that e^{δ}, known as Gelfond's constant, is transcendental.

Gödel, Kurt (1906–1978). Perhaps the most important of all logicians, Gödel was born and worked in Austria and Czechoslovakia but came to Princeton during early World War II. His most famous work is known as Gödel's Incompleteness Theorem, which had profound implications for the logical foundations of mathematics. He also proved that the Continuum Hypothesis could not be disproved within the standard axioms of set theory.

Hermite, Charles (1822–1901). This French mathematician made important contributions to number theory and algebra. He spent his professional life at the École Polytechnique and in the Faculty of Sciences of Paris and proved that e is transcendental in 1873. His technique was used by Lindemann to prove that π is transcendental.

Hilbert, David (1862–1943). Born in Prussia, this German mathematician was one of the most broadly accomplished and widely influential mathematicians in the late 19th and the 20th century. He spent most of his professional life at the University of Göttingen, a top center for mathematical research. His presentation in 1900 of unsolved problems to the international Congress of Mathematics is considered to be one of the most important speeches ever given in mathematics. He was a vocal supporter of Georg Cantor's work and presented the Continuum Hypothesis as the first problem on his list in 1900.

Kronecker, Leopold (1823–1891). This German mathematician made contributions in number theory, algebra, and analytic ideas of continuity. As an analyst and logician, he believed that all arithmetic and analysis should be based on the integers and, thus, did not believe in the irrational numbers. This put him at odds with a number of colleagues and, especially, the new ideas of Cantor in the 1870s.

Lambert, Johann (1728–1777). The son of a poor tailor, Lambert was a German mathematician, astronomer, and physicist. In 1761, he gave the first proof that π is irrational. He studied geometry and the origins of the solar system, spending the last 10 years of his life under the sponsorship of Frederick II of Prussia.

Leibniz, Gottfried (1646–1716). The son of a philosophy professor, Leibniz was highly influential as a mathematician, scientist, and philosopher. He discovered calculus independently of Newton and created the notation still in use today. He created the binary number system that is the basis of the computer and was the first to use the term *transcendental*.

Lindemann, Ferdinand von (1852–1939). A German mathematician and son of a language teacher, Lindemann is best known for his 1882 proof that π is transcendental. While a professor at the University of Königsberg, he supervised the Ph.D. thesis of David Hilbert.

Liouville, Joseph (1809–1882). This French mathematician worked in many fields but is perhaps best known for his proof of the existence of transcendental numbers, given in 1844. He constructed actual examples, and a special class of transcendental numbers is now called Liouville numbers.

Napier, John (1550–1617). A Scottish mathematician, physicist, and astronomer, Napier is best known for inventing the logarithm. He was the first to reference the number e and encouraged the use of the decimal point.

Oresme, Nicolas (1323–1382). Perhaps one of the most original thinkers of his time, Oresme was a French philosopher, mathematician, and scientist. Though he may have been the son of peasants, he became highly educated, ultimately serving as chaplain and advisor to the king of France. He was the first to consider numbers raised to fractional exponents and considered many other innovative ideas that presaged mathematical advances many centuries ahead of his time.

Pythagoras (c. 569 B.C.E.–c. 507 B.C.E.). Though best known for the theorem about right triangles that bears his name, Pythagoras had a much broader influence on mathematics and scholarship in general. Born on the Greek island of Samos, he moved to what is now southern Italy and founded a religious and scholarly community called the Brotherhood. Because they left no written records, knowledge about these Pythagoreans comes from later sources, including Plato and Aristotle. The Brotherhood considered numbers the basis of all reality; Pythagoras is called the "Father of Number Theory." Together with the Jains in India, the Pythagoreans were the first to study numbers as abstract objects, opening the door to mathematics as an intellectual and creative pursuit.

Riemann, Bernhard (1826–1866). A major figure in mathematics during the mid–19th century, Riemann made important contributions to analysis, geometry, and topology. Calculus students everywhere know of the Riemann integral. His Ph.D. advisor was Gauss, and he spent his brief career at the University of Göttingen. His conjecture about the distribution of primes, called the Riemann Hypothesis, is one of the most important unsolved questions in mathematics today.

Schneider, Theodor (1911–1988). This German mathematician, who taught at the University of Göttingen, is best known for his 1935 solution to the seventh of Hilbert's questions posed in 1900. The question was also answered independently in 1934 by Aleksandr Gelfond; thus, the result is called the Gelfond-Schneider Theorem. The Gelfond-Schneider Theorem can be used to show that e^δ is transcendental.

Stevin, Simon (1548/49–1620). A Flemish mathematician and engineer, Stevin wanted to bring about a second age of wisdom in which all earlier knowledge could be rediscovered. He discovered many fundamental results in physics and geometry and was a strong advocate for the adoption of the decimal system for numbers and coinage. He may have been the first to consider the number line.

Weierstrass, Karl (1815–1897). A German mathematician at the Technical University of Berlin, Weierstrass made many important contributions to calculus and analysis. Known as the "Father of Modern Analysis," he formalized the work of Bolzano and Cauchy to construct fundamental definitions still used in calculus today. He was a strong supporter of Georg Cantor.

Bibliography

Burger, Edward B., and Michael Starbird. *Coincidences, Chaos, and All That Math Jazz: Making Light of Weighty Ideas*. New York: W. W. Norton & Company, 2005 (general). This highly readable book for the general public describes many abstract mathematical ideas in concrete terms without the usual cryptic mathematical notation. Included are several chapters on infinity.

———. *The Heart of Mathematics*, 2nd ed. Emeryville, CA: Key College Publishing, 2004 (text). This very accessible college textbook for non–science students covers a wide range of mathematical topics, including Fibonacci numbers, prime numbers, irrational numbers, and infinity.

Burton, David. *The History of Mathematics: An Introduction*, 6th ed. New York: McGraw-Hill, 2005 (text). Although this textbook is written for college mathematics and math education majors, there is a large body of material that is accessible to a more general reader.

Cajori, Florian. *A History of Mathematical Notations*, Vol. 1. Chicago: Open Court Publishing Company, 1928 (advanced monograph). An advanced, nearly encyclopedic treatise on the history of mathematical notation and symbols.

Dauben, Joseph W. *Georg Cantor: His Mathematics and Philosophy of the Infinite*. Princeton: Princeton University Press, 1990 (general/advanced). This detailed and highly researched biography includes perspectives on Cantor's life, mathematics, philosophy, and theological viewpoints.

Gazale, Midhat. *Number: From Ahmes to Cantor*. Princeton: Princeton University Press, 2000 (general). This historical survey of the development of numbers focuses on the key individuals whose ideas have shaped the concept of number.

Hellman, Hal. *Great Feuds in Mathematics: Ten of the Liveliest Disputes Ever*. Hoboken: John Wiley & Sons, 2006 (general). This book chronicles some of the famous feuds between mathematicians throughout history. Of interest to this course is the intellectual battle between Cantor and Kronecker.

Ifrah, Georges. *The Universal History of Numbers: From Prehistory to the Invention of the Computer*. Hoboken: John Wiley & Sons, 2000 (general). A comprehensive treatment of the development of counting and numbers from prehistoric times to the present; full of details but written for the nonexpert.

Kaplan, Robert. *The Nothing That Is: A Natural History of Zero*. New York: Oxford University Press, 1999 (general). A philosophical and historical discussion of zero written by a mathematics teacher who suggests several ancient origins for this fundamental notion.

McLeish, John. *Number*. New York: Bloomsbury, 1991 (general). A history of numbers that explores the cultures in which various number developments took place, as well as the impact these advances in number then had on the cultures themselves.

Seife, Charles. *Zero: The Biography of a Dangerous Idea*. New York: Penguin, 2000 (general). A history of zero that includes numerous cultural contexts in which zero played a role.

Answers to Selected Questions to Consider

Lecture Three, Question 1:

67, 34, 76, 20

Lecture Four, Question 2:

$2017 = (2 \times 10^3) + (0 \times 10^2) + (1 \times 10^1) + (7 \times 10^0)$.

$1101_2 = (1 \times 2^3) + (1 \times 2^2) + (0 \times 2^1) + (1 \times 2^0) = 8 + 4 + 1 = 13$ in base 10.

$212_3 = (2 \times 3^2) + (1 \times 3^1) + (2 \times 3^0) = 18 + 3 + 2 = 23$ in base 10.

Lecture Five, Question 1:

The proper divisors of 28 are 1, 2, 4, 7, and 14; because $1 + 2 + 4 + 7 + 14 = 28$, 28 is perfect.

Observe that $1184 = 2 \times 2 \times 2 \times 2 \times 2 \times 37$; thus, the proper divisors of 1184 are 1, 2, 4, 8, 16, 32, 37, 74, 148, 296, and 592; $1 + 2 + 4 + 8 + 16 + 32 + 37 + 74 + 148 + 296 + 592 = 1210$.

Observe that $1210 = 2 \times 5 \times 11 \times 11$; thus, the proper divisors of 1210 are 1, 2, 5, 10, 11, 22, 55, 110, 121, 242, and 605; $1 + 2 + 5 + 10 + 11 + 22 + 55 + 110 + 121 + 242 + 605 = 1184$.

Thus, 1184 and 1210 are amicable.

Lecture Five, Question 2:

1	14	8	11
15	4	10	5
12	7	13	2
6	9	3	16

Lecture Six, Question 1:

The table below shows that summing the squares of consecutive Fibonacci numbers yields the sequence 2, 5, 13, 34, and so on. We recognize this as a list of every other Fibonacci number starting with 2. (This result can be proved in general.)

F_n	$(F_n)^2$	
1	1	
1	1	$1 + 1 = 2$
2	4	$1 + 4 = 5$
3	9	$4 + 9 = 13$
5	25	$9 + 25 = 34$
8	64	$25 + 64 = 89$
13	169	$64 + 169 = 233$
21	441	$169 + 441 = 610$
34	1156	$441 + 1156 = 1597$

Lecture Seven, Question 1:

Look at the number $2 \times 3 \times 5 \times 7 \times \ldots \times 1{,}000{,}000 + 1$. Dividing this number by 2 or 3 or 5 and so on up to 1,000,000 will always give a remainder of 1; thus, no prime less than 1,000,000 divides this number. Although it is enormous, it is still a natural number; thus, it must either be prime or be factorable into primes, and therefore there must exist at least one prime number greater than 1,000,000.

Lecture Eight, Question 1:

We suppose $\sqrt{3}$ is rational and work toward a contradiction. If $\sqrt{3}$ is rational, then $\sqrt{3} = \dfrac{m}{n}$ for some integers m and n. Thus, $3 = \dfrac{m^2}{n^2}$, and $3n^2 = m^2$. Recall that every natural number can be written uniquely as a product of primes. Note also that 3 is prime and that 3 must appear an even number of times in the prime factorizations of m^2 and n^2; in that case, however, the equation $3n^2 = m^2$ would have an odd number of 3s dividing the left side and an even number of 3s dividing the right side, which is impossible. Thus, our original assumption must have been faulty, which means that $\sqrt{3}$ is irrational.

Lecture Eight, Bonus Challenge:

Unlike 2 and 3, 4 is not prime; rather, $4 = 2^2$. Thus, in the attempted proof, the equation $4n^2 = m^2$ becomes $2^2 n^2 = m^2$. No contradiction is reached because there will be an even number of 2s dividing both sides of this equation.

Lecture Nine, Question 1(a):

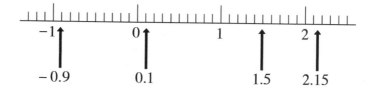

Lecture Nine, Question 1(b):

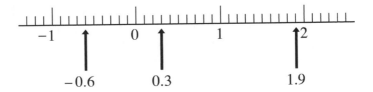

Lecture Nine, Question 2(a):

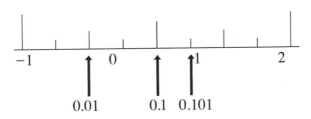

Lecture Nine, Question 2(b):

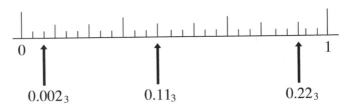

Lecture Ten, Question 1:

Only the second number, 3.7878787878…, is rational. It is rational because its decimal expansion repeats 78 forever and, thus, is periodic. The remaining three numbers are irrational. The number 0.10110111011110… will never be periodic because the number of 1s increases after each 0. The number 0.123456789101112… will never be periodic because its decimal digits are created by listing the natural numbers in increasing order, a list that never repeats. Finally, the number selected at random will never be periodic by the argument explained in the lecture.

Lecture Eleven, Question 1:

The numbers 0 and $\frac{1}{3}$ are in the Cantor set because they can be written in base 3 using only 0s and 2s: $0 = 0.000..._3$ and $\frac{1}{3} = 0.1_3 = 0.222..._3$. The numbers $\frac{1}{2}$ and $\frac{4}{5}$ are not in the Cantor set. This is easiest to see using the geometric view of the Cantor set. The number $\frac{1}{2}$ lies in the middle third of the interval from 0 to 1 and, thus, is removed. The number $\frac{4}{5}$ lies in the middle third of the interval from $\frac{2}{3}$ to 1 and, thus, is also removed. (We see this last fact by noticing that the middle third of the interval from $\frac{2}{3}$ to 1 lies between $\frac{7}{9}$ and $\frac{8}{9}$, then observing that $\frac{7}{9} = 0.777...$, $\frac{4}{5} = 0.8$, and $\frac{8}{9} = 0.888...$; thus, $\frac{4}{5}$ lies in this interval.)

Lecture Twelve, Question 1:

The expressions $5x^3 - 6x + 17$ and $\frac{3}{2}x^5 - 0.6x^2 + x - \frac{27}{13}$ are polynomials. The other two expressions are not polynomials. One contains the variable x under a square root sign; the other involves the variable in a quotient.

Simplified numbers: $25^{3/2} = \left(\sqrt{25}\right)^3 = 5^3 = 125$, $8^{2/3} = \left(\sqrt[3]{8}\right)^2 = 2^2 = 4$.

Lecture Thirteen, Question 1:

First recall that 360° equals 2π radians: the circumference of a circle of radius 1. Then, because 45° is $\frac{1}{8}$ of 360°, it equals $\frac{1}{8}$ of 2π radians, or $\pi/4$ radians. Similarly, because 60° is $\frac{1}{6}$ of 360°, it equals $\frac{1}{6}$ of 2π radians, or $\pi/3$ radians. Finally, because 30° is $\frac{1}{12}$ of 360°, it equals $\pi/6$ radians.

Lecture Fifteen, Question 1:

The number $\sqrt[3]{2}$ is a solution to the equation $x^3 - 2 = 0$. When we substitute $\sqrt[3]{2}$ in place of x, we get $\left(\sqrt[3]{2}\right)^3 - 2 = 2 - 2 = 0$, which is a valid equation.

Lecture Sixteen, Question 2:

$$i^2 = \left(\sqrt{-1}\right)^2 = -1$$

$$i^4 = \left(\sqrt{-1}\right)^4 = \left(\sqrt{-1}\right) \times \left(\sqrt{-1}\right) \times \left(\sqrt{-1}\right) \times \left(\sqrt{-1}\right) = (-1) \times (-1) = 1$$

$$i^5 = i^4 \times i^1 = 1 \times i = i$$

$$i^{25} = i^{24} \times i^1 = i^4 \times i^4 \times i^4 \times i^4 \times i^4 \times i^4 \times i = \left(i^4\right)^6 \times i = 1^6 \times i = i$$

$$i^{1,000,000} = \left(i^4\right)^{250,000} = 1^{250,000} = 1$$

Lecture 17, Question 1:

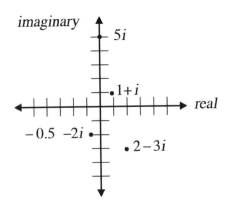

Lecture Eighteen, Question 1:

$$|4|_2 = |2 \times 2|_2 = \frac{1}{2 \times 2} = \frac{1}{4};$$

$$|12|_2 = |2 \times 2 \times 3|_2 = \frac{1}{2 \times 2} = \frac{1}{4};$$

$$\left|\frac{1}{6}\right|_2 = \frac{1}{|6|_2} = \frac{1}{|2 \times 3|_2} = \frac{1}{\frac{1}{2}} = 2;$$

$|0|_2 = 0$ by definition;

$$\left|\frac{24}{25}\right|_2 = \frac{|24|_2}{|25|_2} = \frac{|2 \times 2 \times 2 \times 3|_2}{|5 \times 5|_2} = \frac{\frac{1}{2 \times 2 \times 2}}{1} = \frac{1}{8}.$$

Lecture Nineteen, Question 1:

$$|5 - 0|_5 = |5|_5 = \frac{1}{5}$$

$$|25 - 0|_5 = |25|_5 = |5 \times 5|_5 = \frac{1}{5 \times 5} = \frac{1}{25}$$

$$|25 - 5|_5 = |20|_5 = |4 \times 5|_5 = \frac{1}{5}$$

Lecture Nineteen, Question 2:

Following the 3-adic example in the lecture, we claim that $5 + 5^2 + 5^4 + 5^8 + 5^{16} + 5^{32} + \ldots$ is a 5-adic irrational number.

Lecture Twenty-One, Question 2:

To construct a decimal number that we know will not be on the list, we will choose digits to ensure that our new number differs from the first number on the list in its first digit, differs from the second number on the list in its second digit, differs from the third number on the list in its third digit, and so on. To simplify our choices, let us say that we will choose a 9 if the diagonal digit is *not* a 9 and a 0 if the diagonal digit *is* a 9; thus, the first seven decimal digits of our special number are: 0.9909090... .

By its construction, we know that this number differs from each number on the given list in at least one decimal place and, thus, cannot appear anywhere on the list.

Lecture Twenty-Two, Question 1:

There are 16 elements in the power set. In other words, the set {♣, ᵃ, ©, ¨} has 16 subsets:

{ }

{♣}, {ᵃ}, {©}, {¨},

{♣,ᵃ}, {♣, ©}, {♣,¨}, {ᵃ, ©}, {ᵃ, ¨}, {©, ¨},

{♣, ᵃ, ©}, {♣, ᵃ, ¨}, {♣, ©, ¨}, {ᵃ, ©, ¨},

{♣, ᵃ, ©, ¨}.

Lecture Twenty-Three, Question 1:

Suggest that she consider the power set of her collection. Cantor's Theorem guarantees that it will have a larger cardinality than her original set.

Notes

Notes